A TEXTBOOK OF RADAR

A TEXTBOOK OF
RADAR

A Collective Work by the Staff of
THE RADIOPHYSICS LABORATORY
C.S.I.R.O., AUSTRALIA

Edited by
E. G. BOWEN
O.B.E., PH.D., M.SC.

Second Edition

CAMBRIDGE
AT THE UNIVERSITY PRESS
1954

CAMBRIDGE
UNIVERSITY PRESS

University Printing House, Cambridge CB2 8BS, United Kingdom

Cambridge University Press is part of the University of Cambridge.

It furthers the University's mission by disseminating knowledge in the pursuit of education, learning and research at the highest international levels of excellence.

www.cambridge.org
Information on this title: www.cambridge.org/9781316509654

First edition by Angus and Robertson, Sydney 1947
Second edition revised 1954
First paperback edition 2015

A catalogue record for this publication is available from the British Library

ISBN 978-1-316-50965-4 Paperback

PREFACE TO THE FIRST EDITION

THE Radiophysics Laboratory of the Council for Scientific and Industrial Research was established in 1939 as a result of consultations between the United Kingdom and Australian Governments. It was set up as a centre of radar research and development for the Allied Forces in the Pacific Area.

During the war the subject matter of radar grew to large proportions as a result of the activities of many research laboratories in Great Britain, the United States, Canada and Australia, and a field of scientific knowledge came into being which is unknown outside the group of scientists and engineers who were directly involved. The relaxation of military secrecy now makes it possible to publish this material.

Throughout the war there occurred among the Allied countries the freest exchange of scientific and technical data and this aided the activities of the Radiophysics Laboratory considerably. From the nature of the work and the magnitude of the research and development effort which occurred it is difficult to make adequate acknowledgement to the many laboratories and individuals who contributed. It is particularly difficult to acknowledge the work of the pioneers and those who were responsible for the developments in England which did so much to achieve the preparedness of Britain's defences before the outbreak of war. We wish to acknowledge that much of the information contained in the present volume was obtained in the first place from the work of our colleagues in Great Britain, the United States and Canada, and every effort has been made to give credit where it is due by referring to written work which is known to us. In the majority of cases the references are to documents which were given only a limited circulation and which may still be difficult of access.

PREFACE

We regret that this procedure is so inadequate both from the point of view of the reader and as a means of acknowledging original work.

We cannot omit paying tribute by name to the Telecommunications Research Establishment of the Ministry of Aircraft Production, the Radar Research and Development Establishment of the Ministry of Supply, and the Admiralty Signals Establishment in Great Britain ; also to the Radiation Laboratory of the National Defense Research Committee, the Naval Research Laboratory and the Bell Telephone Laboratories in the United States. Their efforts have been outstanding in the field of radar and their work has been freely drawn upon in compiling this book. We acknowledge also the help that has been given to the Radiophysics Laboratory by many other Government, Service and industrial organisations in Australia, and in particular we express our gratitude to the Amalgamated Wireless Valve Company for permission to publish Chapter III of this volume.

Although the names of individual members of the Staff of the Radiophysics Laboratory appear as authors at the head of each chapter, the work described in this book was contributed to by a much larger group of scientists and engineers. Many other members of the staff have also assisted materially in its preparation. We wish to express our appreciation to all those who are not mentioned by name but whose efforts helped to make this book possible.

PREFACE TO THE SECOND EDITION

THE subject-matter of radar falls into two fairly distinct parts: the basic principles of the subject and their application to practical use.

The basic principles are described in the first sixteen chapters of this book. In preparing the second edition the only changes which have been necessary are a number of small corrections and some minor alterations to the sense. An important improvement has, however, been made in the references to original work. Due to the early date at which the first edition was written the references were often to papers which could not easily be found by the majority of readers. These have now been altered to material which is readily available in the literature.

The last three chapters, dealing with the practical applications of radar, have been completely re-written. The description of military radar has been compressed to cover only the main types; a new account is given of the civil uses of radar and a chapter has been added on the very extensive applications of radar technique to the physical sciences which have taken place in recent years.

E. G. B.

Sydney
March 1952

CONTENTS

(ix)

LIST OF PLATES

CHAPTER I

INTRODUCTION

RADAR can be described as the science of locating distant objects by radio. Many of the techniques used are as old as radio itself ; others have been developed under forced draught to meet the urgent demands of war and it is these which will be described in this book.

Radar depends on two fundamental processes neither of which is essentially new—the use of short pulses for the measurement of distance and the scattering of radio waves by material bodies.

The first use of radio frequency pulses was in 1925 when Breit and Tuve[1] used them as a means of investigating the reflecting regions of the upper atmosphere. They radiated pulses of approximately 1/1000th second duration in a vertical direction, which after reflection at the ionosphere were received on a receiver adjacent to the transmitter. The time of transit of the radio waves was observed and, their velocity of propagation being known, a direct measurement was obtained of the equivalent height of the ionosphere. In the years which followed this became the standard method of conducting such investigations in nearly every country in the world.

The scattering of radio waves was first observed by Hertz during his classical experiments in 1886. On many occasions since, the suggestion has been made that such scattering might be used for the location of obstacles but it was never demonstrated in practice. The next reported observation of interest was by a team of British Post Office engineers[2] who in 1932 observed the reflection of radio waves scattered from an aircraft in flight.

---

[1] G. Breit and M. Tuve, " A test of the existence of the conducting layer," *Physical Review*, Vol. 28, September, 1926, p. 554.
[2] F. E. Nancarrow, A. H. Mumford, F. C. Carter and H. T. Mitchell, " The further development of transmitting and receiving apparatus for use at very high radio frequencies," Post Office Radio Report No. 223, June, 1932.

It was the combination of this observation with the use of pulses for measuring distance which, in the years between 1930 and 1940, gave rise to the remarkable techniques which have since become known as radar.

Historical Development

These events were part of the normal process of scientific research, the results of which were freely published and discussed throughout the world. But when their vital importance was realised, a cloak of secrecy fell on the work and obscured its subsequent history. It is known, however, that the broad path of scientific development leading to radio location was followed in England, America, France, Italy and Germany, and in many of these countries the years 1934 to 1936 saw the culmination of experimental work and the emergence of embryonic radar systems.

In England it is interesting that radar was not so much the result of a happy inspiration as the consequence of a deliberate and carefully planned policy. In 1935 there was apathy in the land and the danger of an impending air attack went unheeded except by a small minority of people. Prominent among the few who recognised the danger was Sir Henry Tizard, who formed a committee of scientists to investigate the problem of the defence of Great Britain against air attack. They studied all forms of defence—fighter aircraft, anti-aircraft guns and the passive methods of air raid precautions—and arrived at one outstanding conclusion : that unless there was at least twenty minutes warning of the approach of enemy aircraft, practically all forms of defence were useless. Tizard and his committee were therefore able to postulate the basic problem, and once it was stated in a simple form, the solution was soon forthcoming from another source. Putting together his experience of the pulse method of investigation of the ionosphere and his knowledge of the observation of the reflection of radio waves from aircraft, Watson Watt (now Sir Robert Watson Watt) was able to suggest the radio method which not only provided the necessary warning but gave precise location as well. Following this proposal, Watson Watt and a small team of co-workers succeeded in receiving pulsed echo signals from a distant aircraft for the first time in June, 1935.

In America[3] the search for methods of obstacle detection began much earlier than in Britain and there is record of work being done at the Naval Research Laboratory, Anacostia, in 1922. As a result of these efforts, pulsed radar detection equipment appeared in the United States at about the same time as in England. A remarkable feature of these efforts was that in spite of the fact that there was no communication between the participants, there was striking similarity in the final equipment. The first American air warning equipment for use on board ship operated on 200 megacycles per second, a frequency identical with that of a British ground-based equipment. The aerial system consisted of the same number of elements mounted on a similar reflecting mattress, while the transmitter and receiver were almost indistinguishable in each case.

War Application

The first application of radar in war was as a defensive weapon. Because radio waves can propagate farther than visible light, at night or through adverse weather conditions, the primary function of radar was to deny the enemy that element of surprise which is normally a potent factor in offensive operations. With later development, the wider potentialities of radar were recognised and it finally proved as effective in offensive operations, on land, on board ship and in the air. Its applications can best be described in terms of the objects from which echoes are obtained when radar equipment is used in various locations.

As shown in Fig. 1, a radar set on the ground obtains echoes

[3] " Radar—A report on science at war," released by the United States Joint Board on Scientific Information Policy, August, 1945.

from distant aircraft at moderate or high altitude. This is its air warning function, detection of the approach of hostile aircraft being used to alert the defences and to scramble fighter

Figure 1.—The reception of echoes from ships and aircraft on a ground radar set.

aircraft. It is followed by the active defensive phase during which other radar equipments fix the position of individual hostile craft more accurately and use the information for the control of fighter aircraft, for searchlight direction and control of gunfire against seen or unseen targets. In addition, ground radar obtains echoes from ships at sea, the information being used for control of fire from coastal batteries.

The use of radar on board ship is similar to that on land, as shown in Fig. 2. For tactical reasons the division of functions is into air warning, sea search and fire control, the last being against both surface and air targets. In addition, the

Figure 2.—The reception of echoes from other ships, aircraft and land targets on a shipboard radar set.

ship's radar equipment obtains echoes from nearby coastlines, a feature which has proved useful for navigation in coastal waters.

Perhaps the most varied uses of radar are in aircraft, because echoes are received from a greater variety of targets, as shown in Fig. 3. Echoes are received from other aircraft in the sky, enabling air interception to be carried out at night and in certain cases, air-to-air blind firing. Ships or surfaced submarines, in fact anything protruding from the surface of the sea, can be detected and attacked by blind bombing methods. Coastlines and hills provide massive echoes which

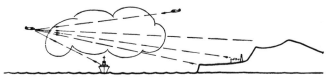

Figure 3.—Airborne radar equipment can receive echoes from other aircraft, from ships or surfaced submarines, the ground or built-up areas.

are detected at extremes of range and have proved invaluable for navigational purposes. Finally, built-up areas may be recognised by the larger signals which are reflected from them compared with those from open country, and attacked without the pilot seeing them by eye.

The first great test of radar in war came during the Battle of Britain. The Hurricanes and Spitfires of the Royal Air Force Fighter Command were superior to the aircraft against them and their pilots were without equal ; but their numbers were small against a huge attacking force. The use of radar so enabled them to conserve and direct their available strength that the attack was finally beaten back with disastrous losses to the enemy.

There followed the Battle of the Atlantic which was notable for the use by the enemy of every technical trick to an extent which exceeded that in any other phase of his war effort. Initially enemy submarines had the advantage. They were free to cruise at will and at considerable speed on the surface at night and were obliged to submerge only during the hours of broad daylight. Soon after hostilities commenced radar changed this set of conditions in a startling manner, since it provided patrol ships and aircraft with means of spotting submarines at night or in any weather. The submarines were forced to remain submerged for long periods to preserve themselves from crippling attacks and this so reduced their mobility that they were prevented from making effective attacks on our shipping. So passed the first great wave of shipping losses and the submarine menace was temporarily reduced to a low level.

The submarines collected their forces and came back with a countermeasure which again gave them virtual freedom of the seas. It was the use of listening receivers which gave warning of the approach of radar-equipped ships or aircraft. Shipping losses rose to a new peak, but for reasons not in any way connected with submarine warfare, the solution was ready in the form of microwave radar detection equipment which had been developed for other purposes. Not only did it have a performance superior to that of earlier equipment but it could not be received on the submarines' listening gear. The installation of the new equipment was carried out with all possible speed and it was instrumental in restoring the ascendancy to our ships and aircraft. Shipping losses were once more reduced while enemy submarine losses mounted to a greater height than at any other period. The waves of measure and countermeasure continued with the scales weighted heavily in the Allies' favour, until in 1943 a situation existed in which there were a greater number of submarines at sea than at any other time, but the intensity of our patrols and attacks was so great that shipping losses were negligible. The " one single weapon " to which Hitler attributed the failure of his submarine warfare was at once an admission of the superiority of Allied equipment and a tribute to the scientific way in which it was used.

Finally in the air offensive operations which were of such importance in the later stages of the war both in the European and Pacific theatres, radar was effective in overcoming the limitations of weather which at one time threatened to cripple the scale of our effort. The number of occasions on which aircraft could leave their base in clear weather and find their target also clear was surprisingly small ; in Europe this would have restricted useful air operations to some six days a month. By providing aids to navigation, new target finding methods and blind bombing techniques of startling precision, radar unshackled the bonds of cloud and weather and allowed the final attack to be launched at its full intensity.

The Special Features of Radar

The features of radar which make it outstanding are its ability to measure distance, its use of radio frequencies which propagate over considerable distances, and the extraordinarily sensitive methods it incorporates for detecting minute quantities of electromagnetic energy.

The measurement of distance is accomplished by determining the time of transit of a radio wave from its point of origin to the target and back again. The velocity of propagation being known, the distance to the target can then be determined in miles. The precision of the observation depends on the timing accuracy and our knowledge of the velocity of propagation of electromagnetic waves in the lower atmosphere. Using refined methods, the timing may be as good as 1 in 10^9 (or a few feet in a million miles), but the final accuracy is at present limited to 1 in 10^5 due to variations in atmospheric propagation. In practice it has been found relatively easy to measure distance to an accuracy of a few hundred yards, while for special military purposes a precision of $\pm$ 20 yards has been achieved over a distance of 200 miles. It is now known that this can be improved by a further factor of 3 or 4.

The frequencies which have come into widespread use for radar purposes are in the region between 100 and 10000 megacycles per second, within which band atmospheric attenuation is negligible. The range of detection therefore extends to the line of sight which in the case of high flying aircraft is considerable. This line of sight performance is almost invariant, but is sometimes improved by abnormal refraction around the earth's surface.

The sensitivity of radar receivers is extraordinary. The pulse transmitted power averages about 10^5 watts, while the received energy from a target at maximum range may be as little as 10^{-14} watts. The overall operating efficiency is therefore 10^{-19} and it is a great tribute to the pioneers of radar that they persisted in their efforts to attain apparently impossible ends.

The techniques which have made these achievements possible are sometimes new and original, sometimes an extension of those common in radio practice. The first tendency was to press the upper limit of frequency of operation well beyond that found in radio. This process was assisted by the fact that operation under pulsed conditions allowed existing valves to operate at excess voltage levels and still with low dissipation. As a result, many early radar sets were designed to operate at frequencies up to 200 megacycles per second when the highest frequency in use for radio purposes had scarcely gone beyond the 50 to 75 megacycles per second region.

The demands of war for still more refined techniques to combat a new threat or meet a new tactical situation accelerated this process until it culminated in the development of entirely new methods of operating in the centimetre wave region. In this field the resonant magnetron stands out as perhaps the most brilliant single invention since radar itself. It made power outputs of 10^6 watts available on wavelengths of a few centimetres, whereas some years previously 1 watt was a notable achievement. Associated with it was the klystron, a modified version of which played a great part in all manner of radar receivers.

Other interesting developments arose from the necessity of displaying the received information to the radar operator in an easily readable form. In ordinary radio practice, signals are reproduced aurally. In radar, the small time intervals involved—pulse durations of millionths of a second and measurement of echo delay times of a thousandth of a second—are beyond the scope of human senses, and instrumental aids must be invoked. The instrument most used for the purpose was the cathode ray tube, on the face of which distance and bearing or height of the target was displayed in pictorial form. The plan position indicator gave simultaneous distance and bearing indications and stands out as the most elegant of a number of brilliant conceptions. It was followed by others such as the range-height display which in meteorological work can give results of great practical importance and incidentally of surpassing beauty.

Those displays are being continually improved and as picture resolution increases, it is becoming possible to appreciate outlines of a target on the radar screen. With further development, precise details of form and configuration will ultimately be revealed. The process is leading inevitably to complete radar vision in which pictures of surrounding objects equal to those seen by eye will become possible through any weather conditions.

Further Applications

Until recently applications of radar have been predominantly of a military character but the techniques can also be applied to the peacetime activities of mankind..

Its use as an aid to aviation and marine navigation is obvious and substantial developments in these fields have already taken place. Not so well known, perhaps, is its application to the study of fundamental meteorological processes. Its ability to "see" raindrops and ice and snow particles in regions normally inaccessible to visual or direct observation is opening up new methods of investigating the physics of the formation of cloud and rain, whether occurring naturally or stimulated by artificial means.

The military methods of blind bombing which depended on the exact location and control of aircraft from points at a home base revealed that the accuracy with which the position of an aircraft could be fixed was often in excess of that with which points on the ground were then known by the ordinary methods of survey. A substantial re-survey of parts of the European theatre was therefore necessary before this type of bombing could be employed to full effect. The methods have a direct peacetime use in the survey of undeveloped or inaccessible terrain and in the determination of the figure of the Earth.

The precision attainable in the measurement of distance by radar means is very high and it was the use of such methods over accurately surveyed baselines which first drew attention to the fact that the velocity of electromagnetic waves is some 16 kilometres per second greater than the previously accepted value. Two recent determinations, one of which is based on the use of radar techniques, have substantiated this higher value (299,792 kilometres per second) for the velocity of propagation in free space. Activity in many other branches of physical science is similarly being stimulated as the application of radar methods of experiment and instrumentation becomes more widely known.

The successful reception of radar echoes from the Moon by the United States Army Signal Corps in January 1946 is noteworthy as the first step in sounding the universe, a step only made possible by the extraordinary instrumental developments which are an essential part of radar. The application of the same general techniques has opened up an entirely new field of astrophysical investigation, the study of the Sun and the stars by means of the radio-frequency radiation of thermal and electro-dynamic origin which they emit. This is the province of Radio Astronomy, a new branch of science which has already produced something of a revolution in astronomical ideas and which may well lead to a reconsideration of the philosophy upon which our theories of the universe are based.

CHAPTER II

FUNDAMENTALS

1. Components of a Radar Set

THE components of a simple radar set are shown in schematic form in Fig. 4. The sequence of events originates in the

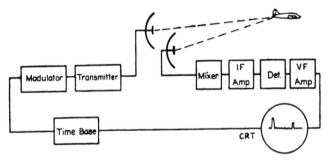

Figure 4.—Components of an elementary radar set.

modulator which produces a series of short DC pulses at the chosen recurrence frequency. These are converted into radio frequency pulses in the transmitter and radiated into space from the transmitting aerial. Some of the energy leaks back to the receiving aerial whence it is conveyed to the receiver, amplified and made to cause a vertical deflection of the cathode ray tube spot. This first signal is known in radar terminology as the "direct pulse." The greater part of the radiated energy travels outwards and that part impinging on hills, ships, waves, aircraft etc. is reflected, giving rise to echo signals in the receiver which arrive with delay times corresponding to the time of travel of the wave to and fro. For clarity in presentation only one echo is shown in Fig. 4.

In order to distinguish the various returned echoes, a horizontal deflection of the cathode ray tube spot, the time base, is superposed giving the characteristic trace depicted in the figure. This trace is repeated each cycle so that it appears stationary to the eye. The range of the target is then obtained by measurement of the displacement of this echo signal from the main pulse. The bearing is obtained by rotating the aerial in azimuth to give maximum signal strength.

Fig. 5 shows schematically a second equipment which represents a more modern type of set. It has a single aerial

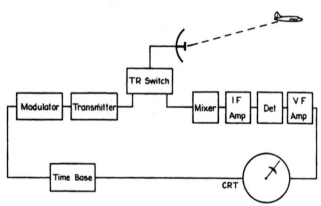

Figure 5.—Components of a typical radar set.

for transmitting and receiving and an elegant form of display known as the Plan Position Indicator (PPI).

In order to transmit and receive on the same aerial, a special duplexing component known as the TR switch is used. This consists of an extremely quick-acting switch which disconnects the receiver during the transmitted pulse and cuts off the transmitter for the remainder of the time.

In the PPI display the time base is radial and starts from the centre of the tube. Instead of causing a deflection, an echo causes the spot to brighten at the appropriate range. The aerial is made to rotate continuously and the time base rotates in synchronism with it. As the beam comes on to a target, the spot brightens, traces a small arc of a circle and disappears again as the beam goes off the target. In this way each echo gives rise to an elongated bright spot at the appropriate range and azimuth. In order to give a more or less permanent pattern on the face of the cathode ray tube, a long-delay phosphor is used which has a decay period of the same order as that of the aerial rotation. The PPI display is most spectacular when used in an airborne radar equipment looking at the ground below. A picture of the countryside is then drawn on the screen as shown in Plate I.

The arrangements shown are not exclusive ; timing may originate in a unit other than the modulator, the aerial may be required to search in elevation as well as in azimuth, and so on. But the two figures do represent arrangements of real radar equipments and serve to illustrate the interrelation of radar components with a minimum confusion.

2. The Echoing Process

The fundamental equation for the intensity Φ of a wave at a distance d from a transmitter of power P and aerial gain G in a given direction is

$$\Phi = \frac{PG}{4\pi d^2}.$$

When a wave is incident on a target a certain amount of energy is intercepted. This is partly absorbed and partly scattered, but for most radar targets the scattering predominates. The directional pattern of the scattered radiation is quite complex and varies radically with slight changes in orientation.

We specify the backward scattering or echoing power of an object in terms of its " echo cross section " σ for a particular orientation and polarisation of incident and scattered waves. If the object is in a plane wave of intensity Φ_o, the intensity of the scattered wave in the backward direction Φ_s at a large distance d is given by

$$\Phi_s = \frac{\Phi_o \sigma}{4\pi d^2}. \tag{1}$$

Consequently we may associate $\Phi_o \sigma$ for the echo wave with PG for the primary wave and consider $\Phi_o \sigma$ as the product of two factors, total power scattered and directional gain of the scattered power.

In the case of a non-absorbing object large with respect to a wavelength the total scattered power is $\Phi_o S$ where S is the projected area. The case of a large sphere is particularly instructive because it scatters equally in all directions, so that $\sigma = S$. Table 1 gives values of σ for some common objects.

Objects which, for their size, have unusually high values of σ are tuned half wave dipoles and large flat plates normal to the direction of propagation. The former depend on intercepting a large amount of energy relative to their size ; the latter on concentrating the energy intercepted. The enormous echo from a large flat surface is restricted to a very small range of angles about the normal. If, however, a ray is reflected in turn from three planes at right angles, the internal corner of a cube, the emergent ray is returned in the same direction as the incident ray and the arrangement gives a large echo over a considerable range of angles.

In military practice the targets of most interest are aircraft, ships, and topographical features. In each case these are large and of complex shape so that the directional diagram is also complex. If the target moves, this lobe pattern changes continually causing the received echo to undergo deep fading which is characteristic of the target.

Reported measurements of the mean values of σ for aircraft show great diversity. However a value of from 1 to 10 square metres, independent of wavelength in the range 3 to 200 centimetres, yields sensible results when used to calculate the performance of radar equipments.

In the case of ships, practical problems are complicated by the fact that the ship extends over a region of grossly varying field. An empirical treatment[1] based on the hypothesis that, on the average, the ship structure is complex enough to be treated as a " white body " in the optical sense, appears to give a fairly good approximation. At close ranges where the echoes are not reduced below those for free space, values of σ

---

[1] M. V. Wilkes and J. A. Ramsay, "A theory of the performance of radar on ship targets," *Proc. Camb. Phil. Soc.*, Vol. 43, 1947, pp. 220-31.

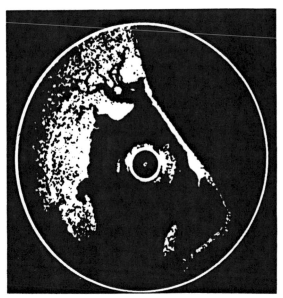

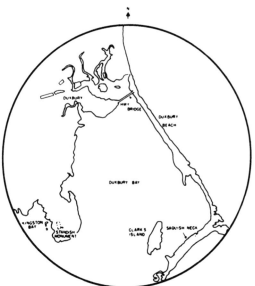

Duxbury Bay, Massachusetts, U.S.A. Height of aircraft, 2,000 feet ; radius of display, 3 nautical miles (ground range) (see p. 13).

[*By courtesy of Radiation Laboratory, Massachusetts Institute of Technology.*

PLATE I.—PPI display in an aircraft over a coast line.

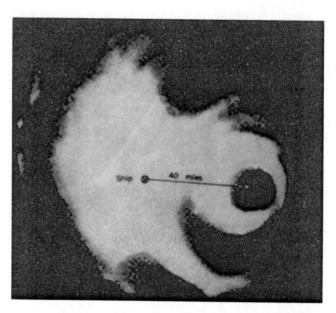
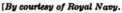

(a) 10-centimetre radar PPI display. Rain associated with a typhoon in the Pacific.

[By courtesy of Royal Navy.

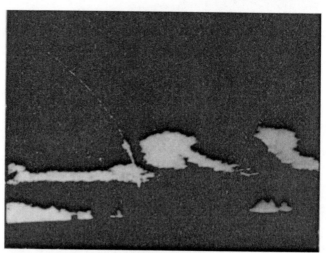

(b) 3-centimetre radar range-height display of strata of heavy cloud, presumed containing large drops, near Cambridge, Massachusetts, U.S.A. The towering structures to the right of the picture lie between 6 and 8 miles out.

[By courtesy of Radiation Laboratory Massachusetts Institute of Technology.

PLATE II.—Examples of rain or cloud echoes.

see p. 37]

TABLE 1—ECHO CROSS SECTION σ OF PERFECTLY CONDUCTING GEOMETRICAL OBJECTS

Object	Size	Orientation	Range of Validity	σ
Dipole (unloaded)	$\lambda/2$	Parallel to electric field	—	$0{\cdot}86\lambda^2$
Sphere	radius a	—	$a \gg \lambda$	πa^2
Curved surface	Radii of curvature ρ_1, ρ_2	Measured where ray normal to surface	ρ_1 and $\rho_2 \gg \lambda$	$\pi\rho_1\rho_2$
Flat plate	Area S	Normal to ray	Dimensions $\gg \lambda$	$\dfrac{4\pi S^2}{\lambda^2}$
Cylinder	Length l, Radius a	Axis normal to ray	Dimensions $\gg \lambda$	$\dfrac{2\pi}{\lambda} al^2$

of about 10^4 and 10^5 square metres for destroyers and battleships respectively have been found. These are enormously greater than those for aircraft.

Echoes from topographical features, hills, buildings etc. are normally encountered in profusion on all radar sets. In all except sets intended for navigational purposes they con⌐ stitute a nuisance. At oblique incidence echoes from land are much greater than from water so that a coastline presents a well defined boundary. Over flat country the echoes returned from cities frequently stand out above others, and this fact was utilised in navigational radar equipments during the bombing of Germany.

3. The Free Space Radar Equation

To derive this equation we require certain aerial relations which are given a more complete treatment in Chapter VIII.

A useful figure of merit for the sensitivity of an aerial is its gain G. For a given polarisation, it is defined as the ratio of the power required by an isotropic radiator to produce a certain field at a distant point, to the power required by the aerial. A similar definition exists for a receiving aerial and the gain is identical for receiving and transmitting. If losses are negligible, it depends only on the directional diagram and is then equal to the directivity.

A similar quantity, useful in the case of a receiving aerial, is the " absorption cross section " A for the given direction and polarisation. If the aerial is placed in a field of intensity Φ and it delivers power P to a matched load, then A is defined by

$$A = P/\Phi. \tag{2}$$

For any aerial

$$G = \frac{4\pi}{\lambda^2}A. \tag{3}$$

It is usual to define G and A for the direction of maximum radiation.

For broadside aerials large with respect to the wavelength and uniformly illuminated in phase and amplitude over their entire surface, A is equal to the area independently of the wavelength. In practice, due to non-uniform illumination, A is always less than the area, the factor being about 0·6 for common aerials and dropping to very low values for aerials designed to have peculiar diagrams.

For aerials of similar design, equation (3) indicates that the gain for a fixed aperture is inversely proportional to the square of wavelength. In the case of broadside aerials this is associated with a beam width given roughly by λ/a radians where a is the width of the aperture in the relevant direction.

Now let

P = transmitter power delivered to aerial (watts),

p = received power delivered to receiver (watts),

p_o = minimum detectable signal (watts),

G = aerial gain (G_T transmitting, G_R receiving),

A = aerial absorption cross section (A_T transmitting, A_R receiving) (square metres),

Φ_o = intensity of primary wave at target (watts per square metre),

Φ_s = intensity of echo wave at receiving aerial (watts per square metre),

σ = target echo cross section (square metres),

d = target distance (metres),

d_o = ultimate range on particular target (metres).

In free space the primary wave intensity at the target is given by

$$\Phi_o = \frac{PG_T}{4\pi d^2}$$

the scattered wave intensity at the receiving aerial by

$$\Phi_s = \frac{\sigma\Phi_o}{4\pi d^2}$$

and the power delivered to the receiver by

$$p = A_R\Phi_s.$$

Hence

$$p = \frac{A_R\sigma PG_T}{(4\pi d^2)^2}.$$

It is convenient to use this equation to specify ultimate range

d_o of a radar equipment for a particular target. Then d becomes d_o, p becomes p_o, and in addition we use equation (3) to eliminate either A or G, giving

$$d_o{}^4 = \frac{P}{p_o} \cdot \frac{A_R A_T \sigma}{4\pi\lambda^2} \text{ or } \frac{P}{p_o} \frac{A^2\sigma}{4\pi\lambda^2} \qquad (4)$$

or

$$d_o{}^4 = \frac{P}{p_o} \frac{G_T G_R \lambda^2 \sigma}{(4\pi)^3} \text{ or } \frac{P}{p_o} \frac{G^2\lambda^2\sigma}{(4\pi)^3} \qquad (5)$$

where in each case the second equation refers to the usual single aerial case.

These are the basic free space radar equations.

It is noteworthy that, because of the twice repeated divergence of energy in radar, d_o varies as the fourth root of P/p_o. Consequently great increases in power pay small dividends in increased range. Aerial aperture is a more important range controlling factor, d_o being proportional to the linear dimensions for similar aerial types.

The variation of ultimate range with wavelength requires careful consideration. When the limit to aerial gain is set by the physical size of the aerial, equation (4) applies and d_o is inversely proportional to $\sqrt{\lambda}$. In certain cases however the aerial is required to have a specified polar diagram so that G, not A, is fixed. Equation (5) then applies and d_o is, in contrast, directly proportional to $\sqrt{\lambda}$.

These equations suggest that a useful figure of merit for the sensitivity of a radar set is the calculated value of d_o for unit echo cross section. It is particularly apt because the effective echo cross section of an aircraft appears to be of the order of one square metre, so that values of d_o calculated in m.k.s. units give approximately correct ranges to aircraft.

Numerical Values

In these equations all quantities concerning the radar equipment are measured and specified except for the minimum detectable signal p_o. This depends on the noise generated in the receiver and on display factors which usually permit detection of a signal about equal to the receiver noise power referred to the input terminal. Table 2 gives these quantities

TABLE 2—CHARACTERISTICS OF TYPICAL RADAR SETS

	Air Warning Types (Land Based)				Airborne Types		
Wavelength (centimetres)	150[1]	150[1]	10	3	150	10	3
Peak power (kilowatts)	10	150	700	80	10	100	35
Receiver noise power (watts)[2]	2×10^{-14}	2×10^{-14}	2×10^{-13}	2×10^{-13}	1×10^{-13}	4×10^{-13}	7×10^{-13}
Aerial gain	100	100	10,000	17,000	2	340	1,300
Aerial size (feet)	10×20	10×20	25×8	10×3	Dipoles	2·5 (diam.)	1·5 (diam.)
Calculated ultimate range on a target with echo cross section of 1 square metre (miles)	30	60	130	50	3	12	9
Observed range on bomber aircraft (miles)	120	180	180	60	4	10	8
Observed range on fighter aircraft (miles)	70	180	100	40	3	6	3

[1] Observed ranges nearly doubled by earth reflections.
[2] Variations due chiefly to different bandwidths associated with different pulse lengths.

for a few typical radar sets, p_o being estimated in the manner indicated above, together with calculated values of d_o and observed ultimate ranges of aircraft.

Coverage

For a stationary aerial and a given target, the ultimate range in different directions varies owing to the directivity of the aerial. The locus of points at which the target gives a just detectable signal is a surface in space known as the " coverage diagram " of the system.

This diagram is geometrically similar to the polar diagram of the aerial expressed in field strength, not power. This is readily seen since, in a direction in which field is reduced by the factor a in Fig. 6, echo field is reduced in the ratio a^2 and

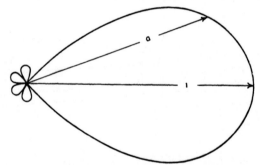

Figure 6.—Coverage diagram of an equipment. This is geometrically similar to the polar diagram of the aerial expressed in volts.

echo power in the ratio a^4. Consequently range is reduced by the factor $(a^4)^{\frac{1}{4}}$ which equals a. The coverage diagram is simply a special contour of a family which gives equal echo strengths and also equal primary fields at the target.

It is clear that the requirement of a specific coverage diagram for a given target defines the aerial gain and hence for a given wavelength the power ratio P/p_o. From equation (5) it follows that this ratio is inversely proportional to λ^2. This result is at first surprising but is related to the greater area and absorption cross section of longer wave aerials for the given gain.

This power ratio is not exactly related to volume or cross

sectional area of a coverage diagram but a very useful approximation exists in the case of a so called " rectangular " aerial. In this case

$$\frac{P}{p_0} \propto S^2 \qquad (6)$$

where S is the area of the section of the coverage diagram cut by a plane through the aerial parallel to one or other side.[2]

This concept, presented for a stationary aerial, can be extended to cover aerials which scan by inserting factors to cover the loss of sensitivity arising from the scanning process. A useful application is to the case of an air warning radar set with an aerial which rotates about a vertical axis. In order to detect aircraft at appreciable angles of elevation the vertical section of the diagram must be wide. If it is too wide the aerial gain and hence ultimate range will be unduly decreased. Equation (6) indicates the actual relationship, S being the vertical section of the coverage diagram. If, for example, the vertical coverage is to be everywhere doubled by adjustment of the vertical aerial diagram alone, the horizontal cover or range will be halved in order to maintain the vertical section constant. Alternatively if we wish to maintain the ultimate range by increase of power, the required increase will be proportional to S^2, or 4 times.

4. Normal Propagation

We shall now consider the departures from free space propagation which occur in practice due to the presence of the earth.

The earth gives rise to a ground reflected wave which reacts with the direct wave and produces an interference pattern in the region above the horizon. Below the horizon there is a shadow region into which some energy passes by diffraction. The field above the horizon may be evaluated fairly accurately from " ray theory " but this method fails in the shadow region. Fortunately radar signals, unlike one-way signals, are usually beyond the limit of detection in this latter region so that most cases can be treated without invoking the full wave theory.

The ray theory concept is that the field at a distant point is the resultant of a direct ray and one or more rays reflected

[2] The diagram has the form $f_1(\theta) . f_2(\phi)$, where θ and ϕ are angles with respect to the perpendicular sides of the aerial.

from the earth according to the laws of geometrical optics. In the case of transmission over a smooth surface such as the sea there is a single reflected ray as indicated in Fig. 7.

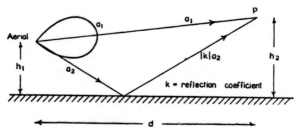

Figure 7.—Direct and reflected rays giving rise to interference phenomena above the horizon.

As the point P is raised the path difference increases and the field goes through a succession of interference maxima and minima. If the relative amplitudes a_1 and a_2 of the rays leaving the aerial are equal, then the maxima are of magnitude $1 + |k|$ and the minima $1 - |k|$ relative to the free space fields where k is the reflection coefficient. In the case of

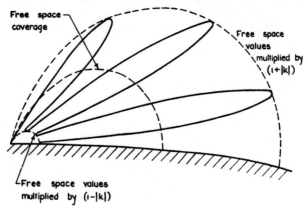

Figure 8.—Normal coverage diagram in presence of the earth.

the sea k is very nearly equal to -1 for both vertical and horizontal polarisation at small angles of elevation. The vertical coverage diagram is therefore modified in the way indicated in Fig. 8, a series of lobes being formed with the first minimum on the surface. The range of detection in the middle of the lobes is almost doubled but at the expense of gaps in coverage.

Referring again to Fig. 7, if atmospheric refraction and earth curvature are neglected, the path difference Δ between the direct and reflected rays for h_1 and h_2 much less than d, is given by

$$\Delta = \frac{2h_1h_2}{d}.$$

For a reflection coefficient of -1, the resulting field strength for one-way transmission is modified by the factor $2 \sin 2\pi h_1 h_2/\lambda d$ which has a zero at the surface, and for low heights increases linearly with height. In this region, writing the angle for its sine, the modified form of the basic radar equation (4) becomes

$$d_o^4 = \frac{P}{p_o} \left(\frac{4\pi h_1 h_2}{\lambda d_o}\right)^4 \frac{A^2\sigma}{4\pi\lambda^2}$$

or

$$d_o^8 = \frac{P}{p_o} \frac{(4\pi)^3 h_1^4 h_2^4 \cdot A^2\sigma}{\lambda^6}.$$

It is of interest that d_o is proportional to $(P/p_o)^{\frac{1}{4}}$, $h_1^{\frac{1}{2}}$, $h_2^{\frac{1}{2}}$ and to $\lambda^{-\frac{1}{4}}$. For constant aerial gain, equation (5) applies and d_o is proportional to $\lambda^{-\frac{1}{4}}$. It is not permissible to neglect earth curvature except at rather short range, so that these equations are restricted in application. The theory for a spherical earth has been discussed by Jaeger.[3] It is found that as for a flat earth, the range of detection at low heights is greatly enhanced by the use of microwaves and by increasing aerial height.

The above treatment is applicable to propagation over the sea or smooth ground. Propagation over irregular ground is very complex.

Normal Refraction

So far we have neglected effects due to refraction. These occur even in a normal atmosphere as the refractive index decreases with height and so tends to bend rays downward.

---

[3] Jaeger's work is unpublished. See *Radio Wave Propagation* (N.D.R.C. Report), Academic Press, New York, 1949, pp. 53-63.

If the gradient of refractive index with height is uniform, its effect can be treated elegantly by a method due to Schelling, Burrows and Ferrell,[4] in which the coordinate system is transformed so that rays travel rectilinearly and the effective radius

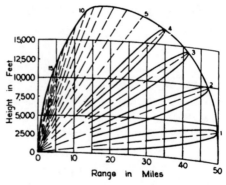

Wavelength :– 150 centimetres, Aerial height :– 100 feet

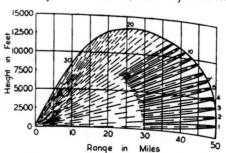

Wavelength :– 10 centimetres, Aerial height :– 33 feet

Figure 9.—Typical shipboard radar coverage diagrams.

of the earth is modified.   This is usually a good approximation using an effective radius increased in the ratio 4/3.   There is evidence that for low heights over the sea, especially in warm climates, a factor of 2/1 gives better results.

In Fig. 9 are shown vertical sections of typical shipboard radar coverage diagrams on 1·5 metres and 10 centimetres

[4] J. C. Schelling, C. R. Burrows and E. B. Ferrell, " Ultra-short wave propagation," *Proc. I.R.E.*, Vol. 21, March, 1933, pp. 427-63.

from which are apparent the outstanding features of increased number of lobes and decreased height of the bottom lobe with reduced wavelength. Land-based air warning equipments employ larger aerials, and ranges of detection of aircraft from 70 to 200 miles are usual.

5. Superrefraction

We have already mentioned that the normal atmosphere, because of its greater density near the ground, causes some bending of rays around the earth. This causes an increase of signal at points near to and beyond the horizon. It is quite common for this effect to be accentuated markedly by the presence of low lying masses of optically dense air. Under these conditions, ships, low flying aircraft, and land are often seen at distances up to one or two hundred miles by radar sets which have normal ranges of perhaps 20 miles. The phenomenon is known as superrefraction. The converse condition in which signals from low lying objects are reduced occurs but is less common.

Superrefraction has been studied extensively and a theory due to Booker[5] gives a good qualitative explanation of the phenomena. Such confirmatory measurements as have been made do not show good quantitative agreement, however, and it is not clear whether this is due to the paucity and complexity of the data or to the omission of an essential factor such as heterogeneity of the atmosphere.

The refractive index of air for most radar wavelengths differs from that for light in that water vapour is much more refracting in the former case. The collapse of dielectric constant of water from 81 at long wavelengths to about 1·7 at light wavelengths does not commence till near the absorption band at 1·2 centimetres. Consequently humidity plays a very important part in superrefraction.

[5] H. G. Booker and W. Walkinshaw, "The mode theory of tropospheric refraction and its relation to waveguides and diffraction," Physical and Royal Meteorological Societies' Report, 1947, pp. 80-127.

The index of refraction of moist air n is given by the semi-empirical formula

$$n = 1 + \left\{ \frac{79P_a}{T} + \frac{380,000e}{T'^2} \right\} 10^{-6}$$

where P_a = partial pressure of dry air (millibars), e = partial pressure of water vapour (millibars), T = temperature (degrees Kelvin). We shall tentatively use ray theory and note that bending of rays is caused by gradients of refractive index. Quantitatively, for nearly horizontal rays the curvature ψ of the ray can be shown to equal the vertical gradient of refractive

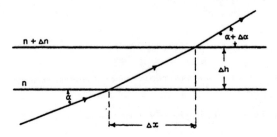

Figure 10.—Passage of a ray through horizontally stratified dielectric.

index as follows. Referring to Fig. 10 which depicts a ray passing through a horizontally stratified medium, Snell's law gives

$$\frac{\cos (\alpha + \Delta\alpha)}{\cos \alpha} = \frac{n}{n + \Delta n}.$$

For small values of α and n approximately unity this gives

$$1 - \Delta n \doteq \frac{1 - \frac{1}{2}(\alpha + \Delta\alpha)^2}{1 - \frac{1}{2}\alpha^2} \doteq 1 - \alpha\Delta\alpha .$$

Substituting $\Delta h/\Delta x$ for α and rearranging gives

$$\frac{\Delta n}{\Delta h} = \frac{\Delta\alpha}{\Delta x} = \psi$$

as stated in words above.

If we transform coordinates so that the earth is flat a ray will have a net curvature ψ_1, away from the transformed earth, given by

$$\psi_1 = \frac{dn}{dh} + \frac{1}{a}$$

where a is the radius of the earth. This transformation is illustrated in Fig. 11. We could, from measurements of

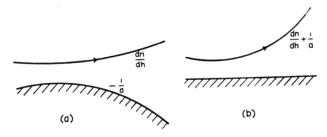

Figure 11.—Transformation to a flat earth. The symbols indicate curvatures away from the earth.

temperature and humidity over a range of heights, calculate n for each height, deduce dn/dh, and by adding $1/a$ determine whether a ray was being bent towards or away from the earth. The standard procedure now adopted is to introduce a new quantity, the "modified refractive index" M, defined by

$$M = \left[(n - 1) + \frac{h}{a} \right] 10^6,$$

where h is the height above an arbitrary datum. Modified refractive index is plotted against height. The slope dM/dh of this line then defines the relative ray curvature ψ_1, since

$$\frac{dM}{dh} = \left(\frac{dn}{dh} + \frac{1}{a} \right) 10^6 = 10^6 \times \psi_1.$$

In the formula defining M, the 1 and the 10^6 are introduced in order to give M values convenient for plotting (M is in typical cases about 300). In order to clarify the effect of positive and negative gradients of M, Fig. 12 illustrates ray paths for the two simple cases of constant M gradient, (a) positive, (b) negative. If mean values of the variation of temperature and humidity with height are taken, as in the specification of the NACA[6] standard atmosphere, it is found that M varies approximately linearly with height, as in Fig. 12 (a), up to a considerable height. The gradient dM/dh is about $+ 35$ per 1000 feet and the effective earth's radius is increased to 4/3

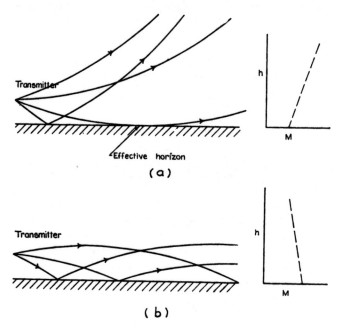

Figure 12.—Ray paths and M-h curves for constant dM/dh.
(a) dM/dh positive. (b) dM/dh negative.

of its value neglecting refraction. As in the case without refraction there is a definite tangent ray and no ray can, on this restricted theory, penetrate into the shadow region beyond the radio horizon.

The case of Fig. 12 (b) is in marked contrast in that rays are being continually bent towards the earth and no shadow is

[6] W. S. Diehl, " Standard atmosphere—tables and data," National Advisory Committee for Aeronautics, Technical Report No. 218, 1925.

formed. This negative gradient of M cannot persist to great heights because the refractive index at the surface only exceeds unity by a small amount and cannot decrease below unity. Nevertheless it is only necessary for the radar set to be within the region of negative gradient for nearly horizontal rays to be prevented from escaping. This is illustrated in Fig. 13.

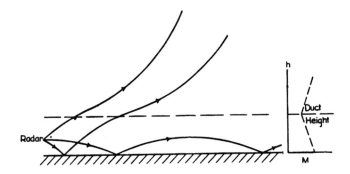

Figure 13.—Ray paths and M-h curve for simple duct with base on surface.

This phenomenon, which does not depend on the precise shape of the M curve but only on the location of the transmitter within a region of negative M gradient, is known as trapping. The region of negative gradient is known as a duct. A duct acts as a waveguide in guiding energy along its length from a source within it.

We have however omitted an important factor in treating the phenomenon from the point of view of geometrical optics. The theory is invalid if the structure is not large with respect to the wavelength. In this case the wavelength concerned is the vertical distance between successive points of equal phase, λ cosec α, where α is the angle of elevation. For nearly horizontal rays this can be large and in consequence situations showing negative M gradients will only produce substantial trapping if they extend over an adequate height for the wavelength concerned. The required value depends on detailed structure, but as a guide, 50 feet at a wavelength of 10 centimetres, and 750 feet at 150 centimetres are effective. Ducts which are not thick enough for the wavelength behave as leaky waveguides.

The case considered so far is that of a duct on the surface. Under certain conditions, the shape of the M curve may be complex giving rise to an elevated duct as illustrated in Fig. 14.

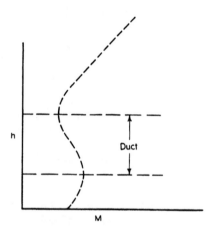

Figure 14.—M-h curve for elevated duct.

A great many measurements have been made of humidity and temperature of the air in the first few thousand feet above ground using aircraft, kites and tethered balloons and attempts have been made to relate these to radio measurements. In general it has been shown that the existence of pronounced superrefraction is associated with regions of negative dM/dh and there is good qualitative agreement between theory and experiment. On a quantitative basis there still remains some doubt but the subject is made very difficult by the usual complexity of a meteorological phenomenon. Normal conditions are illustrated in Fig. 15 (a), and conditions associated with moderate superrefraction in Fig. 15 (*b*). The modification to a vertical coverage diagram which results from superrefraction is shown diagrammatically in Fig. 16.

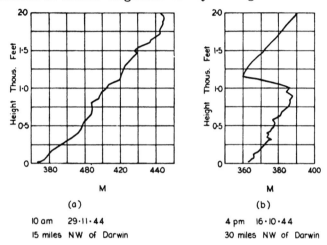

(a)

10 am   29·11·44
15 miles NW of Darwin

(b)

4 pm   16·10·44
30 miles NW of Darwin

Figure 15.—Experimental M-h curves over sea near Darwin. (*a*) Typical normal atmosphere associated with onset of north-west monsoon. (*b*) Example of elevated duct (causing superrefraction) associated with sea breeze.

[*From C.S.I.R. Radiophysics Laboratory, Report RP 260, 1945.*]

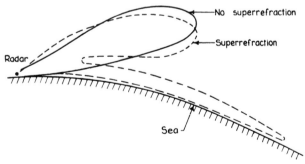

Figure 16.—Modification during superrefraction to bottom lobe of vertical coverage diagram of 1·5 metre radar due to surface duct of moderate strength. Higher lobes are not seriously modified.

Superrefraction occurs frequently in many parts of the world. It is caused when meteorological conditions are such that relatively cold or damp air underlies warm dry air. The

situation which has received most attention following the
pioneer work of Booker and Woodward[7] is the off-shore streaming of relatively hot dry air over the sea. Evaporation takes
place and if the air was originally warmer than the sea it cools
and is convectively stable, so that a blanket of cool humid
air is formed at the surface. This condition is most frequently
developed when trade winds blow from a desert over the
ocean, and spectacular examples are known to occur off the
coasts of the Sahara Desert, Arabia and north-west Australia.

A second important coastal mechanism giving rise to superrefraction in temperate latitudes is that known as " the
coastal front " by Australian meteorologists. It is essentially
the sea breeze mechanism (bringing in sea air under the continental air) engendered by a considerable contrast in temperature between land and sea. When well developed off a continental coast the effect can persist throughout both day and
night. It occurs when the motion of the continental air mass
brings hot dry air to the coast so that the general meteorological
situation is similar to that for off-shore streaming described
above, but differing in that the local winds are on shore and
not off shore. This mechanism is predominant on temperate
coasts, as for example off the southern half of Australia in
summer. Both these mechanisms show a diurnal variation
with an afternoon and evening maximum.

Over land, superrefraction can be as pronounced as at sea.
It occurs at night and is due to the cooling of the land and
lower levels of the atmosphere by radiation. It occurs principally during fine cloudless nights and the detailed mechanism
is complicated by evaporation from or deposition of dew on
the ground and has not been extensively studied. It is of
some interest that such a radiation inversion can cause superrefraction over the sea as it blows outward.

    [7] H. G. Booker, "Elements of radar meteorology: How weather and
climate cause unorthodox radar vision beyond the geometrical horizon,"
J. Instn. Elect. Engrs., Vol. 93, Part IIIA, 1946, pp. 69-78.

These three mechanisms tend to give ducts of the order of 1,000 feet in height and tend to influence radar propagation similarly in the range of wavelengths from 3 to 150 centimetres.

Over the sub-tropical oceans of the world in the region of the trade winds a further mechanism is operative. The air is blowing into regions receiving increasing amounts of solar energy. This energy is chiefly effective in evaporating water. Consequently a layer of humid air is formed over the ocean. When the wind is weak the layer of excess humidity is very shallow ; when strong, increased turbulence increases the thickness of the layer. The height of the duct so formed is roughly proportional to wind speed and is about 40 feet for a 10 knot breeze. These ducts cause pronounced super-refraction but only at short wavelengths. Radar sets on 3 and 10 centimetres show marked effects more particularly if the aerial is low so that it lies in the duct.

Under normal atmospheric conditions echoes from ships or land are seen on powerful surface radar equipments up to distances of 20 to 40 miles. Superrefraction which increases these distances to 100 to 200 miles is common, occurring near warm temperate coasts perhaps one tenth of the total time. Less frequently and only in favourable localities much more spectacular increases are observed. Thus 1·5 metre air warning sets near Darwin, Australia, report echoes from the coast of Timor, 300 to 500 miles distant, several times a month.[8] A similar equipment near Broome in north-west Australia reported echoes in November, 1944, from the coast of. Java 900 to 1,100 miles away.

Intense superrefraction also occurs in the Arabian Sea. Echoes on 1·5 metres from a number of places on the coast

---

[8] F. J. Kerr, "Radio superrefraction in the coastal regions of Australia," *Aust. J. Sci. Res.* A, Vol. 1, 1948, pp. 443-63.

from Aden to Karachi have been obtained from Bombay at distances up to about 1,500 miles.[9]

Ranges of detection of ships which are normally of the order of 20 miles are, similarly, frequently increased to 100 miles. It is unusual for a range greater than 200 miles to be reported but an extreme range of 700 miles has been reported from Bombay.

Normal ranges of detection of aircraft at medium heights are not so severely limited by the shadowing effect of the earth as those of surface targets, and range increases due to super-refraction are relatively less. The usual effect is an increase of range on low flying aircraft to 100 or 200 miles and it is suspected, but not proved, that this is associated with some reduction of range to high flying aircraft. Occasional increases of range are reported greatly beyond the normal even for high flying aircraft. Thus in February, 1944, the air warning station at Geraldton, West Australia, followed a Catalina flying boat, outward bound from Perth to Colombo, almost continuously out to a distance of 800 miles.[10]

It is noteworthy that all the above outstanding increases of range of detection have been obtained on 1·5 metres, but it is not clear whether this should be attributed to specially favourable propagation or to the particular distribution of radar equipments from which reports are available.

6. Atmospheric Absorption

Absorption in air at ordinary radio frequencies is negligible but at the highest frequencies used in radar, this is not so.[11]

[9] H. G. Booker, "Elements of radar meteorology: How weather and climate cause unorthodox radar vision beyond the geometrical horizon," *J. Instn. Elect. Engrs.*, Vol. 93, Part IIIA, 1946, pp. 69-78.
[10] "Abnormal ranges on aircraft in western area," R.A.A.F. Report to Radiophysics Laboratory, March, 1944.
[11] J. H. Van Vleck, "The absorption of microwaves by (a) oxygen and (b) uncondensed water vapour" (two papers), *Phys. Rev.*, Vol. 71, 1947, (a) p. 413, (b) p. 425.

Two atmospheric gases, oxygen and water vapour, show absorption which may be regarded as due to extreme infra red absorption bands associated with the magnetic dipole moment of oxygen and the electric moment of water.   Fig. 17 gives

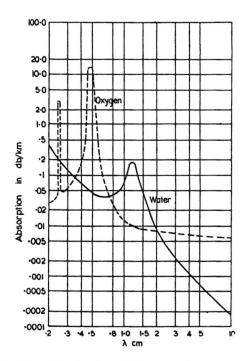

Figure 17.—Absorption in atmospheric gases (one way attenuation).
(a) Water vapour of partial pressure 7·6 millimetres of mercury.
(b) Oxygen in a normal atmosphere of total pressure 76 centimetres of mercury.

theoretical attenuation figures for a typical atmosphere.   The attenuations are proportional to the partial pressures of the respective gases and are not rapidly varying with temperature.

The attenuation will be highest in the tropics and may reach about 0·1 decibels per kilometre at a wavelength of 3 centimetres and 0·5 decibels at 1·2 centimetres, the centre of the water vapour absorption band.

The atmosphere may also carry solid or liquid particles in

the form of fog, cloud, rain etc. and, as in the case of light, these particles cause scattering and absorption. The absorption is trivial in the case of the longer radar wavelengths but is large in the one centimetre region. Fig. 18 gives estimated

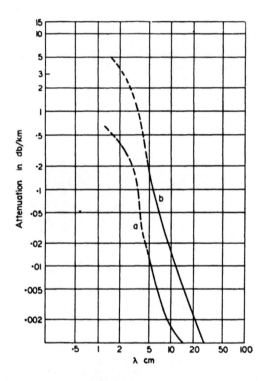

Figure 18.—Attenuation, one way, due to rain. (a) Moderate rain (6 millimetres per hour of known drop size distribution). (b) Rain of cloud burst proportion (43 millimetres per hour of known drop size distribution).

values for moderate and extremely heavy rain. Values for cloud are much less. A rough generalisation is that attenuation due to rain becomes appreciable at a wavelength of about 10 centimetres, that due to cloud at 1 to 3 centimetres.

For drops which are small with respect to the wavelength the absorption is proportional to the total water content per unit volume.

7. Atmospheric Scattering

In a manner analogous to that in which light is scattered by fine dust particles, radar waves are scattered by rain drops.[12] The amount of scattering from heavy rain is adequate to produce echoes on powerful 1·5 metre sets and is very large on microwave equipments. Excepting for very large drops at the shortest wavelengths, the echo power reflected from a single drop obeys Rayleigh's law, being proportional to D^6/λ^4 where D is the diameter. Consequently for uniform rain of density N drops per unit volume, the total energy returned is proportional to ND^6 which may be considered as the product of total mass of water per unit volume and mass of a single drop. It is therefore sensitive to drop size whereas attenuation is not.

If on the other hand we vary wavelength the echo intensity does not usually show a simple dependence on $1/\lambda^4$ because we usually simultaneously vary a number of other controlling factors, such as aerial gain, pulse length, etc.

Because echoes are adequately intense and absorption not excessive, microwave radar equipments can be used to observe the structure of rain clouds in a way which is not possible by eye owing to the heavy attenuation of light. Plate II (a) shows the distribution of heavy rain echoes obtained during a typhoon. It is a photograph of the PPI display of a medium power 10 centimetre radar set on a ship located about 40 miles from the " eye " of the typhoon which is clearly visible.

Plate II (b) shows not a plan view but an elevation of a less spectacular cloud formation obtained on a 3 centimetre equipment with an RHI (range-height-indication) display.

[12] J. W. Ryde, "The attenuation of centimetre radio waves and the echo intensities resulting from atmospheric phenomena," *J. Instn. Elect. Engrs.*, Vol. 93, Part IIIA, 1946, pp. 101-3.

CHAPTER III

THE MAGNETRON

1. Early Magnetrons

FOR many years prior to 1946 the magnetron was studied and used as a generator of high frequency oscillations. Its construction followed the well known lines of a cylindrical anode surrounding a filamentary cathode, and it operated in an axial magnetic field. In many cases the anode was split into two or four segments, alternate segments being connected together in the latter case.

Such devices were used to excite oscillatory circuits of the Lecher wire type connected between filament and anode, or between anode segments, and power was generated at wavelengths between one metre and one centimetre approximately.

It was known that there was more than one mechanism whereby oscillations were produced, and most interest centred on the generation of power at the shortest wavelengths where the principle of operation was least understood. Much ingenuity was used and there were many variations in design ; but it cannot be said that startling results were forthcoming in the range of wavelengths around 10 centimetres in which we are at present interested.[1]

Some typical results obtained at the shortest wavelengths will now be noted. Rice,[2] using a single anode magnetron, produced 3 watts at a wavelength of 4·8 centimetres, the efficiency being only 1 per cent. Cleeton and Williams[3] produced very low powers at wavelengths of 1·87, 1·22, and

---

[1] A useful review of early work, with a bibliography, is given by A. F. Harvey, *High Frequency Thermionic Tubes*, Chapman and Hall, London, 1943.

[2] C. W. Rice, "Transmission and reception of centimetre radio waves," *Gen. Elec. Rev.*, Vol. 39, 1936, pp. 363-68.

[3] C. E. Cleeton and N. H. Williams, "The shortest continuous radio waves," *Physical Rev.*, Vol. 50, November, 1936, p. 1091.

0·64 centimetres. Linder[4] with two different designs working at about 10 centimetres, produced powers of 2·5 watts (12 per cent efficiency), and 13 watts (20 per cent efficiency).

In view of developments to be described we shall not dwell on pre-war work on the magnetron except to glance at the design used by Rice, which is shown diagrammatically in

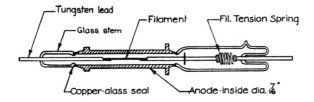

Figure 19.—Single anode magnetron—as used by Rice (1936).

Fig. 19, and the anode used by Linder, Fig. 20. The Rice magnetron was designed so that the anode formed part of a coaxial transmission line resonator, while the anode of the Linder magnetron was split so as to form part of a two-wire transmission line resonator. These were both attempts to

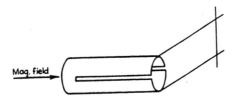

Figure 20.—Magnetron anode system—as used by Linder (1939).

resolve the conflict between high power which requires large dimensions, and short wavelength which requires the opposite. We will now see how this problem was finally solved.

[4] E. G. Linder, " Description and characteristics of the end-plate magnetron," *Proc. I.R.E.*, Vol. 24, Part I, 1936, p. 633-53 ; and " The anode-tank-circuit magnetron," *Proc. I.R.E.*, Vol. 27, 1939, pp. 732-38.

2. The Resonator Magnetron

The modern high-power magnetron was developed in the Physics Department of the University of Birmingham, with the co-operation of the General Electric Company (Wembley), as part of a nationally inspired programme of work on wireless waves of a frequency of 3000 megacycles per second and greater. We shall briefly describe this very remarkable achievement, and in doing so we cannot do better than quote from Oliphant's foreword to the report which describes the original work.[5]

" An examination of the very recent literature and of first principles showed that the only hope for producing efficient generators was to combine generator and circuit in a single unit, and that the circuit should be one of high efficiency made from the best possible electrical conductors. Accordingly, a programme was drawn up for detailed investigation of the velocity modulation methods which had just been published using efficient ' rhumbatron ' resonators, while a paper by Rice in the General Electric Review for 1936 which describes a transmission line type of magnetron, encouraged us to investigate the possibility of introducing rhumbatron technique to improve the circuit of the magnetron. A very superficial examination of existing magnetron devices showed that circuits were made of highly resistant materials, and in a form where radiative and resistive damping seriously reduced the efficiency. Plans were drawn up of possible trial apparatus and the work was entrusted to Dr. Randall and Mr. Boot. They found it was not at all easy to transform the existing types of resonator rhumbatrons for use in the magnetron, where a cylindrical symmetry was desirable, and with considerable insight decided to try the less efficient cylindrical form of resonator, which was at once successful. The general ideas formed at that time with regard to mode of action and general construction have been a little modified by subsequent work, though what was rather a crude laboratory instrument

[5] J. T. Randall, H. A. H. Boot and S. M. Duke, " Magnetron development in the University of Birmingham," Co-ordination of Valve Development Report, Department of Scientific Research and Experiment, Admiralty, 1941.

was transformed through the experience of the Research Laboratories of the General Electric Company into a valve of great simplicity which was easily manufactured and was reliable enough for service use."

The first valve produced is shown in Plate III (a). The anode system with its six cylindrical resonators was machined out of a solid copper block, the block itself with metal end-plates forming the main valve envelope. The resonator size was determined on the basis of the Hertzian dipole ring for which $\lambda = 7 \cdot 94d$, where d is the diameter of the ring. Power was taken out through a coaxial seal terminating in a coupling loop in one of the resonators. The cathode was a pure tungsten filament 0·75 millimetres in diameter. This arrangement provided an immediate solution to the pressing high-power magnetron problem. It introduced high Q resonant circuits of the right size and permitted high anode dissipation. Glass bombardment by electrons escaping from the ends of the anode did not exist.

This magnetron, running on the pumps, worked for the first time on February 21st, 1940. Approximately half a kilowatt of power at 10 centimetres was generated, the power input being 4 kilowatts. This was a tremendous advance over the performance of any previous magnetron. The anode voltage was 8,000 and the magnetic field 0·11 webers per square metre (1100 gauss). The anode block dimensions were : resonator diameter 1·2 centimetres, anode diameter 1·2 centimetres, slot width 0·1 centimetres, and slot length 0·1 centimetres.

The next step was the construction of sealed-off tubes, which was made possible by the use of the General Electric Company gold-ring seal to be described later. An indirectly heated, oxide coated cathode was also added and this made a material improvement in efficiency. Developments were rapid and large numbers of magnetrons of various operating frequencies and powers were made, and used by the Armed Services in subsequent years. Production and development of these tubes was carried out in laboratories and factories throughout the English-speaking world, including Australia. Information from other countries is at present meagre, but there is evidence of advanced Russian work in this field.[6]   It is also known that much work as yet unreported in detail was done in Japan and Germany.   Magnetrons designed in those countries used resonator systems somewhat similar to the one described above, but mounted in large glass envelopes.   They performed badly, and were difficult to construct.[7]

3. Theory of Magnetron Operation

Before describing further developments, let us see if we can form some picture of how these magnetrons work.   Following a useful discussion by Slater[8] we shall consider electron motions in the simple case of a plane magnetron with suitably arranged electric and magnetic fields.   Space charge effects will be ignored and initial electron velocity will be taken as zero.

Referring to the conditions of Fig. 21, simple analysis

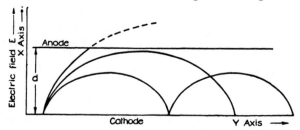

Figure 21.—Electron paths in steady crossed electric and magnetic fields.

shows that the electron paths are cycloids traced out by a point on a circle of radius $R = mE/eB^2$ rolling along with angular velocity $= Be/m$ and with linear speed E/B, where e and m

[6] N. F. Alekseev and D. D. Malairov, "Generation of high-power oscillation with a magnetron in the centimetre band," *Proc. I.R.E.*, Vol. 32, March, 1944, pp. 136-39.

[7] A study of the literature, including patents, shows that some of the ideas embodied in the Birmingham magnetron, including the association of cavity resonators with a number of segments, had occurred to previous workers at one time or another.   Credit must be given to the Birmingham group, however, for the development and construction of the first practical high-power magnetron from which has sprung a whole series of sturdy valves operating over wavelengths from 1 to 50 centimetres, and generating peak powers up to several megawatts.

[8] J. C. Slater, "Theory of the magnetron oscillator," Massachusetts Institute of Technology Radiation Laboratory, Report, August, 1941; J. B. Fisk, H. D. Hagstrum, P. L. Hartman, "The magnetron as a generator of centimetre waves," *Bell Syst. Tech. J.*, Vol. 25, 1946, pp. 167-348.

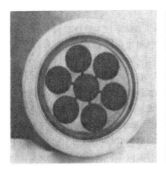

(a) *Left.*—Machined copper anode used in first Birmingham magnetron.
Right.—The first Birmingham magnetron assembled.
[*By courtesy of Professor Oliphant, Physics Department, University of Birmingham.*

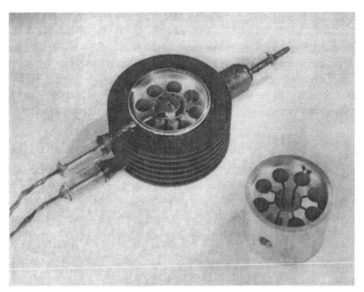

(b) Typical modern magnetron type CV76. Peak power output 400 kilowatts, with an efficiency of 50-60%, at a wavelength of 10 centimetres.

PLATE III.

are the charge and mass of the electron, E is the electric field strength and B is the magnetic flux density. Three paths are indicated for different values of E/B^2.

It is noteworthy that at the cusps near the cathode plane the kinetic energy of the electron is zero, whereas at the top of the path the work done by the field on the electron appears as kinetic energy. It is also useful to remember that with respect to axes moving along at speed E/B, the coordinates of the electron's position at any time can be written

$$X' = R \cos\omega t, \qquad\qquad Y' = R \sin\omega t.$$

Suppose that there is superimposed on the steady electric field a harmonic electric field of a period equal to $2\pi/\omega$ the period of rotation of the electron. The electron paths can be computed by direct solution of the equations of motion, but all we require here is a qualitative idea of these paths which can be obtained by thinking of the effects of the sinusoidal field on the simple harmonic components of the electron's motion. When the field is in phase with those components, the rotational motion of the electron is built up as the field does work upon it, while a field opposite in phase will slow down the rotational motion. Picturing only the first few cycles of either case, the paths shown in Fig. 22 are obtained.   The energy exchanges

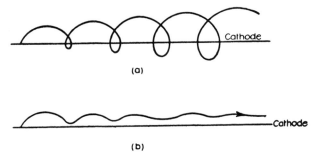

(a)

(b)

Figure 22.—Electron paths in crossed electric and magnetic fields, with added sinusoidal electric field.

are important : in the first case the electron absorbs energy from the alternating field, in the second the electron gives up energy to the alternating field.   These are of course the extreme cases ; electrons emitted at other times either receive, or give up energy more slowly. Thinking for a moment of the second case which is the more favourable for yielding energy to the alternating field, we see that the electron finishes up with its drift speed E/B only, its rotational motion having been lost by the action of the alternating field. It is moving at a distance R from the cathode. Hence substituting from above, the work done on it by the steady field is,

$$EeR = Ee \cdot \frac{mE}{eB^2} = m\,\frac{E^2}{B^2}$$

which equals twice the kinetic energy of drift $\frac{1}{2}m(E/B)^2$. Thus in a favourable case, the electron is able to give up to the alternating field one half of the energy extracted from the steady field.

There is one vital point, however, about paths of the first type which must not be overlooked. All such paths cross the cathode because of the increase of speed due to the gain of energy from the field. They will therefore strike the cathode after one period and have no further opportunity of absorbing energy.

From this simple case we can now understand why a single anode magnetron can be expected to oscillate. Remembering that electrons are emitted from the cathode in all phases, we see that there are two groups. Roughly half the electrons absorb energy from the alternating field, but are removed after one period and thus do not have much damping effect on the alternating-field generator. The remainder, which can yield up to the alternating field about half of the energy they have received from the steady field, will clearly tend to sustain oscillations once they commence.

It should be noted that the direction of the superimposed field in the plane of the orbits is not significant. It must not be forgotten that the " spent " electrons which have delivered energy to the field must be removed before they can absorb energy again by building up their rotational motion and that only oscillations of a frequency roughly equal to the natural electron frequency $\omega = Be/m$ have been considered.

A magnetron working in this way will generate power at a wavelength $\lambda = 2\pi c/\omega = 2\pi mc/eB$. We have therefore $\lambda B = 10\cdot 6 \times 10^{-3}$ where λ is in metres and B in webers per square metre ; hence if λ is in centimetres and B in gauss, $\lambda B = 10,600$.

This relation will be found to be approximately satisfied in the case of the simpler single and split-anode cylindrical magnetrons used by the workers mentioned earlier.

The Multi-segment Cylindrical Magnetron

Turning now to the electrode system used in the Birmingham magnetron, the system becomes cylindrical instead of plane and the electric field is complicated by the presence of the transverse fields across the slots.   Exact tracing of the electron

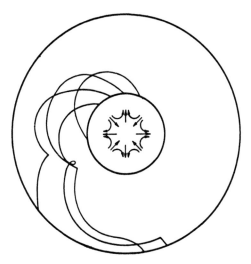

Figure 23.—Electron orbits in a magnetron with 8 segments.   The wavelength is 10 centimetres, cathode radius 3 millimetres, anode radius 8 millimetres, anode potential 10 kilovolts, AC potential 5 kilovolts, magnetic flux density 0·13 webers per square metre (1300 gauss), assuming a linear potential.   Arrows show the phase of the electric field at the cathode at the instant of emission.

orbits is difficult even if space charge be neglected, but it can be shown that the electrons move around the system in cycloid-like paths, and can both absorb energy from and deliver it to the alternating fields which are produced across the slots by the resonators. When the drift speed of the electrons has the right relation to the period of the resonators so that electrons pass each slot in about the same phase, we find that the energy-

Figure 24.—Orbits of Fig. 23 as seen from rotating co-ordinates.

absorbing electrons are removed, while the energy-yielding electrons move out and strike the anode with very little residual energy. This represents an important change from the simple case, when the spent electrons remained in the middle of the discharge space to absorb energy. In the multi-segment magnetron no such problem arises, and some of its increased efficiency may be attributed to this fact.

Efficiency is also increased by the fact that the residual energy of some electron orbits is much below 50 per cent of the work done on them, making the " orbit efficiency " high. These are the orbits with cusps close to the anode. In the plane case maximum orbit efficiency is 50 per cent.

As a practical example, Figs. 23 and 24 show orbits for an eight-segment magnetron, calculated by Slater by numerical integration of the equation of motion. There has as yet been no mention of space charge effects, but what is believed to be a fair approximation to take care of this has been made in computing the orbits shown (Slater,[8] p. 118).

The orbits are of familiar shape. Three of them have a chance of giving energy to the field ; the fourth begins to absorb energy, but is lost before a complete cycle occurs. The energy calculations show that of the work done on these four electrons, only 60 per cent appears at the ends of the orbits, and 40 per cent is available for maintaining oscillation.

The grouping of orbits shown in Fig. 24 gives an interesting physical picture of the cloud of electrons rotating as a toothed wheel with the speed of the travelling voltage wave at the anode. This picture turns out to be useful in the understanding of space charge effects.

Much work[8,9] has been done on the theory of the multi-segment magnetron, investigation having been carried out on :

(a) The behaviour of the multi-cavity resonator system, its modes of oscillation and the fields it produces in the working space.

(b) Electron concentrations and paths in the working space under the influence of fields due to the resonators and space charge.

The problem has to be split up in this way to make headway in the face of its complexity and numerical integration must be resorted to at most stages.

A brief consideration of modes of oscillation is given in the next section and we will now consider problem (b). Here the greatest success has come from a " self-consistent field " method of approximation. In Slater's words, the problem is : " Given a certain boundary condition in the form of a field at the anode surface, what must be the electric field in the anode-cathode region, in order to produce the motion of the electrons which will in turn set up the space charge necessary to produce the field ? " The method is to make a reasonable guess at a field and compute orbits and space charge. Then

[9] A series of reports of work done by teams at Manchester and Leeds Universities, led respectively by Dr. Hartree and E. C. Stoner, has been issued as Co-ordination of Valve Development Reports by the Admiralty Department of Scientific Research and Experiment, numbered Mag. 1, 2, 3 etc., commencing August, 1941.

by deriving a field from this space charge (with perhaps some smoothing), the orbits are re-computed and so on. Convergence of assumed and derived fields gives the " self-consistent field " solution. Examples of such a computation with a discussion of its problems are given by Hartree and others.[9] As a starting point, use is often made of the space charge disposition which has been employed with some success in the single-anode magnetron. It should be noted that a complete calculation will give figures for emission and anode currents as well as energy exchange.

While some success attended the efforts of those working on this difficult problem, Slater's opinion in 1941 was that there was at that time no completely satisfactory solution of the space charge problem. He pointed out however that whatever uncertainty there was in the region close to the cathode, the electron orbits were generally little changed by variation in the assumed space charge disposition. If we recall Fig. 24, we may say that we do not know all about the hub of the rotating toothed wheel but we are fairly certain that the orbits in the teeth approach reality, and this is where most of the energy exchanges take place. A good investigation of magnetron operation may therefore be made without a serious consideration of space charge at all, and this Slater has done. He has been able to specify a number of points for practical magnetron design. These are briefly noted in the next sub-section.

Magnetron Design

Because of the difficulty of the theory, most magnetron design has proceeded empirically guided only by general theoretical conclusions. Slater deduces that for efficient operation:

(a) The drift speed of the electron must be such that the electrons pass each slot in about the same phase.

(b) The magnetic field should be above "cut-off." This is of course to give the useful electrons an orbit with at least one cusp, with a good chance of giving up their energy.

(c) The distance between slots should be roughly the same as the anode-cathode distance. This is to allow a reasonable tangential AC field component all over the working space.

None of these conditions is at all critical and it is not surprising to find that a magnetron will work over a large range of voltages, fields, and currents.

Equations describing (a) and (b) above for a multi-segment cylindrical magnetron are:

$$V = \frac{300\pi}{n\lambda}(\gamma_a^2 - \gamma_c^2) \left\{ B - \frac{0\cdot 0106}{n\lambda} \right\} \times 10^6 \qquad (1)$$

$$V = 22,200\, B\, \frac{(\gamma_a - \gamma_c)^2(\gamma_a + \gamma_c)^2}{\gamma_a^2} \times 10^6 \qquad (2)$$

where V is the steady anode potential, B is the magnetic flux density, λ is the magnetron wavelength, γ_c and γ_a are the anode and cathode radii; and n is the "mode number" or number of repeats of the voltage around the anode.

Hartree,[10] also, produced from other considerations an equation differing slightly from (1). The "Hartree diagram," well known in the literature, shows the cut-off curve plotted from equation (2); and a series of straight lines plotted from his equation for various values of n.

These equations have proved very useful as guides to magnetron design. They only go part of the way, of course, because questions of cathode emission, anode current, anode and cathode heating, and resonator dimensions must also be considered if a magnetron is to be designed *ab initio*. They can be used for "scaling"—the means whereby, given a successful magnetron, designs for higher or lower frequency operation are produced.

It is an interesting fact that most practical magnetrons whose dimensions were chosen by considerations of resonator size and manufacturing technique, require such high voltages and currents that pulse operation is necessary to prevent overheating of anode and cathode.

[10] H. A. Boot and J. T. Randall, "The cavity magnetron," *J. Instn. Elect. Engrs.*, Vol. 93, Part III A, 1946, p. 935.

4. Modes of Oscillation

It did not take early users of multi-segment magnetrons long to discover that even the most carefully made valves could oscillate at more than one frequency. It is easy to see why this can happen if we reflect that there may be 8 resonant circuits with some degree of coupling between each, so that a number of modes of oscillation is expected on general grounds. A simple example will show the type of " moding " which can exist. Fig. 25 gives a view of the segments of an

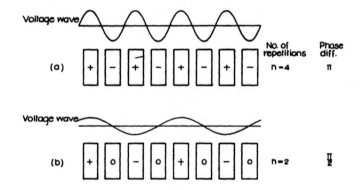

Figure 25.—Moding in an 8-segment magnetron. Segments shown " unwrapped," as seen from the cathode.

8 resonator magnetron which has been " unwrapped " as seen from the cathode. Two possible modes are shown, the charges on each segment being indicated by signs. Another way of describing these conditions is by giving the voltage wave for each case ; it may be considered as travelling in either direction, or as a stationary wave.

Mode (a) is described as the " π mode " or $n = 4$ mode, π being the phase difference between adjacent segments and n the number of repetitions of the wave pattern around the anode. The $n = 2$ mode is also shown and it is clear that there are modes of this type corresponding to $n = 1, 2, 3$ and 4 for the 8 segment magnetron. No further elaboration of this complex question will be made here. An important point, however, is that even with perfectly made anodes, these modes give rise to different wavelengths which unfortunately are closely grouped—hence the difficulty in operation.

Strapping

The above analysis indicated to early workers with the Birmingham magnetron that some measure of mode selection could be attained by means of physical connection between segments.   For instance if alternate segments of Fig. 25 are strapped electrically, the $n = 4$ mode would appear to be favoured at the expense of mode $n = 2$.   These ideas were tested and found to be along the right lines, by " cold block " experiments in which the resonant behaviour of the anode

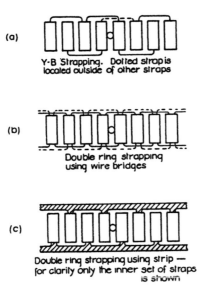

(a)

Y-B Strapping. Dotted strap is located outside of other straps

(b)

Double ring strapping using wire bridges

(c)

Double ring strapping using strip — for clarity only the inner set of straps is shown

Figure 26.—Methods of strapping. Segments are shown " unwrapped " and the position of the coupling loop is indicated by a small circle.

block was examined by injection of a steady signal.   Strapped magnetrons proved to be very free from " moding " troubles, i.e. it was difficult to produce energy at more than one frequency and the efficiency of operation was higher. A slight drawback was that strapping increased the change of frequency with load impedance.

Fig. 26 shows strapping methods which have been used in practice, unwrapped magnetron segments being shown in each case and the position of the output coupling loop indicated by a small circle. Strapping is by wire or strip bridges mounted on the ends of the anode segments. Magnetrons strapped according to these methods have been shown to operate with high efficiency in the π mode (or $n = 4$ for eight-hole blocks).

The exact effects of strapping have been investigated thoroughly by Slater[11] and it appears that the presence of straps produces frequency separation of modes and a distortion of the field pattern of unwanted modes, which tends to make operation in the unwanted mode difficult. One point which has arisen and been found useful in practice is that a break in the strapping to give " incomplete strapping " (of which Y-B strapping is an example) helps in suppressing modes near the one desired. Incomplete strapping is always arranged symmetrically with respect to either the output loop or the cathode support. Such breaks exist in the strapping of magnetrons to be described later.

An important practical point is that strapping permits some degree of frequency adjustment during assembly. The block with straps assembled, is fed with a signal of the desired frequency and the whole tuned to resonance by bending the straps. This feature is much appreciated by manufacturers who have a rigid frequency tolerance to satisfy.

5. General Characteristics

Operating Conditions : Characteristic Diagrams

As there is no theoretical guide to the precise operating conditions of a magnetron, a series of measurements is usually made and presented in the form of a characteristic diagram. The coordinates are anode voltage and anode current, and

[11] J. C. Slater, " Resonant modes of the magnetron," Massachusetts Institute of Technology Radiation Laboratory, Report 43-9, August, 1942.

contours of constant power output, efficiency, frequency, and
magnetic field may be drawn.   An example is shown in Fig. 27.
On this diagram peak readings of volts, amperes and kilowatts
are given, as this characteristic has been taken under the pulse
conditions which we have seen are necessary for high power
magnetrons.

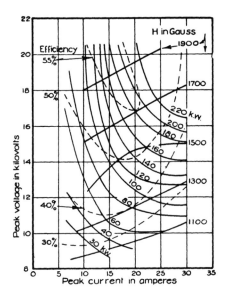

Figure 27.—Magnetron characteristics showing the relation between peak
voltage and peak current for various magnetic fields. Contours of
constant power output and efficiency are shown.

The characteristic diagram gives a good picture of magnetron
behaviour, and enables suitable operating conditions to be
fixed.   It also shows the DC resistance offered to the source
of pulse voltage, which information is often required for pulse
network design.   Discontinuities in any of the contours are
matters of importance as they expose regions of instability
which may result in " double moding."   The example given
is free from such defects.

Frequency Stability

Although the frequency of these magnetrons is nominally
fixed, we find that there is nearly always a smooth change of
frequency with magnetic field and peak current, as one would
expect from space charge movements and the associated
variations of effective dielectric constant in the working space.
Such changes are usually exposed by the characteristic diagram, and with proper control of operating conditions need
not cause trouble except in connection with the RF spectrum
referred to below.

Change of frequency with block temperature is almost
inevitable and in practice imposes a warm-up period during
which tuning adjustments may have to be made.

RF Spectrum

The energy emitted by a magnetron extends over a range
of frequencies, and the most desirable state of affairs exists
when it is concentrated in one narrow band. Excitation of
more than one mode, which unfortunately is always possible
due to some peculiarity in design, manufacture, or operation,
results in a serious loss of useful RF energy in most radar
systems. With only one mode excited, RF bandwidth depends
both on the magnetron itself and on pulse shape and duration.
For further discussion of this important point see, for example,
Collins.[12]

Effect of Load Changes on Frequency and Power Output.

Another operating difficulty may now be referred to. Particularly with magnetrons delivering full power it is found
that the frequency of oscillation is affected to a disturbing
degree by changes in the impedance presented by the feeder.
Such behaviour is not surprising when we consider that the
internal resonators are closely coupled to the external circuit.

[12] G. B. Collins (Ed.), *Microwave Magnetrons* (Massachusetts Institute of
Technology, Radiation Laboratory Series, Vol. 6), McGraw-Hill, New York,
1948, pp. 54-6, 66-9, 83-92. 135-40.

Fortunately, it is possible to minimise these effects by a proper choice of mean load impedance, and an excellent method of achieving this has been put forward by Collins,[13] who has made a systematic experimental study of the loading of a number of types of magnetron. The magnetron under test was fed through a slotted section with a voltmeter probe as used in normal impedance measuring gear, into a transformer and load which were capable of presenting a large range of impedance to the valve. With a fixed anode current, measurements of power output, frequency and sometimes anode voltage were made for various impedances which were noted implicitly in terms of standing wave ratio and distance of the voltage minimum from some reference point.

These data were plotted on a Smith chart,[14] giving one form of what is now called a Rieke diagram. An example is given in Fig. 28. The polar coordinates (r, φ) are given by

$$r = |K| = (1 - S)/(1 + S)$$
$$\text{and } \varphi = 2\theta$$

where $|K|$ is the absolute value of reflection coefficient, S is the voltage standing wave ratio, and θ is the distance of the voltage minimum from the reference point expressed in electrical degrees.

This chart then shows very clearly the effects of variation of load impedance on both frequency and power output. A feature of great significance which is common to many types of magnetrons is that the region of highest power output is near the region of greatest frequency change. This requires some compromise in choice of working point. The choice may be assisted by putting on the chart other information such as RF bandwidth, and regions of sparking, over-heating, or frequency instability. For the diagram shown, a good working point would lie near the centre, where powers of 60 to 70 watts would be obtained. In this region a change of standing wave ratio from 1·0 to 1·5 would produce a frequency change of $\pm$ 5 megacycles per second at the most.

[13] G. B. Collins (Ed.), *Microwave Magnetrons* (Massachusetts Institute of Technology, Radiation Laboratory Series, Vol. 6), McGraw-Hill, New York, 1948, pp. 40-2, 316-20.

[14] See Chapter VI, section 6.

Rieke diagrams may of course be drawn on any suitable impedance chart.   There is no need to stress their importance to the user who requires to minimise frequency variations.

The problem of frequency dependence on impedance is also being attacked from the other direction.   Attempts are being made to construct magnetrons which show no such frequency

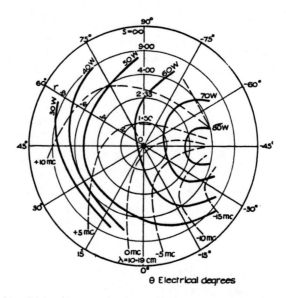

Figure 28.—Rieke diagram showing the effects of varying load impedances, on frequency and power output of a magnetron as measured by standing wave ratios and displacement of the voltage minimum.

criticality when delivering maximum power output, the method of attack being to insert a highly resonant cavity between the load and the tube.

6. General Features of the Practical Resonator Magnetron

In order to get a concrete idea of the practical fruits of development let us examine a successful type, the British CV76, which was made in very large numbers and saw wide service in the war.

The CV76 Magnetron

The operating conditions of this tube under pulse conditions are : wavelength, 10 centimetres ; magnetic flux density 0·23 webers per square metre (2300 gauss) ; peak anode current,

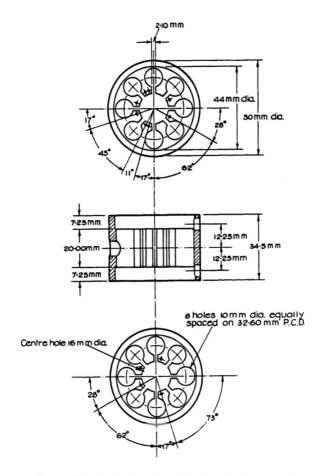

Figure 29.—Anode block used in type CV76 magnetron.

40 amperes ; peak power output, 400 kilowatts ; efficiency, 50 to 60 per cent ; pulse duration, 0·5 microseconds ; and

pulse repetition frequency 500 cycles per second. The life is of the order of 1,000 hours.

The tube with end-plates removed is shown in Plate III (b), along with the anode block. This block contains the resonant circuits and the anode cavity as in the early Birmingham magnetrons, and forms the main valve envelope. A drawing of the block is given in Fig. 29 and in view of the tolerances required, it offers an interesting problem in machining. The first blocks of this kind were turned and jig-bored, but in quantity production the body and the centre anode hole are turned, and the eight concentric holes for the resonant circuits drilled with a jig. The slots are cut in a shaping or slotting machine with proper indexing fixtures, or are broached. The side holes for the output and filament leads are jig drilled and reamed or tapped.

The mode-locking straps are bridges of copper wire, peened into holes drilled in the segments. The strapping used is that referred to above as Y-B strapping and aims at π mode excitation.

The cathode is mounted centrally on the two tungsten filament leads which are sealed to the body by the use of

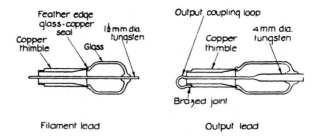

Figure 30.—Arrangements for filament and output leads of a type CV76 magnetron.

feather-edged copper thimbles, glass-copper seals, and glass-tungsten seals. The complete filament lead assembly is shown in Fig. 30, along with the similarly constructed output lead.

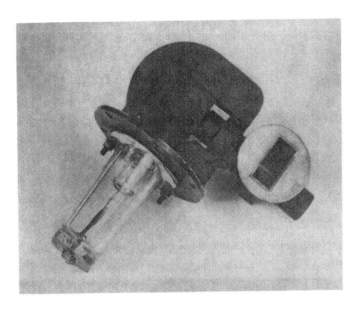

(a) The Western Electric Magnetron type 725.

(b) The Raytheon Magnetron type HK7 (see p. 64).

PLATE IV.

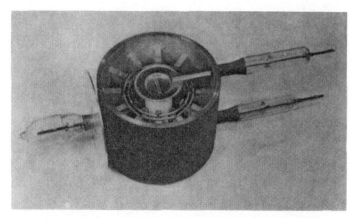

(a) Partial assembly of AWV Company magnetron type AV20 (see p. 66).

(b) An internal view of the 1·25-centimetre "rising-sun" magnetron type 3J21 (see p. 66).

[From "Radar Systems and Components", Members of the Technical Staff of Bell Telephone Laboratories, D. van Nostrand Co. Inc., 1949.

PLATE V.

The cathode is of nickel of 6 millimetres outside diameter and 0·5 millimetres wall thickness, and is fitted with end-plates. It is coated with the conventional emission mixture of double barium-strontium carbonates which on subsequent exhaust treatment give an oxide coating similar to that widely used on receiving-tube cathodes. This cathode is indirectly headed by an internal spiral of tungsten wire.

In assembly, the filament and output leads are fitted into the anode block and brazed into position. The cathode is then welded onto the filament leads. The end-plates, which are flat copper discs turned to fit the ends of the anode block, are either brazed on, or attached by the extremely useful General Electric Company (Wembley) device of a ring gasket of gold wire. In this latter method, the body with its assembled seals, and plates, and gold rings, is clamped between plates and transferred to the exhaust bench where the gold gasket forms a vacuum tight seal during the baking part of the exhaust process. The brazing processes require the use of a protective gas and chemical washing at various stages in order to keep the interior of the tube clean enough for the exhaust to be effective.

The exhaust process is carried out on a pumping system of the usual kind employing a diffusion pump backed by a mechanical pump. An exhaust schedule includes about an hour's bake at 500 degrees centigrade, followed by cathode activation and the passage of some space current—steady DC or pulsed with magnetic field. A getter is not generally used.

The external fittings are then soft soldered to the exhausted valve. These are the cooling radiator and the output thimble sleeve. In operation the valve is forced-air cooled, an air flow of the order of 50 cubic feet per minute being used. Forced air cooling is in universal use for high power magnetrons—indeed this is one of the advantages of the method of construction initiated at Birmingham.

Magnetron Cathodes

Magnetron cathodes raise several problems of great interest. High power magnetrons demand emission of the order of

10 peak amperes per square centimetre at a wavelength of 10 centimetres, and as much as 30 amperes per square centimetre at 3 centimetres. Those figures are much higher than those normally accepted for oxide cathode design (0·5 to 1·0 ampere per square centimetre). Oxide cathodes are used universally. Their high efficiency made them attractive in the first place and they have subsequently given excellent service. Two factors may contribute to such high ratings :

(a) " Back bombardment " is inseparable from magnetron operation, and oxide cathodes at normal working temperatures show marked secondary emission ; factors of 5 and greater are possible.[15]

(b) Pulse operation allows long recovery periods between short active periods.

It is interesting to note that back bombardment results in dissipation of heat at the cathode which allows most magnetrons to run with their heaters switched off once they are started. On the other hand it presents a problem of adequately cooling the cathode in small high-power magnetrons. In the CV76, therefore, the cathode is made of heavy gauge nickel (0·020 inches) welded at one end to a heavy end-plate which is itself welded to one of the filament leads so that as much conduction of heat as possible takes place.

Experience has shown that a cathode poor in emission can lead to frequency instability and sparking. This has been investigated at the M.I.T. Radiation Laboratory (by Coomes, 1944),[16] where DC pulse emission in the absence of a magnetic field was correlated with magnetron performance. Poor emission, measured in this way, was accompanied by poor performance and provided a useful test for the development of better cathodes.

Much work[17] has been done to produce long life cathodes,

[15] M. A. Pomerantz, "Magnetron cathodes," Proc. I.R.E., Vol. 34, 1946, pp. 903-10; and G. B. Collins, Microwave Magnetrons,[13] pp. 517-9.

[16] E. A. Coomes, "The pulsed properties of oxide cathodes," J. Appl. Phys., Vol. 17, 1946, pp. 647-54; and G. B. Collins, Microwave Magnetrons,[13] pp. 379, 505-6.'

[17] G. B. Collins, Microwave Magnetrons,[13] pp. 503-39; and J. B. Fisk, H. D. Hagstrum, P. L. Hartman, "The magnetron as a generator of centimetre waves," Bell Syst. Tech. J., Vol. 25, 1946, pp. 167-348.

particularly in higher frequency magnetrons. It was realised that perhaps the most predominant cause of short life was destruction of small areas on the cathode surface by sparking, which produced intense local heating. The use of nickel mesh on the cathode surface was introduced. The mesh was packed with carbonates which provided a reservoir of oxides, and by improving the conduction of heat away from a hot spot, minimised the effects of the arc. More recent developments along these lines include the use of porous metal sintered layers instead of mesh and the use of a mixture of carbonates and powdered metal (e.g. nickel), with porous layers and with mesh.

The use of end-plates on cathodes is universal. Their purpose is to confine the electrons to the active part of the valve. They can cause trouble if they give rise to electron emission, whether primary or secondary, and at times measures have to be taken to deal with this.[18]

The Output Transformer

A resonator magnetron sets a problem in microwave power transmission, since the power generated in small cavity resonators must be extracted and delivered to a feeder. Starting from a loop on one cavity, the most practicable step is to use a short length of evacuated coaxial line of small diameter. That shown on the CV76 is typical. The small line is connected to a large coaxial or waveguide feeder and as the proper load impedance must be presented to the magnetron an impedance-matching transformer is necessary. A transformer for low-power magnetrons is shown diagrammatically in Fig. 31. The slugs are adjusted to produce maximum power. Another type of output transformer for direct connection to waveguide is constructed so that the inner conductor of the coaxial output protrudes into a waveguide and acts as an aerial.

[18] G. B. Collins (Ed.), *Microwave Magnetrons* (Massachusetts Institute of Technology, Radiation Laboratory Series, Vol. 6), McGraw-Hill, New York, 1948, pp. 379, 537-9.

Magnetrons will be described later which typify more recent developments in which a fixed transformer is built into the tube itself and adjusted in the factory so that direct connection may be made to the feeder.

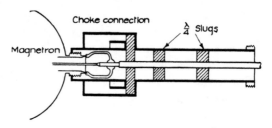

Figure 31.—Coaxial output transformer, for a low-power magnetron.

Much work has been done on magnetron output transformers (for example of this see Collins[19]). Besides presenting a suitable load to the magnetron for which they are designed, they must be mechanically satisfactory and adjustments, if provided, must not be critical.

7. Examples of Magnetron Design

Further information on the design of resonator magnetrons will be found in the literature,[20] together with details of other developments such as tunable magnetrons and so-called "packaged" versions which incorporate a built-in permanent magnet.

In order to convey some idea of the trends in modern design, a description follows of the salient features of a number of magnetrons which have come into widespread use.

A 3-Centimetre Magnetron. Western Electric Type 725

The Western Electric Type 725 is a highly successful 3-centimetre magnetron developed by the Bell Telephone Laboratory.

[19] G. B. Collins, *Microwave Magnetrons*,[13] pp. 481-98.
[20] "Cavity magnetrons," *Electronics*, Vol. 19, 1946, pp. 126-31; J. B. Fisk, H. D. Hagstrum, P. L. Hartman, "The magnetron as a generator of centimetre waves," *Bell Syst. Tech. J.*, Vol. 25, 1946, pp. 167-348; W. E. Willshaw, L. Rushforth, A., G. Stainsby, R. Latham, A. W. Balls, A. H. King, "The high-power pulsed magnetron. Development and design for radar applications," *J. Instn. Elect. Engrs.*, Vol. 93, Part III A, 1946, pp. 985-1005.

Its general appearance is shown in Plate IV (a) and its performance is as follows : wavelength, 3·2 centimetres ; magnetic flux
density, 0·55 webers per square metre (5500 gauss) ; peak
anode current, 10 amperes ; peak anode potential, 12 kilo-

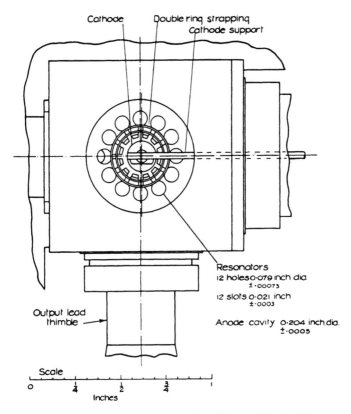

Figure 32.—The internal construction of the Western Electric Magnetron
type 725.

volts ; peak power output, 40 kilowatts ; efficiency, 30 to 40
per cent ; duty cycle, 0·001.

A drawing of the anode block is shown in Fig. 32, and
attention is drawn to the dimensions of the resonators and
anode hole. It is obvious that a formidable manufacturing
problem is presented by this minute block, yet this' valve was

put into mass production during the war. Double ring strapping is used at each end to ensure π-mode excitation. The cathode, which is indirectly heated by an internal heater, is turned from nickel rod and has a covering layer of nickel mesh impregnated with the usual oxides.

This magnetron has a built-in output transformer which is adjusted and fixed at the factory so that direct coupling can be made to rectangular waveguide with the assurance both of good power output being passed into a reasonably terminated guide, and good frequency stability.

The output transformer is shown diagrammatically in Fig. 33 from which it will be observed that the output lead, which

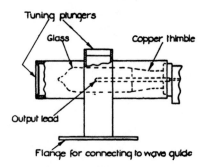

Figure 33.—Output transformer of type 725 magnetron.

acts as an aerial in the waveguide, is vacuum enclosed. This is a simple arrangement which avoids corona losses at the lead and allows one vacuum seal to be dispensed with.

Finally it is noteworthy that very thorough mechanical protection is afforded the fragile copper-glass seals by both output transformer and glass housing round the filament leads. This is a feature of great practical importance.

A High Power 10-Centimetre Magnetron : Type HK7

The HK7 is a high power magnetron which has been made in quantity by the Raytheon Manufacturing Company, and its construction is shown in Plate IV (b) and Fig. 34. Approximate performance figures are :   wavelength, 10 centimetres ; magnetic flux density, 0·22 webers per square metre (2200 gauss) ; peak anode current, 40 amperes ; peak anode potential,

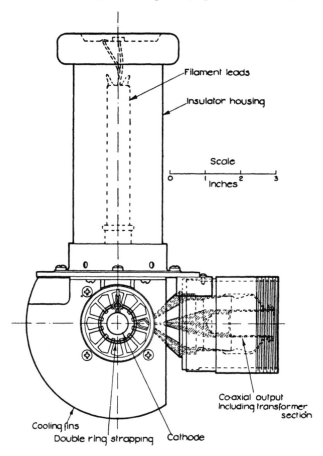

Figure 34.—Construction of the Raytheon Magnetron type HK7.

30 kilovolts ; peak power output, 750 kilowatts ; efficiency, 60 to 65 per cent.

Recent work on a valve with the same basic design as this one has produced power outputs of no less than 3·5 megawatts peak.

Points of interest in the design are : *The coaxial output* which is built to carry high power and incorporates a transformer section which gives a proper match to a 50-ohm line. Mechanically, this output connection is excellent and the glass work is well protected. *The resonators* whose peculiar shape is associated with the interesting method of constructing the anode block from stamped-out sheets approximately 0·060 inches thick, a number of which is piled together and brazed to form a solid whole. This method avoids fine machining of each block and is therefore well suited to mass production. *The filament leads* which are very long to prevent flash-over. They are also provided with corona rings and are well protected by a robust insulator housing.

A 25-Centimetre Magnetron : Type AV20

Plate V·(*a*) shows a partial assembly of an Australian magnetron which was designed in the Physics Department, University of Melbourne, and produced by the Amalgamated Wireless Valve Company. It is of the vane type, the resonator assembly being made up of copper vanes brazed to a copper shell. Double ring strapping is used, the strapping rings being formed by copper wire bridges. The design as a whole is orthodox and permits the valve to be made entirely from local materials. In its final form it is fitted with cooling fins and a filament lead housing.

A 1·25-Centimetre "Rising-Sun" Magnetron: Type 3J21

Plate V (*b*) shows some details of the 3J21 magnetron which uses the "rising-sun" anode block. This system of anode cavities gives good mode separation without straps, and provides a practical construction for the minute blocks necessary for magnetrons operating at very short wavelengths. The cavities are made out of a solid block of copper by a single hobbing operation. The waveguide output coupled to the resonators by a   window" transformer should also be noted.

CHAPTER IV

TRIODE POWER OSCILLATORS

ALTHOUGH easily the most outstanding valve developed for radar purposes has been the resonant magnetron, triodes have nevertheless played an important part as oscillators in pulse transmitters. Their contribution has been on the longer wavelengths associated with radar and in this role they are unlikely to be replaced. Since the development of triodes for very high frequency operation was not a radar development and since they are already described fully in the literature, this chapter concerns itself only with those aspects of valve design which are peculiar to pulsed operation. These are discussed first in relation to the electrical design of the valve itself and then from the point of view of the reaction of this design on the geometry of the valve and its associated circuits.

1. Electrical Design

It is fortunate that pulsed operation not only requires but permits high peak powers to be obtained from the triode valves, often at frequencies higher than is possible in continuous wave operation.

Three interrelated factors enter into production of high powers ; the cooling of the valve, the applied DC voltages and the emission of the cathode. A direct result of operating the valve over only a short portion of the total time is that the mean power to be dissipated is reduced much below the peak value, the factor being generally of the order of 1,000. Hence a given valve can handle peak powers many times greater than its normal rating, provided always that it can be made to generate these higher peaks of power. In order to increase the power output, the anode voltage and/or current must be increased. Even with conventional designs, it is possible to increase the voltage as much as 4 times and the current by a factor of perhaps 2, giving a power increase of 5 or 6 times.

This factor can be increased even more by modification of the valve design. A good vacuum and the elimination of all unnecessary insulating material within the valve ensure that high voltages will cause little internal trouble. High voltages can be tolerated external to the valve because the ions associated with incipient corona have time to disperse or can be forcibly dispersed by air blast during the relatively long rest periods between pulses. Full use can be made of this by pulsing the power supply to the valve anode—by so-called anode modulation.

It was originally thought that tungsten filaments were necessary for operation at high anode voltages, and attempts to increase the emission current were then limited by grid heating. Fortunately it was found that oxide coated cathodes could withstand quite high anode voltages under pulsed conditions and a tremendous gain in emission followed immediately. An additional advantage was the reduction in filament power which decreased the total power loss to be dissipated ; because it is continuous, the filament load always represents a considerable portion of the total.

Even with the great increase in peak power output, no very special demands are made by the cooling requirements which can always be met by the provision of radiator fins and an air blast. The overall result is that in spite of the high peak power available the valve is quite small in size, a further advantage for very high frequency operation.

2. Physical Design

The main effort put into the design of valves for radar purposes has been in the direction of increasing the upper limit of frequency and of increasing their efficiency near this limit. The main factor limiting the high frequency operation of triodes is electron transit time which in turn is related to the size of the valve and the applied potentials. It has been shown above that pulsed operation permits a decrease in size and an increase in applied voltage both of which raise the frequency limit.

The need for improved efficiency has resulted in designs which have developed along the following lines. At the desired frequencies, the interelectrode capacities within the valve represent impedances which are very much smaller than those considered optimum in low frequency practice. In the first place they can give rise to high currents and therefore high losses, and then they require small tuning inductances which are not easily realised. The capacities should therefore be kept small and of such a construction that they can be connected to the remainder of the circuit by leads of low inductance. For the same reasons no more circuit elements than are absolutely necessary for proper operation as an oscillator should be added.

A design procedure[1] is available for determining the circuit elements which should be added to give the desired frequency, output and control conditions. It remains to find a suitable physical form for these elements. The principal element is always inductive and the inductor which has the greatest physical size for a given value of inductance is of the transmission line type, which may take the form of a twin or coaxial line or at the highest frequencies may be a radial line or cavity. The valve structure must permit a simple connection to such elements and it has therefore taken the concentric form which is well illustrated by the NT99 valve (Plate VI (a)). This valve is capable of a peak power output of 40 kilowatts at frequencies in the vicinity of 600 megacycles per second ; the anode voltage is 8,000 and the filament heating power is 40 watts. It is typical of many produced by the Research Laboratory of the General Electric Company, Wembley, which played an outstanding part in the development of such valves for radar purposes.

[1] O. O. Pulley, " Ultra high frequency oscillators," C.S.I.R. Radiophysics Laboratory, Report RP 92, 1941.

An alternative construction popular at lower power levels is the plane-element type with disc seals of low inductance. It is exemplified by the CV90 and the GL446 described in Chapter XI.

In both these types the inductance of the lead from the electrode to a point external to the valve where the circuit can be connected is kept as low as possible. This allows a high proportion of the inductive tuning element to be external to the valve and hence under control. Known artifices for increasing the size of the external circuit such as the use of a line length greater than half a wavelength are permissible but seldom satisfactory.

The relatively simple method of increasing power output by the use of several valves in parallel or push-pull is much used and has a bearing on valve design because of the length and hence impedance involved in the inter-connections. The design procedure quoted above shows that in push-pull opera-

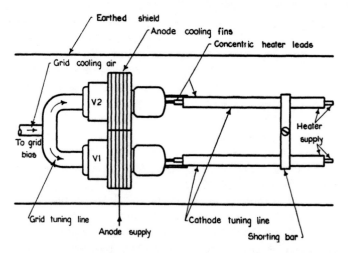

Figure 35.—Typical schematic arrangement of a push-pull oscillator for 500 megacycles per second.

tion the impedance between one pair of corresponding electrodes should preferably be zero or very small, and this can be achieved in a valve such as the NT99 by strapping the two

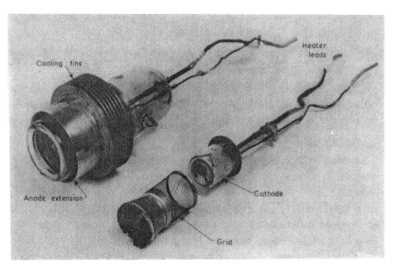

(a) Valve type NT99 (General Electric Company, Wembley) showing the concentric construction used for direct connection to concentric line tuning devices.

(b) 200 megacycles per second push-pull oscillator, showing details of valve mounting, cooling arrangements and connections to grid and filament tuning lines (see p. 71).

PLATE VI.

anodes directly together. The grid and cathode circuits then become part of a twin transmission line system. To prevent radiation they should be shielded by an extension of the anode system.

Fig. 35 shows schematically the arrangement of a push-pull oscillator suitable for operation at about 500 megacycles per second while Plate VI (b) illustrates the construction of a transmitter for a somewhat lower frequency. In the latter the grid and cathode tuning lines run backwards from the valve assembly while the valves themselves are mounted on a closed box of insulating material through which the cooling air is blown.

Coupling to the load can be effected by tapping off either the grid or cathode lines, the latter usually being more convenient. Loop coupling can also be used and in some cases is necessary in order to isolate the oscillator from its load.

Many variations in arrangement, construction, etc. have been described in published literature or can readily be devised to suit particular conditions.

CHAPTER V

MODULATORS

RADAR transmitters are required to radiate pulses of radio frequency for periods from 0·1 to 20 microseconds repeated at rates up to 2,000 pulses per second. Triode oscillators are used for transmitter frequencies up to about 1,000 megacycles per second, while for higher frequencies magnetrons are generally employed. The usual method of modulation is to apply a high voltage of rectangular waveform to the anode of the oscillator if it is a triode, or to the cathode if a magnetron. Provided the build-up and decay times are short compared with the duration of the pulse, the resultant current waveform and radio frequency envelope are similar to the applied pulse. This method of modulation corresponds to anode modulation of radio telephone practice and, as explained in Chapter IV, it allows tubes to be run at higher anode voltages and consequently with greater power output than is normal. Grid modulation of triode oscillators is seldom used because in this method the anode potential has to be applied continuously and full advantage cannot be taken of the virtues of pulse operation.

Because power is transmitted only during a small fraction of the total operating time, peak powers which are extraordinarily high compared with those of radio communication practice are achieved with moderately sized radar equipment. That fraction of the total time during which the transmitter actually operates is called the " duty cycle," and is equal to the product of pulse duration and repetition frequency. Peak values of quantities such as power can then be converted to average values by multiplying by the duty cycle. For example, a high power air warning set employing a 4 microsecond pulse at a repetition frequency of 250 pulses per second may use a magnetron accepting a peak input power of one megawatt. The duty cycle is 1/1000, so that the mean power input to the magnetron is only one kilowatt.

In practice, the output waveform of a modulator may not have the ideal rectangular shape and the permissible departure depends on the characteristics of the oscillator. A variable anode voltage will cause both magnetron and triode oscillators to change frequency, so that the receiver will not be in tune during some portion of the pulse. In addition magnetrons driven by a varying voltage may jump discontinuously to a different frequency, a phenomenon known as " double moding." This may also be caused by a rapid rise of the leading edge of the pulse and, to prevent it, modulators delivering flat-topped trapezoidal pulses have been proposed.

In the oscillating region, magnetrons have approximately constant voltage characteristics so that any modulator will produce a more rectangular voltage pulse with a magnetron than with a resistive load. The reverse is true for the current pulse.

The techniques described in this chapter have application in fields other than radar, consequently the needs of those who have not previously had access to this information have been kept in mind.

1. Basic Modulator Types

Radar modulators may be divided into two main types. The first applies a high tension supply to the oscillator anode for the duration of the pulse and then disconnects it, as illustrated in Fig. 36. The " on-off " switch is a high vacuum triode or

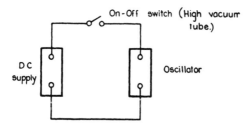

Figure 36.—Basic high vacuum tube modulator circuit.

tetrode which operates under particularly arduous conditions. Alternatively the modulator may be thought of as a power amplifier developing the high voltage pulse across a load, which is the oscillator in Fig. 37. Because of the difficulty of manufacturing suitable switch tubes, and because they operate at high voltage and low current, this type of modulator has been largely superseded. It is referred to as a "high vacuum tube" or "hard tube" modulator.

The second type of modulator uses a pulse forming network containing capacitors, the total energy of which is discharged into the oscillator by a switch which remains closed until after the discharge is complete. The switch must pass from the non-conducting to the fully conducting state in a time which is short compared with the pulse duration, but the only requirement as to its rate of opening is that it should become non-conducting in a time short compared with the interval between pulses. The ability of an open spark in air to perform this operation has led to the adoption of this type of modulator for many applications.

The basic pulse forming network is the loss-free open circuited transmission line, which is charged and connected at one end to a resistance equal to its characteristic impedance. It produces across the resistance a rectangular voltage pulse of magnitude equal to half the voltage to which the line was initially charged. The pulse terminates when the disturbance caused by the closing of the switch arrives back at the load after reflection at the open end. The waveform is determined by the network and not by the switch. The interval between pulses is available for recharging the network so the demand on the power supply is not severe. This type of modulator is referred to as a "line type modulator."

The successful development of voltage step-up transformers capable of handling high power pulses constituted a notable advance in modulator technique. Magnetrons usually operate at inconveniently high voltages of 15 to 40 kilovolts, and the provision of insulation to withstand this or a greater voltage in the power supplies and elsewhere adds considerably to the size of the equipment. With the introduction of "pulse transformers," as they are called, it became practicable to operate the main part of the modulator at the most convenient voltage and restrict the high voltage to the output terminal

A further feature of the use of pulse transformers is that they permit the separation of the modulator[1] from the transmitter, pulses being transferred by means of relatively low voltage coaxial cable. The insertion of a cable does not cause distortion of the pulse if the terminating impedances are equal to the characteristic impedance of the cable. Magnetron input impedances are about 500 to 1,000 ohms while the characteristic impedance of convenient coaxial cable is about 50 ohms so that by connecting the magnetron through a transformer of about 10/1 impedance ratio it is possible to match it to the cable. In the case of a line type modulator the pulse forming network is made equal in impedance to the coaxial cable and only one transformer is employed. A high vacuum tube modulator requires a transformer at each end of the cable since the modulator is itself a high impedance device.

In comparing the two basic modulator types the line type modulator has two salient advantages. It has a simpler switch and when used with a pulse transformer a much lower supply voltage. The high vacuum tube modulator requires a supply voltage some 15 per cent greater than the output pulse voltage, while a typical line type modulator employing DC resonant charging and a 4 : 1 step-up pulse transformer requires a supply voltage only one quarter of the output. The line type modulator can also be made simpler, smaller and more efficient. On the other hand the hard tube modulator[1] is more flexible and is therefore often used for laboratory purposes.

2. High Vacuum Tube Modulators

The basic high vacuum tube modulator circuit shown in Fig. 36 cannot be used in practice because the cathode of the switching tube is at high DC potential. Two practical modifications are shown in Fig. 37 (a) and (b), the first being transformer coupled and the second capacitor coupled. The grid

---

[1] R. H. Johnson, "Hard-valve pulse modulators for experimental use in the laboratory," *J. Instn. Elect. Engrs.*, Vol. 93, Part III A, 1946, pp. 1043-57.

of the switch tube is biased beyond cut-off during the quiescent
period and driven positive rendering the tube highly conducting
during the pulse.

If the pulse applied to the grid is exactly rectangular,
distortion still occurs in the output in the manner illustrated

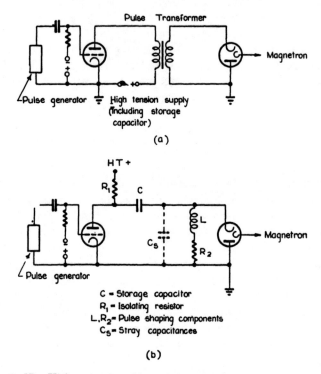

(a)

(b)

C = Storage capacitor
R₁ = Isolating resistor
L,R₂ = Pulse shaping components
Cₛ = Stray capacitances

Figure 37.—High vacuum tube modulator circuits. (a) Transformer
coupled. (b) Capacitor coupled.

in Fig. 38. The initial rise of the pulse is delayed while the
stray capacitances C_s in Fig. 37 (b) are charged through the
switching tube. The voltage falls during the pulse because the
coupling capacitor C is being discharged by the current through
the magnetron and the components L and R_2. At the end of
the pulse the switching valve becomes non-conducting and after
the voltage has fallen about 10 per cent the magnetron also be-

comes non-conducting. The stray capacitances then have to discharge through L and R_2. If the values of these components are properly chosen the discharge can be critically damped. If L is replaced by a high resistance a slow decay results. If L only is used an oscillation with C_s as capacitance is set up. Rapid damping of the oscillation may be obtained by connecting a low impedance diode in parallel with the magnetron with such a polarity that it conducts when the magnetron anode is negative with respect to its cathode. This produces heavy damping while the diode is conducting but wastes no power when the pulse is of opposite polarity. The arrangement suffers from the disadvantage that it adds another major component to the system. Fig. 38 shows the after pulse effects of the various components mentioned.

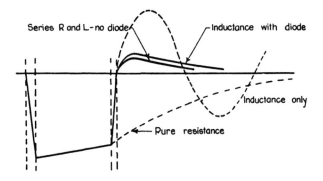

Figure 38.—Magnetron voltage pulse with various damping methods.

The positive pulse required to drive the grid of the switch tube is large and difficulty is experienced in obtaining it from a driving tube of reasonable dimensions. The difficulty arises because, in order to economise in size and power consumption, high power tubes are arranged to conduct only during the pulse. Such tubes give negative pulses to a load in the anode, and to obtain positive pulses it is necessary to couple to the cathode. With cathode coupling a large grid driving voltage, greater than the output voltage, is required. In the widely used " bootstrap " modulator circuit (Fig. 39) this difficulty is overcome by

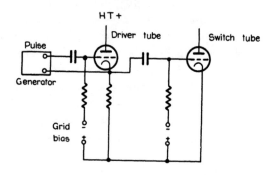

Figure 39.—Bootstrap driver circuit.

connecting a pulse generator between cathode and grid of the driver tube and arranging for the whole of the pulse generator circuits to fluctuate in potential with the driver cathode.

High Vacuum Modulator Tubes

Since the pulse required to drive a typical magnetron is of the order of 30 amperes at 20 kilovolts the switch is required to withstand a similar voltage when non-conducting and to pass a similar current when conducting. The current must be attained with a low voltage drop across the switch tube and it is desirable that the requisite grid swing between the conducting and non-conducting states should be low.

These characteristics are difficult to attain simultaneously in a tube of small size and heater power, and the successful development of such tubes has demanded extremely good high vacuum technique. Table 3 lists the characteristics of some of the tubes used for this purpose.

TABLE 3—CHARACTERISTICS OF SOME TYPICAL HIGH VACUUM
MODULATOR TUBES

Tube Type		Cathode	Cathode power watts	Maximum anode voltage	Peak anode current amperes
NT100	tetrode	oxide	48	12,000	10
829A	tetrode	oxide	14	5,000	10
715B	tetrode	oxide	58	15,000	15
304TH	triode	thoriated	130	18,000	7.5
6C21	triode	thoriated	150	33,000	15

3. Line Type Modulators[2]

The essentials of the line type modulator are the pulse forming
network, the circuit for charging it, and the switch which
discharges the network into the load. Rotary spark gaps,
fixed triggered gaps both open and closed, and thyratrons
have found application as switches, and the comparative ease
with which satisfactory switches can be obtained has, among
other reasons, resulted in the line type modulator almost
completely superseding the high vacuum tube modulator.

Theory of Discharge of a Transmission Line

The development of the pulse forming network followed on
the realisation that a uniform, low-loss, open-circuited transmission line gives a rectangular pulse when discharged into
a resistance equal to its characteristic impedance.

Consider the case of Fig. 40 in which a line of characteristic
impedance Z_o is charged to a voltage V_o and connected to a
resistor R of magnitude Z_o at time $t = 0$. For a short interval
the line behaves as a generator of electromotive force V_o and
internal impedance Z_o applying a voltage $V_o/2$ to the load.
The initial voltage drop from V_o to $V_o/2$ gives rise to a rectangular wave of voltage $V_o/2$ travelling away from the load.
This travelling wave is completely reflected without change of
sign at the open end and travels back to the load, where it

[2] K. J. R. Wilkinson, "Some developments in high-power modulators
for radar," *J. Instn. Elect. Engrs.*, Vol. 93, Part III A, 1946, pp. 1090-112.

reduces the voltage to zero. The system is then completely discharged, the duration of a pulse being equal to twice the time of travel of a wave along the line.

If R is not equal to Z_o, the discharge will reach completion

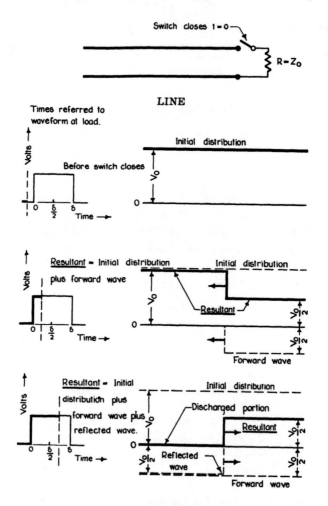

TRAVELLING WAVES

Figure 40.—Discharge of a matched transmission line showing travelling waves.

asymptotically. To find the conditions at any given time the steady voltage originally present and all travelling waves must be superposed.

Let V be the voltage across R. Immediately after closing the switch only one travelling wave of magnitude $V_o - V$ is present. The line current $(V_o - V)/Z_o$ equals the load current V/R so that

$$V = \frac{R}{R + Z_o}\, V_o.$$

This state persists until time $t = \delta$ when the wave reflected from the open end reaches the load. This wave of magnitude $V_o - V$ or $\dfrac{Z_o}{R + Z_o}\, V_o$ is partially reflected at the load, the reflection coefficient being $\dfrac{R - Z_o}{R + Z_o}$, so that in the interval $\delta < t < 2\delta$

$$V = \left(\frac{R}{R + Z_o} - \frac{Z_o}{R + Z_o} - \frac{Z_o}{R + Z_o} \cdot \frac{R - Z_o}{R + Z_o} \right) V_o$$

$$= \frac{R(R - Z_o)}{(R + Z_o)^2}\, V_o.$$

Continuing the calculation, a series of decreasing steps is

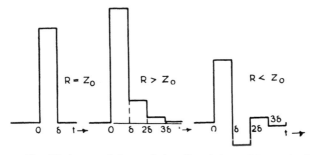

Figure 41.—Discharge of transmission line into resistance loads of different magnitudes.

found, which if $R > Z_o$ are of one sign, or if $R < Z_o$ are of alternate signs as in Fig. 41.

If the load is not a fixed resistance, but has a non-linear voltage-current characteristic independent of the rate at which it is traced out, a stepped waveform will still result. This case, which may be treated graphically, is important in view of the marked non-linearity of magnetron characteristics. Alternatively, the closing of the switch may be regarded as the application of a step-shaped voltage between its contacts. The alternating current impedance of the network which gives a current response of rectangular waveform when such a voltage is applied may be calculated by a well-known theorem.[3] It transpires that a uniform open circuited transmission line in series with a load resistance equal to the characteristic impedance is required.

Networks Equivalent to the Uniform Line[4]

An actual transmission line to produce pulses of the length commonly used in radar (1 microsecond or more) would be of inconvenient dimensions. Hence networks consisting of lumped elements are used to approximate the performance of the open circuited line.

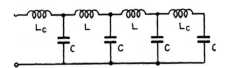

Figure 42.—Low pass filter pulse forming network.

Low Pass Filter Pulse Forming Network. The simplest and most commonly used equivalent network is the low pass filter shown in Fig. 42. Constructional advantages of this network

[3] The current response $A(t)$ of a network to a unit step-shaped applied voltage is related to the corresponding alternating current impedance $Z(j\omega)$ by the equation

$$\frac{1}{aZ(a)} = \int_0^\infty e^{-a\lambda} A(\lambda)d\lambda$$

in which a is real and positive. Assuming a unit current pulse of duration δ for $A(t)$, we have $Z(j\omega) = \frac{1}{2} + \frac{1}{2} \coth j\omega\delta/2$.

[4] E. L. C. White, "The use of delay networks in pulse formation," *J. Instn. Elect. Engrs.*, Vol. 93, Part III A, 1946, pp. 312-4.

are that all capacitors have the same value and the inductors can be wound as a continuous coil on a single former and tapped where required. Two or more of these networks may be connected in tandem to give a choice of pulse duration. The following relations are satisfactory for design purposes

$$Z_o = \sqrt{\frac{L}{C}}$$

$$\delta = 2n\sqrt{LC}$$

where Z_o = characteristic impedance of network at zero frequency (ohms), δ = pulse duration (seconds), L = inductance per section (henries), L_c = end inductance (henries), C = capacity per section (farads), n = number of sections. Empirically it is found that good pulse shapes can be obtained with 3 or 4 sections when L_c/L = 1·1 to 1·2.

For high voltage networks the capacitor and coil assembly is usually placed in a sealed tank with terminals brought out as required. Oil impregnated paper dieletric is most common, although for some low voltage networks special organic-inorganic dielectric capacitors moulded into a solid block have been developed. Plate VII (a) shows some typical low pass filter networks.

Bartlett Pulse Forming Network. Bartlett of the General Electric Company obtained several possible pulse forming networks by expanding the expression for the input impedance and admittance of a uniform open circuited transmission line in various ways. Only the circuit which has found most application will be described here.

The input impedance of the open circuited transmission line of characteristic impedance Z_o in series with its characteristic impedance is

$$Z_{IN} = Z_o + Z_o \coth \frac{j\omega\delta}{2}$$

where ω is the angular frequency and δ is the time taken for a wave to travel to the far end and back. In terms of the partial-fractions expansion of the hyperbolic cotangent, viz.

$$\coth x = \frac{1}{x} + \sum_{n=1}^{\infty} \frac{2x}{n^2\pi^2 + x^2},$$

we have

$$Z_{IN} = Z_o + \frac{2Z_o}{j\omega\delta} + \sum_{n=1}^{\infty} \frac{j\omega\delta Z_o}{n^2\pi^2 - \omega^2\delta^2/4}.$$

The first term is the series resistance Z_o, the second is the reactance of a capacitance equal to $\delta/2Z_o$, and the nth term of the series represents the reactance of a parallel combination of a capacitance $\delta/4Z_o$ with an inductance $Z_o\delta/n^2\pi^2$. Fig. 43 illustrates the equivalent circuit.

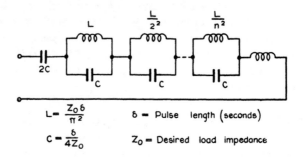

$$L = \frac{Z_o\delta}{\pi^2}$$ δ = Pulse length (seconds)

$$C = \frac{\delta}{4Z_o}$$ Z_0 = Desired load impedance

Figure 43.—Bartlett pulse forming network.

The large storage capacitor $2C$ must withstand twice, and the others $2/n\pi$ times the pulse voltage under normal conditions. If for any reason the transmitter becomes short circuited, however, the voltage across the small capacitors will reach twice the quoted values and spark gaps may be needed for protection. In practice four or five parallel circuits are used and a series inductance added, its value being determined experimentally. This network has the merits that

(a) Pulse forming networks.

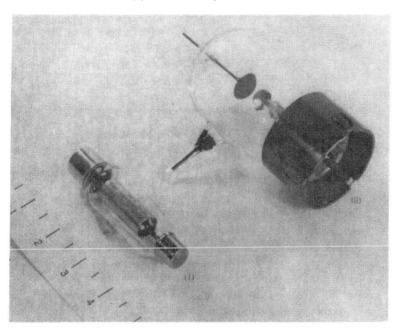

(b) Enclosed fixed triggered gaps.　(i) 1B22.　(ii) CV85 (see pp. 89, 90).

PLATE VII.

facing p. 84]

only one capacitor has to withstand the full voltage and the
pulse forming capacitors are all equal in value. It is therefore
suitable for construction from standard components and is
convenient for experimental purposes. However, these features are not important in a design intended for production,
and the disadvantage that the pulse forming capacitors contribute nothing to the stored energy renders the Bartlett line
inferior to the low-pass filter network.

Special Arrangements for Saving Insulation

The arrangements described above result in a pulse voltage
equal to half the voltage to which the network is charged,
so that the network must be insulated to withstand twice the
voltage of the output pulse. The Blumlein circuit shown in
Fig. 44 gives a pulse voltage equal to the voltage to which

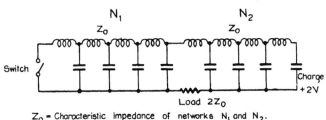

Z_0 = Characteristic impedance of networks N_1 and N_2.

The electrical lengths of N_1 and N_2 are equal.

Figure 44.—Blumlein Circuit.

the lines are charged. The load impedance should equal twice
the characteristic impedance of each of the pulse forming
networks.

Another useful method of economising insulation is to charge
two pulse forming networks in parallel and switch them to
series connection for discharge through the load after the
fashion of the Marx high voltage impulse circuit, as indicated
in Fig. 45. Switch No. 1 operates first and the sudden rise

in voltage fires switch No. 2. The " hold-off " chokes must have sufficient inductance to prevent appreciable current building up in them during the pulse. Arrangements such as these have been superseded by the use of pulse transformers but they may still be found useful for experimental purposes when a transformer is not available.

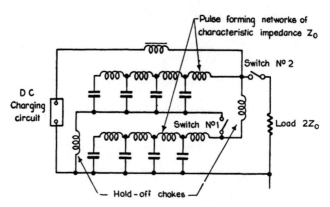

Figure 45.—Parallel charge, series discharge, circuit.

Switches Used in Line Type Modulators[5]

As mentioned earlier the requirements for the switch in a line type modulator are that it should close quickly and remain closed for the duration of the pulse. It should then open in time to permit the charging of the network for the next pulse.

Rotary Spark Gap. The rotary spark gap consists of a set of rotating electrodes which pass close by one or more fixed electrodes. Near the point of closest approach a spark passes which closes the switch. Current ceases on the discharge of the network and the electrodes separate, opening the switch before the voltage again builds up in preparation for the next spark. The repetition frequency is determined by the speed of rotation and the number of electrodes ; for example, a speed of rotation of 3,000 revolutions per minute or 50 per second and ten moving electrodes passing one fixed electrode gives a repetition

[5] F. S. Goucher, J. R. Haynes, W. A. Depp, E. J. Ryder, "Spark gap switches for radar," *Bell Syst. Tech. J.*, Vol. 25, 1946, pp. 563-602.

frequency of 500 cycles per second. Rotary gaps are not suitable for frequencies greater than about 1,000 cycles per second owing to the difficulty of separating the electrodes sufficiently rapidly to eliminate premature sparking while the network is being charged. Two types of construction are used. The first employs short rods mounted on an insulated disc which carries them between two fixed electrodes. The spark then jumps two short gaps in series. The second uses a metal disc to carry the moving electrodes, the disc being earthed by means of a carbon brush or collector ring. These are illustrated in Fig. 46.

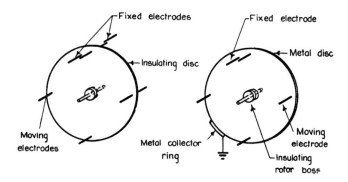

Figure 46.—Two types of rotary spark gap.

The operation of rotary gaps is satisfactory in the range from about 8 to 25 kilovolts. Large currents can flow, the only ill effects being wear of the electrodes. With tungsten electrodes and currents of a few hundred amperes the electrodes require adjustment every thousand hours. Breakdown of a gap is complete in a time of the order of 10^{-8} seconds after its initiation. The residual voltage across the gap is so small that it is difficult to measure and is probably less than 100 volts. Rotary gaps operate effectively in the open air but, unless of low power, are very noisy. It is therefore necessary to enclose them, provision being made for removing the products of the spark (ozone and oxides of nitrogen) from parts which may corrode.

The instant of sparking is irregular, a variation of the interval between pulses of ± 30 microseconds being usual. This makes it necessary to trigger the time base of a radar equipment from the modulator and precludes the use of simple rotary gaps in certain applications which require precise timing.

The ease of construction, simplicity of adjustment, and reliability of the rotary spark gap have contributed largely to the success of line type modulators.

Open Fixed Triggered Gaps. Triggered gaps were developed at the same time as rotary gaps with the object of eliminating moving parts and irregularity in firing. A common type illustrated in Fig. 47 consists of two spherical molybdenum electrodes with a tungsten electrode protruding through a hole

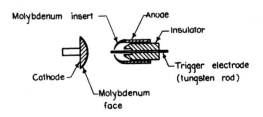

Figure 47.—Open fixed triggered gap.

in the anode. The spacing between the spheres is normally sufficient to prevent a spark occurring. At the desired moment a low power discharge is passed between the trigger electrode and the sphere through which it projects. The resulting ions and ultra violet light cause breakdown of the main gap. After discharge is complete the ionisation decays and the gap is again able to withstand the applied voltage. The switch illustrated is intended for a high power modulator and can handle frequencies from 100 to 2,000 cycles per second. Adjustments are required every 100 hours or so and the life is much less than for a rotary spark gap. The range of voltage for satisfactory operation is restricted but can be extended by blowing or pressurising the gap. A spark gap of this

type developed by Metropolitan Vickers Limited could switch
2 megawatts at 18 kilovolts at repetition frequencies up
to 2,000 cycles per second. The switch fired within 0·1 microseconds of the application of the trigger pulse.

A typical trigger circuit is shown in Fig. 48. A pulse is
applied to the grid of the tube, driving it positive so that
a current builds up in the inductance L. The grid is then

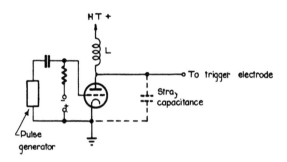

Figure 48.—Trigger circuit.

driven suddenly negative, cutting off the plate current and
producing across the inductance a high voltage pulse which is
applied to the trigger electrode.

Open gaps, both fixed and rotary, if used for airborne
applications, must be housed in an air-tight enclosure to overcome the variations in atmospheric pressure encountered.
This difficulty may be avoided by the use of an enclosed gap.[6]

Enclosed Fixed Triggered Gaps. There are two main types of
enclosed fixed triggered gaps in general use. One type, (ii) in
Plate VII (b) of which the CV85 is typical, is a modification
of the open triggered gap and has the electrodes enclosed in
a mixture of argon and oxygen. Time jitter of the discharge is
of the order of 0·1 microseconds. The other type consists of
two aluminium electrodes enclosed in a mixture of hydrogen and
argon at a little less than atmospheric pressure. Because no

---

[6] J. D. Craggs, M. E. Haine, J. M. Meek, "Development of triggered spark
gaps for high-power modulators," *J. Instn. Elect. Engrs.*, Vol. 93, Part IIIA,
1946, pp. 963-76.

trigger electrode is provided, the tubes must be used two or more in series with the trigger voltage applied at the common point of the tubes. Like the CV85, this tube, the 1B22 (i) in Plate VII (b) handles powers of about 100 kilowatts.

Thyratrons. The characteristics of a thyratron appear to make it ideally suited for application as a line type modulator switch. However, the high values and rapid rates of rise of currents proved beyond the capabilities of the types previously available. More recent work has led to the development of new forms which have satisfactory characteristics but which are difficult to manufacture. Of these, hydrogen thyratrons[7] appear the best and as low power tubes are used extensively. Hydrogen thyratrons have also been designed for power levels up to 15 megawatts. One tube, the 4C35, has a life of 900 hours at 8,500 peak volts and 90 amperes peak current with a 0·5 microsecond pulse recurring 2,000 times per second. The trigger voltage required is 150 volts rising at a minimum rate of 150 volts per microsecond. The time jitter is less than 0·04 microseconds using an AC heater supply and can be made undetectable by the provision of a DC heater supply.

4. Line Type Modulator Charging Circuits

At the cessation of a pulse, a pulse forming network is completely discharged and must be recharged before the commencement of the next pulse. Since the charging period is long compared with the pulse duration, the pulse forming network behaves like a simple capacitor as far as the charging circuit is concerned. Since this capacitor stores the energy for the pulse, its capacitance is given by

$$\tfrac{1}{2}CV_o^2 = \frac{\delta V^2}{R}$$

where C = line capacitance (farads), V = pulse voltage,

    [7] R. H. Wittenberg, "Thyratrons in radar modulator service," *R.C.A. Rev.*, Vol. 10, 1949, pp. 116-33; and H. de B. Knight, L. Herbert, "Development of mercury-vapour thyratrons for radar modulator service," *J. Instn. Elect. Engrs.*, Vol. 93, Part IIIA, 1946, pp. 949-62.

V_o = voltage to which network is charged, R = load resistance (ohms), δ = pulse duration (seconds).

The usual network is equivalent to a simple transmission line so that $V/V_o = \frac{1}{2}$ and $C = \delta/2R$. The average power which must be supplied to the pulse forming network is $\frac{1}{2}f_r C V_o^2$ where f_r is the repetition frequency.

If the pulse repetition frequency is different from the AC mains frequency, the pulse forming network can be charged from a conventional rectifier circuit with filter capacitors. Such a method is referred to as DC charging. If the repetition and mains frequencies are the same or simply related (e.g. half or twice) the filter circuits and often the rectifier may be dispensed with. This is called AC charging. In all cases the components connecting the power supply and the network must be capable of isolating the one from the other during and immediately after the pulse.

DC Charging Circuits

Resistance Charging. The simplest method of charging a pulse forming network is through a resistance connected in place of the inductance in Fig. 49. The efficiency of the system

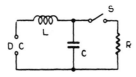

Figure 49.—DC inductance charging—basic circuit.

cannot exceed 50 per cent so it is seldom used except for low power modulators or for experimental purposes.

Inductance Charging. Fig. 49 shows the basic circuit for DC inductance charging, an arrangement which does not suffer from the intrinsic inefficiency of resistance charging. There are three main cases to consider : resonant charging, linear charging and diode charging. In all cases the network is

charged to a voltage equal to twice the DC supply voltage
if losses are neglected. In practice the voltage step-up is
about 1·9 times.

Resonant charging refers to the case in which the resonant
frequency of L and C in the circuit is equal to half the repetition
frequency f_r so that the charging current I_c and network
voltage V_c pass through one half cycle between pulses. Fig. 50
shows the current and voltage waveforms of the charging cycle.

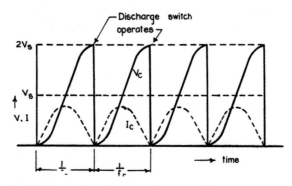

Figure 50.—Voltage and current waveforms in DC " resonant charging."

Since the network voltage remains almost constant for an
appreciable period about the time the switch operates, this type
of charging wave is suitable for use in modulators where the
switch has a time jitter, such as a rotary spark gap. No over-
voltage occurs if the switch fails to operate.

If the resonant frequency f is low compared with half the
repetition frequency f_r, so called " linear charging " results.
The large inductance used tends to make the charging current
constant which in turn charges the network approximately
linearly. Fig. 51 shows typical waveforms obtained with
linear charging. This method allows the pulse repetition rate
to be varied over a wide range without changing the value
of the inductance. Satisfactory results are obtained if the
inductance is made about four times that required for resonant
charging at the lowest repetition frequency. If the switch
should fail to operate, the voltage rises in the manner indicated

by the dotted line in Fig. 51.   If this is excessive the network
can be protected with a spark gap.   A thyratron or fixed
gap should be used for the switch as irregular firing results
in fluctuations of the output pulse voltage.

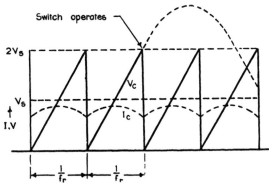

Figure 51.—Voltage and current waveforms in DC ' linear charging."

So called " diode charging " results from the use of a diode
in series with an inductance less than that required for resonant charging.   The network charges in half the resonant
period of L and C, which is less than the interval between pulses,

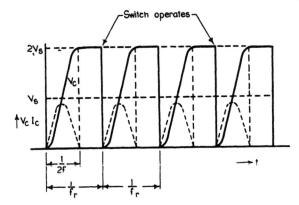

Figure 52.—Circuit and current voltage waveforms in DC "diode charging."

and the diode then prevents the discharge of the network till
the switch operates.   The voltage and current waveform are
shown in Fig. 52.   This circuit has the advantage that it can

handle a wide range of repetition frequencies and, as the voltage is steady prior to discharge, the irregular firing of a rotary spark gap does not produce pulses of irregular amplitude. It suffers from the disadvantage of introducing another high voltage component.

The inductances used in DC charging can be quite readily designed from a knowledge of the incremental permeability of the core material for the currents flowing. The iron core is best utilised without an air gap and at a value of magnetising force of about 4,000 amperes per metre (50 oersteds) at the current peak. The inductance values range from one to several hundred henries depending on the requirements.

DC Power Supplies for Line Type Modulators

Modulators use conventional rectifier circuits to convert AC to DC power, but the problem of the exact calculation of "ripple voltage," or the variation in the voltage to which the pulse forming network is charged, is often difficult. Design information obtained experimentally is given in Fig. 53 for a simple condenser input filter with inductance charging of the network. As given it applies to a half wave rectifier; for a full wave rectifier the supply frequency must be doubled.

If the supply frequency is much less than the repetition frequency, the filter capacitor must be sufficiently large to charge the network several times without its voltage falling unduly. If the supply frequency is higher than the repetition frequency a great reduction in the size of the filter capacitor is possible, since it will receive one or more current impulses from the rectifier during each charging interval of the network. For example, if a 50 cycle full wave rectifier charges a network 500 times per second, a simple calculation shows that the capacitance must be 1,000 times that of the network if the ripple voltage is to be 1 per cent. However, if the supply frequency is about 500 cycles per second, it need only equal

that of the network.   The economy resulting from the use of
a high supply frequency applies also to the rectifier transformer
and other parts of the radar equipment.

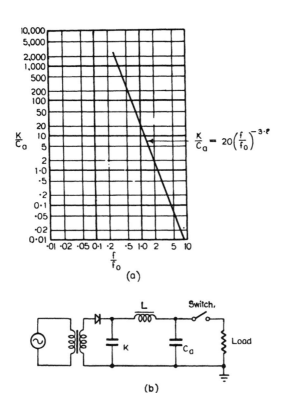

$$\frac{K}{C_a} = 20\left(\frac{f}{f_0}\right)^{-3.7}$$

(a)

(b)

Figure 53.—Ratio of filter capacitance to network capacitance for 1% ripple.

K = filter capacitance
C_a = network capacitance
f = supply frequency
f_0 = repetition frequency.

(a) Experimental relation between C and K. (b) Basic circuit.

Charging from Low Voltage DC Source

If low voltage DC only is available, say from a generator or
battery, the circuit in Fig. 54 enables a pulse forming network

to be charged to any required voltage. The arrangement is essentially the same as for DC inductance charging, except that the inductance used is the transformer leakage inductance. In addition the voltage step-up at the pulse

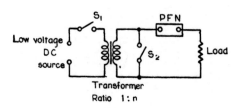

Figure 54.—Basic circuit for DC charging from low voltage source.

forming network is increased by the transformer turns ratio. The switch S_1 is closed during the charging period and opens just before switch S_2 closes to discharge the network into the load. S_1 closes again after the pulse and the cycle is repeated. Switch S_1 is a low voltage and switch S_2 is a high voltage switch. As they must be accurately triggered, thyratrons are the most satisfactory switches for this purpose.

AC Charging Circuits[8]

In the circuits described as AC charging circuits the filter capacitors are omitted, the pulse network being charged by a transformer through a suitable isolating device. The power supply frequency must then be equal to, or simply related to, the pulse repetition frequency. If a repetition frequency of some hundreds of cycles per second is required, it is usual to include a high frequency alternator in the modulator. This alternator may supply power to the whole radar set thus economising in the size of other components. If a rotary spark gap is used, it can be mounted on the alternator

[8] K. J. R. Wilkinson, "Some Developments in high-power modulators for radar," *J. Instn. Elect. Engrs.*, Vol. 93, Part IIIA, 1946, pp. 1090-112.

shaft with provision for correctly phasing the instant of discharge of the network with respect to the output voltage of the alternator.

Fig. 55 shows a practical arrangement in which a diode is used to isolate the pulse forming network at the time of discharge. The figure gives waveforms and indicates the switch

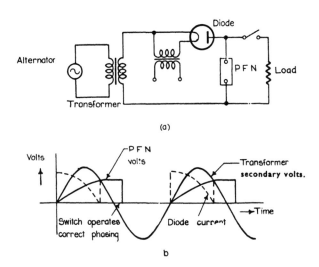

Figure 55.—(a) AC diode charging circuit.   (b) Waveforms.

phasing which keeps the diode inverse voltage as low as possible. This circuit is simple to adjust but since power is supplied to the network during a quarter of a cycle only the root-mean-square current is unduly great and the transformer is not used to the best advantage.

The diode and its filament transformer are high voltage components which can be dispensed with, as in the more efficient arrangement of Fig. 56 (a) which uses an inductance to isolate the network from the transformer. An equivalent circuit is shown in Fig. 56 (b) in which all quantities are referred to the transformer secondary terminals. The inductance L is the equivalent inductance including contributions from leakage in the transformer and the generator. The resistance is usually neglected. The performance may be described in terms of the ratio, k, of the resonant frequency of L and C to the supply frequency.

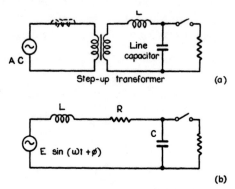

Figure 56.—AC inductance charging circuits. (*a*) Actual circuit. (*b*) Equivalent circuit.

If these frequencies are equal the condition is known as "AC resonant charging." In this case the pulse forming network is charged to a voltage π times the equivalent electromotive force. The switch operates as the electromotive force passes through zero, and the network is completely recharged in one cycle, the process being repeated every cycle. The pulse

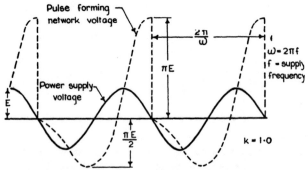

Figure 57.—AC resonant charging voltage waveforms.

network polarity depends on the zero of the applied electromotive force chosen. The transformer secondary carries a direct component of current equal to approximately one third

of the root-mean-square current. Voltage waveforms are
shown in Fig. 57. It is obvious that if the switch fails to
operate large voltages will tend to build up but this may be
prevented by fitting protective spark gaps.

In general the voltage to which the pulse network is charged
will depend on k in the manner shown in Fig. 58 provided the
switch fires at the optimum phase. When $k = 1\cdot4$ the greatest

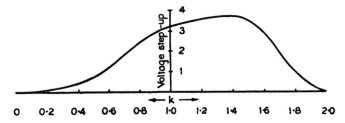

Figure 58.—AC charging. Voltage step-up as a function of k.

step-up occurs, being $3\cdot7$ as against π for the resonant case.
To obtain this step-up ratio the network should be discharged
114 electrical degrees past the zero of the applied electromotive force. Waveforms for this case are shown in Fig. 59.

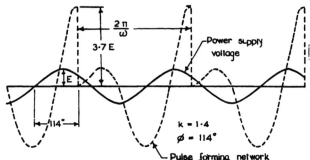

Figure 59.—AC charging waveforms for $k = 1\cdot4$.

A feature of AC charging is the remarkable economy of
components it permits. The accompanying disadvantage is
lack of flexibility in pulse length and repetition frequency,
which is serious in the case of laboratory equipment, but is
often of no importance for service use. Fig. 60 gives a

summary of the main characteristics of the modulator charging circuits discussed above.

Title	Circuit	Typical Waveform	Remarks
D C Resistance Charging	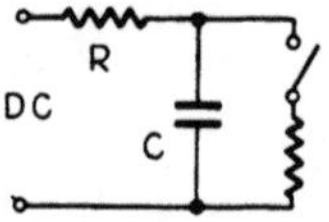		Efficiency low
D C Inductance Charging (a) Resonant (b) Linear	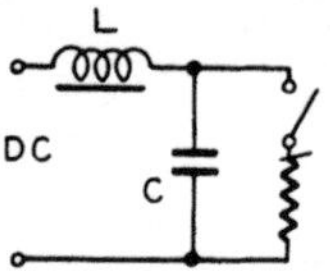	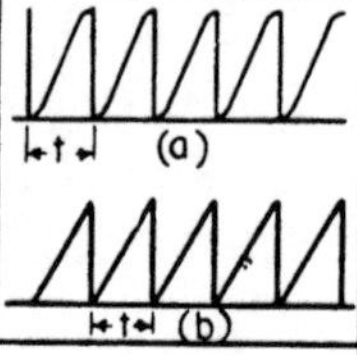	(a) $\pi\sqrt{LC} \doteq t$ (b) $\pi\sqrt{LC} \gg t$
D C Inductance Diode Charging	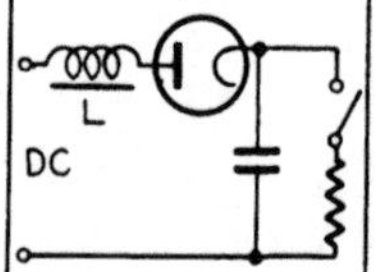	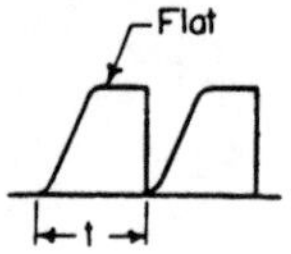	$\pi\sqrt{LC} < t$
D C Transformation Charging	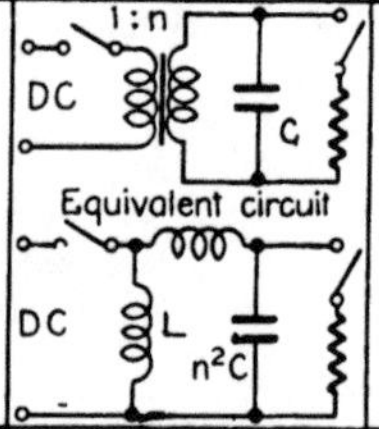		$n\pi\sqrt{LC} < t$ L = Transformer leakage inductance Operated from low voltage DC source
A C Diode Charging	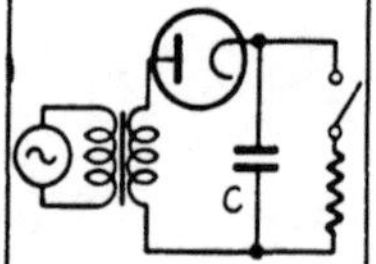	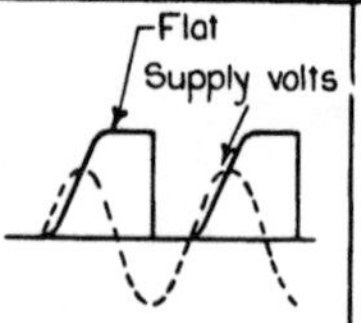	
A C Resonant Charging	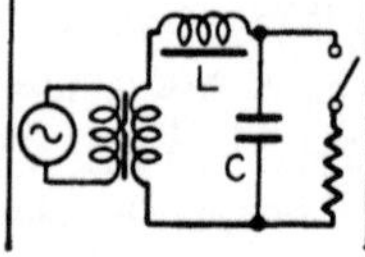	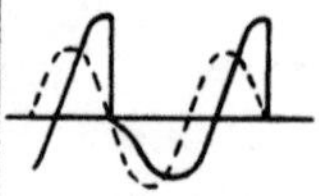	L may be the transformer leakage inductance

Figure 60.—Summary of characteristics of modulator charging circuits.

5. Pulse Transformers

As noted earlier, the introduction of pulse transformers in conjunction with line type modulators permitted a great saving in bulk and weight and also made it possible to separate the modulator from the transmitter.

Pulse transformers may be required to transmit rectangular pulses of the order of 1 microsecond duration and 1 megawatt peak power without significant distortion. A standard input impedance is 50 ohms while the output impedance when feeding a magnetron is about 800 ohms. For a power output of 1 megawatt the input voltage is therefore 7000 and the output 28000 volts. These requirements introduce serious design difficulties both in the reproduction of such short pulses and in the maintenance of adequate insulation.

The problem can be more clearly formulated if we consider the approximate equivalent circuit of a transformer given in Fig. 61. In this circuit all quantities are referred to the primary in the conventional way. The leakage inductance is represented by L_L, the primary inductance by L_p, and C represents the equivalent capacitance across the secondary due to winding capacitances and strays in both the transformer and load. The losses and the more complex stray capacitance distribution which exist in practice are neglected.

If the distortion is not great, the effects due to leakage inductance and stray capacitance may be separated from those due to the primary inductance. The former distort the rise and fall of the pulse causing them to take the form of damped oscillations associated with the circuit R_1, R_2, L_L and C, which is obtained by omitting L_p from Fig. 61.

The primary inductance draws an increasing current through R_1 during the pulse and this causes a fall in the top of the pulse and an overshoot at the end as indicated in Fig. 61. Both types of distortion may be determined quantitatively if the circuit elements are known. To reduce distortion it is necessary to achieve a combination of high primary inductance, low leakage inductance and low winding capacitance. These conflicting requirements have been met by paying close attention to the quality of the magnetic circuit so that physical size can be reduced to a minimum while maintaining the necessary primary inductance. This leads to a reduction of capacitance and leakage inductance but increases the insulation problem.

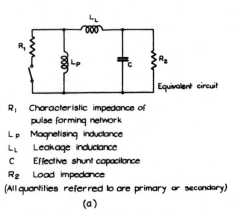

R₁   Characteristic impedance of
        pulse forming network
L p   Magnetising inductance
L_L   Leakage inductance
C   Effective shunt capacitance
R₂   Load impedance
(All quantities referred to are primary or secondary)

(a)

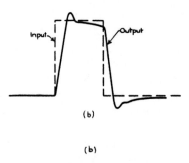

(b)

(b)

Figure 61.—(a) Equivalent circuit of pulse transformer.   (b) Input and output waveforms.

If a core of normal construction were used in a pulse transformer, skin effect would severely reduce its effective permeability. This difficulty is overcome by forming the core from exceptionally thin laminations and by using a core material having a high resistivity which tends to increase the depth of penetration at high frequencies.

In designing a pulse transformer[9] the primary, secondary and leakage inductances are estimated in the manner used for normal transformers. The effective winding capacitance is determined by assuming a voltage distribution on the windings, estimating the capacitative energy associated with significant sections of them, and equating the sum of these energies to $\frac{1}{2}CV^2$ in the equivalent circuit.

Design Procedure

Data required when designing a pulse transformer include pulse duration, primary and secondary pulse voltages and currents, and the admissible peak value of magnetising current (of the order of 10 per cent of the desired pulse current). The design steps follow broadly the sequence set out below :

(a) The best available core material is selected, and suitable peak values of magnetic induction B and magnetic intensity H are chosen using curves obtained under actual pulse conditions.

(b) The core shape is chosen empirically after the core volume is calculated from equation (1) which is obtained by equating the magnetic energy of the core to the energy delivered to it.

$$A\,l \times B\,H \;=\; V_p \delta I_m \tag{1}$$

where A = cross sectional area of core (square metres)
     l = core length (metres)
     B = peak magnetic induction (webers per square metre)
     H = peak magnetic field intensity (amperes per metre)

[9] W. S. Melville, "Theory and design of high-power pulse transformers," *J. Instn. Elect. Engrs.*, Vol. 93, Part IIIA, 1946, pp. 1063-80; N. F. Moody, "Low-power pulse transformers," *J. Instn. Elect. Engrs.*, Vol. 93, Part IIIA, 1946, pp. 311-4; R. Lee, *Electric Transformers and Circuits*, John Wiley and Sons, New York, 1947, pp. 219-57.

V_p = primary pulse voltage

δ = pulse duration (seconds)

I_m = peak magnetising current (amperes).

(c) The number of turns required on the primary and secondary windings is determined from equation (2) and wire sizes are allotted to give suitable current densities.

$$V_p\delta = N_pAB \qquad (2)$$

where N_p is the number of primary turns.

(d) An arrangement of the windings is chosen to fit the core window and to give adequate insulation. The different portions of the windings are interleaved so as to keep leakage inductances and winding capacitances low.

(e) If considered necessary, leakage inductance and winding capacitances can be estimated by methods outlined above and their effect on the output waveform investigated.

Construction

The Core. Core materials combine high resistivity with high permeability at high values of magnetic induction. They are made in laminations about 0·002 inches thick. Typical B-H curves for complete cores when 1 microsecond pulses are used are given in Fig. 62.

In order to obtain a magnetic circuit of the lowest reluctance special precautions must be taken to minimise the effect of joints in the core. Some designs use continuous cores, but special coil winding methods must then be adopted. This difficulty is overcome in one interesting construction in which a continuous core is first wound from strip around a rectangular mandrel and then cut to permit the insertion of coils

wound in the normal way. The core is rejoined after the
faces of the joints have been ground flat and etched to remove
inter-laminar short circuits.

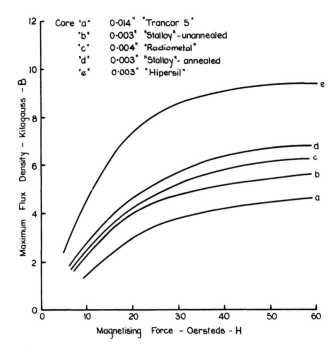

Figure 62.—B-H curves for pulse transformer core materials.

Windings. The winding configuration is determined by the
conflicting requirements of low leakage inductance   which
requires windings in close proximity, and low capacity and
high insulation which require the separation of points differing
appreciably in potential. Separated single layer windings
are used. As most of the heat dissipated in pulse transformers
is due to core losses, very high current densities are used in the
windings viz. 5000 amperes per square inch for continuous
loads and 3000 amperes per square inch for pulse currents.
   A special feature of pulse transformers designed for driving
magnetrons is the use of a bifilar secondary winding which

supplies the magnetron heater current, thus eliminating a separate heater transformer of high insulation.

Insulation problems encountered in pulse transformers are extraordinary and they have been overcome only by close attention to refinements of impregnating techniques. The best transformer oil is used and exceptionally thorough vacuum impregnation and hermetic sealing are essential. Kraft paper is used almost universally for insulation between winding layers and when oil immersed may be stressed up to 300 kilovolts per inch. The oil itself may be stressed up to 500 kilovolts per inch while over the surface of the paper the stress should not exceed 30 kilovolts per inch. Very high voltage stresses between turns are encountered, values of up to 500 volts per turn being quite common. Plate VIII shows photographs of three typical pulse transformers. Plate VIII (i) shows an uncased transformer with a "hipersil" core wound on a mandrel and cut as described on p. 104. Plate VIII (ii) shows a 125 kilowatt 1 microsecond pulse transformer designed for airborne use. The larger transformer shown in Plate VIII (iii) is a laboratory model for 6 microsecond pulses with an output of 2 megawatts 40 kilovolts.

6. Typical Modulator Circuits

High Vacuum Tube Modulator

The circuit of a typical high vacuum tube modulator designed for airborne microwave equipment is shown in Fig. 63. It generates negative pulses of 11 kilovolts and 10 to 15 amperes and of 1 microsecond duration at a repetition frequency which may be varied between 400 and 2000 cycles per second. The pulses are formed at a low power level by discharging a pulse forming network through a small thyratron V_2 which is triggered from an external source. The positive pulse so obtained is amplified by the twin tetrode V_3 which is cathode coupled to the high power switching tubes V_6 and V_7. The whole pulse forming circuit follows the potential of the cathode of V_3, so that this is an example of the so-called " boot-strap "

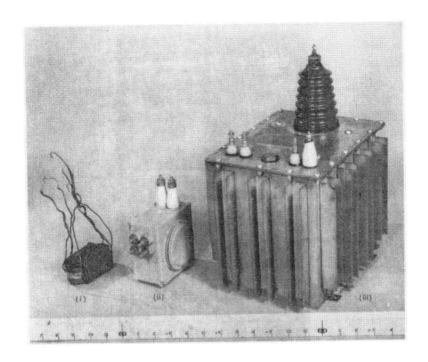

PLATE VIII.—Output pulse transformers.

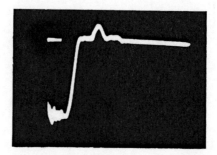

(a) Voltage waveform—modulator discharging into a non-inductive resistor.

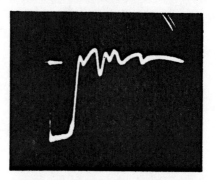

(b) Voltage waveform—magnetron load.

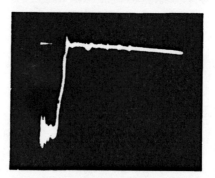

(c) Current waveform—magnetron load.

PLATE IX.—Waveforms for line type modulator (see p. 109).

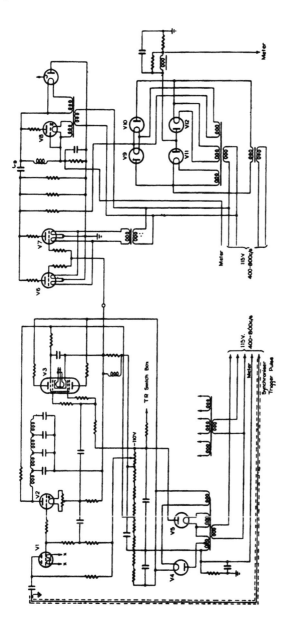

Figure 63.—High vacuum tube modulator circuit developed by Radiation
Laboratory, Massachusetts Institute of Technology.

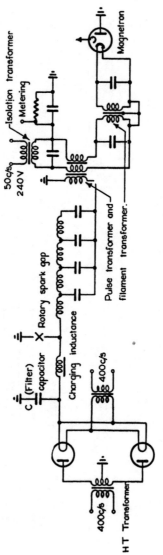

Figure 64.—Line type modulator—resonant charging circuit.

circuit. The system operates on 115 volts 400-800 cycles
per second, requires an input of 1 kilowatt, and has an overall
efficiency of from 10 to 30 per cent. The weight is about
60 pounds.

Line Type Modulator with Rotary Spark Gap

Fig. 64 shows the circuit of a line type modulator designed
for laboratory test purposes. The unit is mounted on a
400 cycles per second alternator set. It uses DC resonance
charging, a rotary spark gap and a pulse transformer. It delivers a pulse of 40,000 volts at 50 amperes or a peak power output of 2 megawatts. Two combinations of pulse length and
repetition frequency are available, 1 microsecond at 500 cycles
per second, and 5 microseconds at 250 cycles per second.

The output waveforms obtained from this modulator are
typical of radar practice and are shown in Plate IX. Plate IX (*a*)
shows the voltage waveform when power is delivered to an
800 ohm non-inductive resistor. The oscillations on top of the
pulse are characteristic. The second oscillogram, Plate IX (*b*),
shows the voltage pulse with a magnetron as load. The flat
top of this pulse illustrates the effect of the constant voltage
characteristic of the magnetron. The third oscillogram,
Plate IX (*c*), that of magnetron current, shows oscillations
which are a little greater than those with a resistive load.

CHAPTER VI

MICROWAVE TRANSMISSION AND CAVITY RESONATOR THEORY

1. Microwave Transmission Lines

THE function of a transmission line is to transfer electromagnetic energy from one point to another with the least loss. The conventional transmission line of two conductors is well known as a means of doing this. Either the conductors are external to each other as in the case of the twin wire line or one completely surrounds the other as for the coaxial transmission line. In the former case the fields spread out into space and loss of energy results at short wavelengths. A coaxial transmission line however confines the field within the outer cylinder and therefore is not subject to attenuation by radiation. In the region of decimetre wavelengths twin wire lines become unusable. The transmission lines used at these and shorter wavelengths are the object of our attention here.

Well into the microwave region the coaxial line continues to be used until accumulating disadvantages limit its further

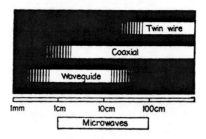

Figure 65.—Field of use of various transmission systems.

application. In addition to the two conventional types of transmission line there is the hollow pipe or waveguide. Unlike the former which find use from the longest wavelengths down, the waveguide does not become practical until short wavelengths are reached. Fig. 65 illustrates the relation between the regions in which the various lines are used. It will be seen that the twin line is of no importance in microwave transmission but that coaxial and waveguide transmission are both met over nearly the whole region.

The word *microwave* is roughly synonymous with centimetre-and-decimetre waves but its real use is to designate certain techniques—consequently definite wavelength limits cannot be set to the microwave region. The techniques symptomatic of microwaves involve the use of completely enclosed apparatus such as waveguides and electromagnetic cavities.

2. Waveguides

A waveguide is simply a hollow conducting pipe through which electromagnetic waves will travel. It differs from the conventional transmission line in having no return conductor and therefore its operation cannot be visualised in terms of current and voltage waves—though the idea of impedance can be retained. To deal with waveguides it is necessary to move the emphasis from voltages and currents to the electric and magnetic fields confined within the guide. Virtually all the field is restricted to the dielectric medium filling the guide ; for alternating fields cannot exist within a perfect conductor and penetrate only slightly into practical conductors at radio frequencies.

Many different field patterns can be propagated along a waveguide or transmission line but with a two-conductor line emphasis falls on one only of these modes, the conventional mode for which the electric and magnetic fields are purely transverse. The existence of possible higher modes of propagation on a twin wire or coaxial line may usually be safely forgotten for a reason which appears below. However in a hollow pipe the conventional or Transverse Electro-Magnetic (TEM) mode cannot exist and attention must be focused on the higher modes of propagation. They are conveniently classed as transverse electric and transverse magnetic according to whether the electric or magnetic field remains purely transverse. Hence their distinguishing feature is the presence of a component of field in the direction of propagation. If longitudinal components of both field types are present the pattern is regarded as a linear combination of transverse electric and transverse magnetic modes and each is treated separately. In general the phase velocities of the components will not be equal and the pattern will change as it travels. Figs. 66 and 67 illustrate fields in coaxial line and circular waveguide. In the former the field components are completely transverse while in the waveguide, although the magnetic

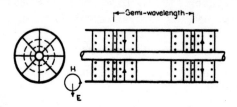

Figure 66.—Conventional mode in coaxial transmission line.

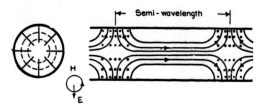

Figure 67.—TM_{01} mode in circular waveguide.

field is similar, the electric lines turn out of the transverse plane and make their way back to the outer wall, having no central conductor on which to terminate.

The property of higher modes which distinguishes them from the conventional mode, is the possession of critical wavelengths. Thus although waves of any frequency can negotiate a two-conductor transmission line, the absence of the conventional mode from a waveguide divides the spectrum into transmitted and non-transmitted frequencies. For example, electro-magnetic waves at optical frequencies will pass down a hollow pipe of moderate size whereas waves at power frequencies will not. These extreme cases are well away from the critical wavelength. Microwave transmission through a waveguide is carried on at a wavelength just shorter than the critical wavelength. Another way of stating this is to say that the waveguide is made just large enough to transmit the wavelength in question.

Each possible field pattern in the waveguide has its own critical wavelength which must not be exceeded if there is to be propagation. A waveguide is usually designed so that it is large enough to transmit in the mode with the longest critical wavelength (the dominant mode) but not large enough to transmit in the next higher mode. If in a particular transverse section of a guide a field pattern is applied which cannot propagate, then the pattern is attenuated exponentially along the guide at a rate depending on how close it is to its critical wavelength. In general the higher the order of a mode the more rapidly it is damped out. Fig. 68 shows how the field

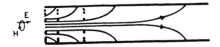

Figure 68.—TM_{01} mode in pipe too small to allow propagation.

in a circular waveguide falls off when a circular magnetic field is applied at the mouth. The mode is the same as that which in Fig. 67 is shown travelling down a pipe of larger size.

It can now be seen why the existence of higher modes is not of importance in conventional transmission lines—for at wavelengths longer than the critical wavelength no energy can be transferred save by the conventional mode. As the longest critical wavelength in a coaxial line cannot exceed the circumference of the outer conductor only at the shortest wavelengths do new considerations arise. The part played by attenuated modes in coaxial lines is one of adjusting boundary conditions. For example the field impressed on the mouth of a coaxial line may be neither circularly symmetrical nor inversely proportional to the radius, yet a short distance along the line the field will be quite pure. This is accounted for by analysing the applied field into an infinite series of orthogonal patterns all of which save one are attenuated to negligible proportions in a distance of the order of a diameter. That pattern which alone can propagate therefore exists in complete purity save in the immediate neighbourhood of discontinuities.

In a waveguide whose surface is perfectly conducting a propagated wave suffers no attenuation, but in a waveguide of finite conductivity the current flow in the walls results in the development of heat which appears at the expense of the energy of the field. As the amount of heat liberated per unit distance at any section is proportional to the power crossing the section, the wave is attenuated exponentially with distance just as waves on conventional transmission lines. This sort of attenuation, which is due to conductor losses, represents a real loss of energy whereas the attenuation in a pipe too small to allow propagation is reactive.

In choosing between the waveguide and coaxial line for a particular application four factors are considered. These are dielectric loss, conductor loss, power handling capacity and mechanical convenience. Without considering details a tendency may be observed for waveguide to replace coaxial line as the wavelength shortens, first for the transmission of high powers and later for local oscillator and signal powers. Thus a decimetre-wave radar may have large air-filled and small polythene-filled coaxial lines for transmitter and receiver respectively, a set operating on short centimetre waves may have waveguide in each case, whereas at frequencies between the extremes it is common to find the transmitter power carried by waveguide and the low powers by flexible polythene cable.

We have seen that there are numerous modes of transmission through waveguides each with its own critical wavelength and phase velocity and that at frequencies higher than the critical frequency there is a damped travelling wave, whilst at lower frequencies there is an attenuated standing wave. To illustrate the derivation of these and other results we consider in detail the rectangular waveguide.

3. Rectangular Waveguide

Solution of Maxwell's Equations

The following theory applies to transverse electric modes in rectangular waveguides. In a passive medium characterised by permeability μ, permittivity ϵ and conductivity g, Maxwell's equations for harmonic electric and magnetic fields are

$$\text{curl } \mathbf{E} = -j\omega\mu\mathbf{H} \tag{1}$$

$$\text{curl } \mathbf{H} = (g + j\omega\epsilon)\mathbf{E} \tag{2}$$

in rationalised m.k.s. units.[1] A factor $\exp(j\omega t)$ is omitted from all field expressions. To specify a medium we may take instead of μ, ϵ and g the derived constants η and σ such that

$$\eta = \sqrt{\frac{j\omega\mu}{g + j\omega\epsilon}} = \mathsf{R} + j\mathsf{X} \tag{3}$$

and

$$\sigma = \sqrt{j\omega\mu(g + j\omega\epsilon)}, \tag{4}$$

where η is the intrinsic impedance of the medium, σ is the intrinsic propagation constant and R and X are the intrinsic resistance and reactance. The propagation constant of a plane wave in the medium is the same as σ whilst η is the ratio of the electric to the magnetic field of a plane wave in the medium. The same advantages are gained by using these derived constants as are gained in transmission line theory by working with characteristic impedance and propagation constant instead of the distributed parameters L, C, R, G.

We desire to find solutions of equations (1) and (2) inside a rectangular pipe extending in the z-direction and bounded by the perfectly conducting planes $x = 0$, a and $y = 0$, b. On these planes there must be no tangential component of electric

---

[1] The rationalised metre-kilogram-second system of units is used throughout this book. For information on this system of units see J. A. Stratton, *Electromagnetic Theory*, McGraw-Hill, New York, 1941. The quantities appearing in Maxwell's equations are measured in the following units :—$\mathbf{E}$ (electric field strength), volts per metre ; $\mathbf{H}$ (magnetic field strength) amperes per metre ; g (conductivity), mhos per metre ; μ (permeability), henries per metre ; ϵ (permittivity), farads per metre. The permeability of free space (μ_0) is $4\pi \times 10^{-7}$ henries per metre ; the permittivity of free space (ϵ_0) is $8\cdot85 \times 10^{-12}$ farads per metre and the intrinsic impedance of free space (η_0) is given by $(\mu_0/\epsilon_0)^{\frac{1}{2}} = 120\pi = 377$ ohms.

field and no normal component of magnetic field.   In Cartesian
coordinates the equations become

$$\frac{\partial E_z}{\partial y} - \frac{\partial E_y}{\partial z} = -j\omega\mu H_x$$

$$\frac{\partial E_x}{\partial z} - \frac{\partial E_z}{\partial x} = -j\omega\mu H_y$$

$$\frac{\partial E_y}{\partial x} - \frac{\partial E_x}{\partial y} = -j\omega\mu H_z$$

(5)

$$\frac{\partial H_z}{\partial y} - \frac{\partial H_y}{\partial z} = (g + j\omega\epsilon)E_x$$

$$\frac{\partial H_x}{\partial z} - \frac{\partial H_z}{\partial x} = (g + j\omega\epsilon)E_y$$

$$\frac{\partial H_y}{\partial x} - \frac{\partial H_x}{\partial y} = (g + j\omega\epsilon)E_z.$$

We shall now assume that a solution exists such that the
electric vector is everywhere parallel to the y-direction i.e.
that $E_x = E_z = 0$. By assuming $E_z = 0$ we are restricting
the solutions to the transverse electric class.   By the further
assumption that $E_x = 0$ we shall be left with a special set of
transverse electric waves.   The equations now become,

$$H_x = \frac{1}{j\omega\mu}\frac{\partial E_y}{\partial z}$$

$$H_y = 0$$

$$H_z = -\frac{1}{j\omega\mu}\frac{\partial E_y}{\partial x}$$

(6)

$$\frac{\partial H_z}{\partial y} = \frac{\partial H_x}{\partial y} = 0$$

$$\frac{\partial H_x}{\partial z} - \frac{\partial H_z}{\partial x} = (g + j\omega\epsilon)E_y.$$

Eliminating **H**,

$$\frac{\partial^2 E_y}{\partial x^2} + \frac{\partial^2 E_y}{\partial z^2} = \sigma^2 E_y \qquad (7)$$

where

$$\sigma^2 = j\omega\mu \, (g + j\omega\epsilon). \qquad (8)$$

From this partial differential equation for the single field component E_y we obtain the solution by separation of variables by substituting $E_y = X(x)Z(z)$. The resulting total differential equations for the sought functions X and Z give

$$X(x) = A \sin hx + B \cos hx \qquad (9)$$
$$Z(z) = C \exp(\Gamma z) + D \exp(-\Gamma z) \qquad (10)$$

where

$$h^2 = \Gamma^2 - \sigma^2. \qquad (11)$$

To determine the integration constants we note that as E_y must vanish for $x = 0, a$

$$X(x) = A \sin \frac{n\pi x}{a}, \, n = 1, 2, .. \qquad (12)$$

and as the terms of Z evidently represent waves travelling in opposite directions we may take $C = 0$ and consider only waves travelling in the positive z-direction. Then if E is the maximum value of the electric field,

$$E_y = E \sin \frac{n\pi x}{a} \exp(-\Gamma z) \qquad (13)$$

where

$$\Gamma = \sqrt{\left(\frac{n\pi}{a}\right)^2 + \sigma^2}. \qquad (14)$$

The propagation constant Γ is equal to $\alpha + j\beta$ where α is the attenuation constant and β the phase constant.

For the remaining field components we have from (6)

$$H_z = -\frac{n\pi E}{j\omega\mu a} \cos \frac{n\pi x}{a} \exp(-\Gamma x) \qquad (15)$$

$$H_x = -\frac{\Gamma E}{j\omega\mu} \sin \frac{n\pi x}{a} \exp(-\Gamma z). \qquad (16)$$

Since E_y is the only component of electric field, the lines of force are normal to the guide faces at $y = 0, b$, and since E_y has been made to vanish when $x = 0, a$, there is no component tangential to the faces $x = 0, a$. It can be seen from the expressions for the field components that H_y is identically zero and that H_x vanishes on the walls $x = 0, a$. The boundary conditions are therefore satisfied.

Propagation Constant and Critical Wavelength

If the medium is loss-free, i.e. $g = 0$, equations (8) and (11) give for the propagation constant

$$\dot{\Gamma} = \sqrt{\left(\frac{n\pi}{a}\right)^2 - \left(\frac{2\pi}{\lambda}\right)^2} \tag{17}$$

since

$$\omega(\epsilon\mu)^{\frac{1}{2}} = 2\pi/\lambda, \tag{18}$$

where λ is the wavelength of a plane wave of angular frequency ω in the medium filling the waveguide.

If λ is small enough to make Γ imaginary then the field equations represent a pattern propagated in the z-direction but if Γ becomes real we have a stationary pattern which attenuates exponentially with distance. The critical wavelength which λ must not exceed if propagation is to take place is given by $\Gamma = 0$. Hence

$$\lambda_c = 2a/n. \tag{19}$$

In the case of propagation we have zero attenuation constant and $\Gamma = j\beta$ where

$$\beta = \frac{2\pi}{\lambda_c}\sqrt{\left(\frac{\lambda_c}{\lambda}\right)^2 - 1} \text{ radians per metre.} \tag{20}$$

In the case of attenuation the phase constant is zero and $\Gamma = \alpha$ where

$$\alpha = \frac{2\pi}{\lambda_c}\sqrt{1 - \left(\frac{\lambda_c}{\lambda}\right)^2} \text{ nepers per metre.} \tag{21}$$

The longest critical wavelength corresponds to $n = 1$ and is twice the length of the side perpendicular to the electric field. Fig. 69 shows the pattern which is obtained when the wavelength is shorter than critical and Fig. 70 shows the

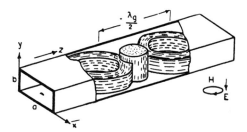

Figure 69.—Field pattern and coordinates for rectangular waveguide.

attenuated field for the same mode. The pattern of Fig. 69 is to be regarded as moving down the waveguide with a velocity $f\lambda_g$ which is somewhat above that of light. At fixed points in the median plane of the guide the electric and magnetic fields will be in phase, reaching their maxima together. At other points there is a phase difference which approaches 90 degrees at the walls $x = 0, a$. The patterns of Fig. 70

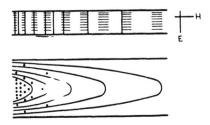

Figure 70.—Attenuated field of the usual mode in a waveguide too small to allow propagation.

are stationary in space but collapse and re-establish themselves harmonically with time. There is a 90-degree difference between the phases of the electric and magnetic fields so that when the electric field is as shown there is no magnetic field and vice versa. This field simply represents a periodic interchange of electric and magnetic energy without any net transfer. To amplify these remarks we note from (13) and (16) that $H_x = (-\Gamma/j\omega\mu) E_y$. Thus when Γ is imaginary the transverse field components are in phase and there is a net flow of energy but when Γ is real the fields are in quadrature and there is oscillation of stored energy only.

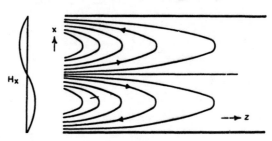

Figure 71.—Attenuated field of the TE_{20} mode in rectangular waveguide.

Fig. 71 indicates the meaning of the integer n which may be described as the number of half-period variations along the x-direction of any field component. The figure shows a field corresponding to $n=2$.

Phase Velocity

The propagation constant determines the wavelength in the guide,

$$\lambda_g = 2\pi j/\Gamma$$

$$= \frac{\lambda}{[1 - (\lambda/\lambda_c)^2]^{\frac{1}{2}}} \tag{22}$$

or in simpler form,

$$\frac{1}{\lambda^2} = \frac{1}{\lambda_c^2} + \frac{1}{\lambda_g^2}.$$

The phase velocity is given by

$$v_p = f\lambda_g$$

$$= \frac{v}{[1 - (\lambda/\lambda_c)^2]^{\frac{1}{2}}}$$

$$= \frac{j2\pi v}{\Gamma\lambda} = \frac{j\omega}{\Gamma} \tag{23}$$

where $v = (\epsilon \mu)^{-\frac{1}{2}}$ is the velocity of a plane wave in the medium filling the guide.

It will be noticed that the wavelength in the guide exceeds the wavelength of a plane wave in the medium and that the phase velocity is greater than that of light.

Group Velocity

Propagation in a waveguide is dispersive, i.e. different frequencies travel at different velocities. The dispersion is normal since the velocity decreases with frequency. A group velocity may be calculated from the usual formula

$$u = v_p - \lambda \frac{dv_p}{d\lambda} \tag{24}$$

giving

$$u = \frac{v^2}{v_p}. \tag{25}$$

Since $v_p > v$ it follows that $u < v$ i.e. the group velocity is always less than the velocity of a plane wave in the medium.

From the field equations we can find the density of energy and flow of power and thus make a direct calculation of the velocity with which energy is transferred through the waveguide. The mean flow of power P down the guide is given by the real part of the flux of the complex Poynting vector Π through a transverse section.

$$P = \text{Re} \iint \Pi \cdot d\Sigma \tag{26}$$

where $d\Sigma$ is an element of area in a transverse section and

$$\Pi = \tfrac{1}{2} E \times H^*. \tag{27}$$

Now

$$\Pi_z = -\tfrac{1}{2} E_y H_x^*$$

$$= \tfrac{1}{2} \frac{\Gamma E^2}{j \omega \mu} \sin^2 \frac{n\pi x}{a}. \tag{28}$$

Hence

$$P = \int_0^a \int_0^b \Pi_z \, dx dy$$

$$= \frac{(Eb)^2}{4j \omega \mu b/a \, \Gamma}. \tag{29}$$

At a point where the electric field strength is E the density of electric energy is $\epsilon E^2/2$. As the electric field in the guide is distributed sinusoidally in two dimensions (the x- and z-directions) with a maximum strength E, the mean electric energy density taken over a section of guide $\lambda_g/2$ long is one quarter this value. Taking account of the equal quantity of magnetic energy we have for the total average energy density,

$$U = \frac{\epsilon E^2}{4}. \tag{30}$$

The energy transport velocity u can be found from

$$u = \frac{P}{abU}, \tag{31}$$

whence

$$u = \frac{\Gamma}{j\omega\mu\epsilon} = \frac{2\pi}{\omega\lambda_g\mu\epsilon}$$

$$= \frac{v^2}{v_p}. \tag{32}$$

The velocity of energy flow is therefore less than the velocity of the wave and is equal to the group velocity.

Attenuation

To find the attenuation due to the finite conductivity of the walls we calculate the power flow into the walls. The attenuation in nepers per unit length is then one half the fraction of the transmitted power which is lost in unit length since $dP/dz = -2\alpha P$. From equation (13) which introduces the definition of attenuation constant we see that the neper, unlike the decibel, measures the attenuation of field or voltage but not power. When the two units are mutually convertible, one neper equals 8·686 decibels.

Assuming that the loss of energy does not appreciably modify the field distribution, the complex power flow into unit length of the walls is

$$\tfrac{1}{2} \int\int E_t H_t^* \, dS \tag{33}$$

where E_t and H_t are the tangential components of the field and dS is an element of area of the walls. The small tangential electric field E_t is given by

$$E_t = \eta H_t \tag{34}$$

where η is the intrinsic impedance of the conductor. Since $\eta = [j\omega\mu(g + j\omega\epsilon)]^{\frac{1}{2}}$ and therefore in a conductor $\eta = (j\omega\mu/g)^{\frac{1}{2}}$ we have

$$E_t = R(1 + j)H_t \tag{35}$$

where R the intrinsic resistance of the conductor is given by

$$R = \left(\frac{\omega\mu}{2g}\right)^{\frac{1}{2}} \text{ ohms.} \tag{36}$$

The intrinsic resistance of a conductor is equal to the resistance to direct current of a square piece whose thickness is equal to the skin depth. Intrinsic resistance increases as the square root of the frequency and is least for good conductors.

We may obtain the skin depth δ from the relation

$$R = \frac{1}{g\delta} \tag{37}$$

giving

$$\delta = \sqrt{\frac{2}{\omega\mu g}}. \tag{38}$$

Skin depth is defined so that the total current in the surface of a conductor, uniformly distributed to a depth δ, would dissipate the same power as the actual exponentially distributed current. Curves showing the skin depth for various conductor materials and frequencies are given in Fig. 106, Chapter VII. The skin depth in a conductor increases as the square root of the wavelength and is least for good conductors.

TABLE 4—CONDUCTIVITIES RELATIVE TO COPPER

Conductor	Ag	Cu	Au	Al	Zn	Brass
g_r	1·06	1.00	0·70	0·61	0·28	0·24
$g_r^{-\frac{1}{2}}$	0·97	1·00	1·19	1·28	1·90	2·04

For copper we have $R_0 = 0 \cdot 00452\lambda^{-\frac{1}{2}}$ ohms and $\delta_0 = 3 \cdot 8\lambda^{\frac{1}{2}} \times 10^{-6}$ metres and for non-magnetic materials whose conductivity relative to that of copper is g_r we have $R = R_0 g_r^{-1}$ and $\delta = \delta_0 g_r^{-\frac{1}{2}}$. Table 4 gives the factor g_r for a few metals.

At a wavelength of 10 centimetres $R_0 = 0.0143$ ohms and $\delta_0 = 1.2 \times 10^{-6}$ metres which is no more than twice the wavelength of light.

The average flow of power W into unit length of the walls is the real part of equation (33), hence

$$W = \frac{R}{2} \int\int H_t H_t^* \, dS. \tag{39}$$

The energy lost per cycle, $2\pi W/\omega$, is

$$\int\int \pi\delta \, (\tfrac{1}{2}\mu H_t H_t^*) \, dS. \tag{40}$$

As $\tfrac{1}{2}\mu H_t H_t^*$ is the density of magnetic energy at the surface of the conductor this result means that in each cycle an amount of energy is dissipated equal to the magnetic energy stored in a skin of thickness $\pi\delta$. An idea of the order of magnitude of the attenuation in a waveguide may therefore be formed by comparing the area of this thin skin ($\pi\delta \times$ perimeter) with the cross-sectional area of the guide leading to

$$\alpha \doteqdot \frac{\pi\delta \times \text{perimeter}}{2 \times \text{area}} \left(\frac{\lambda_g}{\lambda}\right)^2 \text{ nepers per guide-wavelength} \tag{41}$$

or, approximating further

$$\alpha \doteqdot \frac{2\delta \times \text{perimeter}}{\lambda \times \text{area}} \text{ nepers per metre.} \tag{42}$$

These formulae show clearly how waveguide attenuation depends on wavelength, conductor material and size and shape of guide and readily yield its order of magnitude.

Now splitting W into two integrals over the walls parallel and perpendicular to the electric field

$$W = R \int_0^b H_t H^* \, dy + R \int_0^a H_t H_t^* \, dx$$

$$= \frac{-R\pi^2 bE^2}{\eta^2 \Gamma^2 a^2} + \frac{RaE^2}{2\eta^2}$$

$$= \frac{RaE^2}{2\eta^2}\left[1 + \frac{2b}{a}\left(\frac{\lambda}{\lambda_c}\right)^2\right] \tag{43}$$

where the intrinsic impedance of the medium in the guide is

$(\mu/\epsilon)^{\frac{1}{2}}$. The total power flowing through the guide is

$$P = \frac{ab\,\Gamma E^2}{4j\omega\mu}.$$ (44)

Hence the attenuation is

$$\alpha = \tfrac{1}{2}\frac{W}{P}$$

$$= \frac{\pi\delta\left[\dfrac{a}{b} + 2\left(\dfrac{\lambda}{\lambda_c}\right)^2\right]}{a\lambda\left[1 - \left(\dfrac{\lambda}{\lambda_c}\right)^2\right]^{\frac{1}{2}}}.$$ (45)

Thus for a copper waveguide 3 inches by 1 inch internal dimensions this equation gives an attenuation of 0·00253 nepers per metre at a wavelength of 10 centimetres. From the approximate formula (41) we have 0·0026 nepers per metre.

Review

In the preceding paragraphs the properties of a simple waveguide have been worked out. Starting from Maxwell's equations and applying the boundary conditions for a rectangular conducting pipe we have derived a set of solutions [equations (13), (15), (16)] in each of which the electric field is everywhere parallel to the y-direction. Each of these modes has its critical wavelength given by equation (19). At shorter wavelengths than the critical the wave travels with a velocity given by equation (23) and at longer wavelengths equation (21) gives the attenuation constant governing its decay. The group velocity has been calculated from the known frequency-dependent wave velocity [equation (25)] and shown to be equal to the speed at which energy travels in the guide. This speed is always less than that of a plane wave in the medium whereas the wave itself always travels faster than a plane wave. By taking into account the energy spent in heating the walls of the waveguide, the attenuation constant has been found [equation (45)] and a useful approximate result given [equation (41)]. The latter shows that waveguide attenuation per wavelength is of the order of the ratio of skin depth in the conductor to the linear dimension of the cross section.   For a given waveguide the attenuation constant increases as the square root of the frequency but if the size of the waveguide is changed in proportion to the wavelength the attenuation constant increases as the sesquialterate power of the frequency.   Field equations for the circular waveguide are given in section 7.

4. Synthesis of Waveguide Fields

The procedure just described for investigating the properties of a waveguide exemplifies the method of attack on waveguides of other shapes, i.e. Maxwell's equations are written out in a suitable coordinate system and solved for the appropriate boundary conditions.   In the case of the rectangular waveguide carrying modes such as those we have been discussing, an alternative method is available which leads to a useful physical picture of the guided wave.

Consider an infinite conducting plane in a loss-free dielectric, upon which is incident an infinite plane wave whose electric vector is parallel to the plane (Fig. 72).

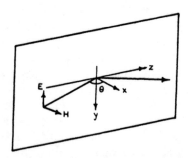

Figure 72.—Plane wave incident on a conducting sheet.

At the plane the electric field is annulled and the reflected wave sets up an interference pattern which moves without change along the plane.

Fig. 73 shows the crests and troughs of the electric field. There is no net flow of energy in the direction normal to the plane, but there is a steady flow parallel to the plane. If x is measured normal to the plane, y parallel to the electric vector and z is the direction in which the energy moves, the only field components are E_y, H_x and H_z. If λ is the wavelength of the incident radiation and θ is the angle of incidence, E_y will

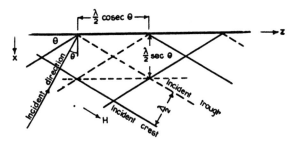

Figure 73.—Interference pattern at a conducting sheet. The electric vector is normal to and the magnetic field parallel to the plane of the diagram.

be zero on the planes $x = 0$, $\tfrac{1}{2}n\lambda \sec \theta$ parallel to the reflecting sheet. Consequently a conducting plane can be introduced into the field at a distance $\tfrac{1}{2}\lambda \sec \theta$ without disturbing the distribution. Moreover if conducting planes are introduced at $y = 0,b$ the boundary conditions will still be satisfied by the existing field.

We are now left with a travelling wave in a rectangular pipe. The field is the sum of two plane waves whose wavefronts make an angle θ with the axis, and which are reflected back and forth down the pipe. The distance measured along the pipe between points in the same phase exceeds the wavelength in the medium and from Fig. 73 is seen to be $\lambda \csc \theta$, hence the phase velocity $v_p = v \csc \theta$. If the width of the pipe $a = \tfrac{1}{2}\lambda \sec \theta$ is kept fixed, then as the wavelength of the radiation is increased the angle θ approaches zero, the phase wavelength approaches infinity and the direction of the component waves approaches normal incidence on the walls. When θ is zero, $\lambda = 2a$; consequently $2a$ is the critical wavelength which λ may not exceed. For very short wavelengths

θ approaches 90 degrees, the phase wavelength approaches the wavelength of a plane wave in the medium, and the direction of the component waves lies mainly along the axis. The energy velocity, as we have seen, is $v^2/v_p = v \sin \theta$ and this is the resolved part of velocity of the plane wave components along the axis. If the amplitude of the plane waves is $\frac{1}{2}E$ then the maximum electric field is E. The power carried by such a plane wave is $E^2/8\eta$ per unit area. If we double this, resolve along the axis and multiply by the area of the pipe, we have for the flow of power, $P = E^2 ab \sin \theta/4\eta = E^2 ab\lambda/4\eta\lambda_g$ which agrees with (44). We also observe that since $\lambda/\lambda_g = \sin \theta$ and $\lambda/\lambda_c = \cos \theta$ it follows that $(\lambda/\lambda_g)^2 + (\lambda/\lambda_c)^2 = 1$. The transverse components of electric and magnetic fields are in a constant ratio independently of the coordinates. For a plane wave $E/H = \eta$; in the waveguide, $-E_y/H_x = \eta/\sin \theta = \eta\lambda_g/\lambda = j\omega\mu/\Gamma$.

Most of the results for the rectangular waveguide have now been obtained by the method of decomposition into plane waves.

5. Nomenclature of Wave Types

To illustrate the rectangular waveguide modes other than those discussed here in detail, Figs. 74, 75 and 76 have been prepared.

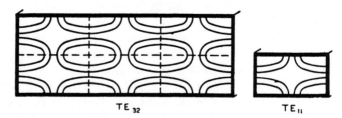

$$TE_{32} \qquad\qquad TE_{11}$$

Figure 74.—Electric lines in transverse section of a waveguide carrying the TE_{32} mode. Conducting partitions may be inserted along the dashed lines without disturbing the field. Each cell then carries a TE_{11} wave.

The TE_{10} mode is the one with which we have been mainly concerned and the subscripts indicate that there is one half-period variation in the x-direction and none in the y-direction. The electric field in the transverse plane of the TE_{32} mode

(Fig. 74) shows three half-period variations in one direction and two in the other and we see that the wave may be regarded as divided into six equal cells in each of which a TE_{11} mode is present. The dividing walls between the cells occur at places where conducting planes could be inserted. To

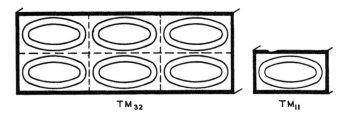

Figure 75.—Magnetic lines of the TM_{32} mode in rectangular waveguide. Each of the six cells carries a TM_{11} wave.

visualise the TM_{32} mode (Fig. 75) one should fill each cell with a TM_{11} wave. As neither of the subscripts of transverse magnetic waves can be zero, all modes in rectangular waveguides are built up from three simple modes only. These are the TE_{10}, TE_{11} and TM_{11}. A grasp of these three modes of propagation, together with the fundamental circular electric mode, will be found sufficient for forming a qualitative picture

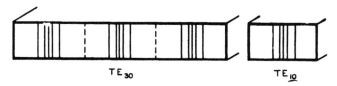

Figure 76.—Electric field of the TE_{30} mode. Each of the cells carries a TE_{10} wave.

of the modes of transmission in waveguides of circular or any other cross-section. To illustrate this Fig. 77 compares corresponding modes in rectangular and circular wayeguides. The correspondence is such that one pattern may be transformed into the other by continuous deformation of the wall. The circular electric mode, which is also shown, has no analogue in rectangular pipes—deformation of the pipe causes lines of force to snap and simple squeezing into a rectangular shape results in a TE_{20} mode.

The nomenclature of wave types in circular pipes is discussed in section 7 under circular cylindrical resonators.

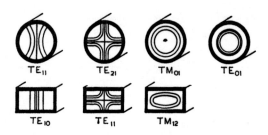

Figure 77.—Corresponding modes in circular and rectangular pipes.

6. Waveguide Impedance Theory

Normalised Impedances in Waveguides

Much of the foregoing matter has been concerned with emphasising the differences between waveguides and conventional transmission lines and these differences must be kept in mind. Yet there is a similarity between the two which is most important in practice for it allows the impedance-measurement techniques in use with conventional lines to be applied to waveguides. If, in a length of waveguide, discontinuities are so remote that the dominant mode exists in virtual purity, then there will in general be unequal waves travelling in opposite directions and giving rise to a standing wave pattern such as might be found on a conventional transmission line. The pattern is sufficiently specified by two real numbers : the standing wave ratio[2] S which is the ratio of the minimum

[2] Standing wave ratio so defined lies between 0 and 1. The reciprocal is sometimes used in which case standing wave ratio ranges from 1 to ∞ . Two advantages are possessed by the present definition : on the Kennelly charts the standing wave ratio of every circle centred on the real axis is readily found as the abscissa of its intersection with that axis, on the Smith chart equal increments in standing wave ratio produce roughly equal increments in the radii of the circles of constant standing wave ratio, thus facilitating interpolation.

to the maximum field and the space phase θ which is the distance in angular measure of a minimum from a fixed reference point. Fig. 78 shows the standing wave on an actual transmission line. The two quantities S and θ may now be used

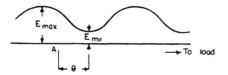

Figure 78.—Standing wave pattern on a transmission line.

to define the impedance at the point A by analogy with the conventional line for which

$$Z = Z_0 \phi \, (S, \theta) \tag{46}$$

where Z_0 is the characteristic impedance of the line. So far the question of the characteristic impedance of a waveguide has not arisen, nor need it here for it will be sufficient to have a definition of the normalised impedance ζ at a point in a waveguide

$$\zeta = \frac{Z}{Z_0} = \phi \, (S, \theta). \tag{47}$$

The actual form of the function ϕ is

$$\phi \, (S, \theta) = \tanh \, (\text{artanh} \, S + j\theta). \tag{48}$$

It will be noticed that in this definition of impedance no ratio of voltages and currents is implied. We have simply defined a function of the observable quantities S and θ, which replaces the customary impedance of conventional line theory, when we are dealing with reflections and impedance matching.

Limitations of the Theory

As an example of the meaning of waveguide impedances, consider a metallic obstacle placed in a transverse plane of a waveguide. Suppose that a generator feeds one end of the waveguide which extends to infinity. Near the obstacle the field will be distorted in such a way that the electric lines terminate normally on the conducting surface. The total field may be regarded as the sum of a number of the possible patterns which can exist in the undisturbed guide. Well away from the obstacle only the dominant mode will be found as no other mode is capable of propagation. Near the obstacle as many additional modes occur as are necessary to satisfy the boundary conditions and they are attenuated exponentially in both directions. The dominant mode will be found travelling in both directions between the generator and the obstacle but beyond the obstacle travelling outwards only. These are the incident, reflected and transmitted waves. As far as the behaviour of the dominant waves is concerned the system may be interpreted as a uniform line having a shunt reactance at the position of the obstacle. If the reactance is negative then the energy stored by the local fields will be predominantly electric.

The last paragraph indicates that the analogy with the conventional line has strict limitations. Two conditions which must be fulfilled are firstly that only one mode can be propagated, and secondly that attenuated modes shall be negligible, in the neighbourhood of the point where impedances are to be measured.

Impedance Charts

The bulk of numerical work with transmission lines is carried out graphically on charts of two kinds, the Kennelly chart, Fig. 79 and the Smith chart,[3] Fig. 80. Each of these consists of two sets of superimposed orthogonal networks, one of S and θ, the other of normalised resistance ($\rho = R/Z_0$) and reactance ($\xi = X/Z_0$). A point on the chart is therefore characterised by a pair of values of S and θ and the components of a complex impedance ζ which satisfies the relation $\zeta = \phi (S, \theta)$.

---

[3] A. E. Kennelly, *Chart Atlas of Complex Hyperbolic and Circular Functions*, Harvard University Press, Cambridge, Mass., 1924 (of this collection the hyperbolic tangent charts are the ones referred to here) ; and P. H. Smith, " Transmission line calculator," *Electronics*, Vol. 12, January, 1939, pp. 29-30.

To construct the Kennelly chart the relation

$$\zeta = \tanh \mu \tag{49}$$

where $\mu = \operatorname{artanh} S + j\theta$ (θ in quadrants), has been plotted on the complex plane of ζ.

By making the transformation

$$w = \frac{\zeta - 1}{\zeta + 1} \tag{50}$$

the Smith chart is obtained. In polar coordinates,

$$w = \frac{1 - S}{1 + S} \exp j2\theta.$$

As the Kennelly chart represents impedances in rectangular coordinates it is useful where lumped resistances and reactances occur and when calculations are carried out in terms of im-

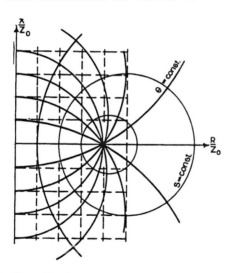

Figure 79.—Kennelly chart used for transmission line calculations.

pedances, as for example when changes in characteristic impedance have to be made. In some experimental work the measurable quantities S and θ may be sufficient for the calculations without at any stage changing to impedances.

Or again highly reactive impedances may be involved. In these cases the Smith chart has advantages for it is easier to plot on the S and θ loci which are radial lines on concentric

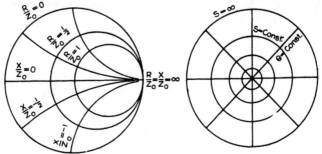

Figure 80.—Smith chart. *Left*—Resistance and reactance circles. *Right*—S and θ loci.

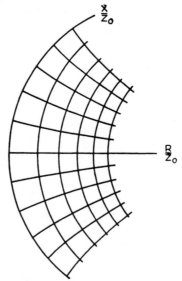

Figure 81.—Passing from the Kennelly to the Smith plane.

circles, and in the second case the chart covers the whole semi-infinite range of values of ζ. In general it is found that one chart usually offers advantages over the other when a particular calculation is to be done and it is necessary to be familiar with both. Fig. 81 suggests a way of picturing the transition from the Kennelly to the Smith chart showing how the infinite half-plane is compressed into a circle.

The graphical charts are not only used for numerical calculations. A good deal of qualitative thinking can be done quickly and easily with their aid which could otherwise hardly be done at all. Examples of this use are given in the next chapter.

Wave Impedance and Power Flow

So far the measurements discussed have been independent of the absolute level of the electromagnetic fields involved. Sometimes however one wishes to calculate the electric or magnetic field when the flow of power is given or to calculate one field component in terms of the other. This requirement introduces the idea of wave impedance which in a waveguide is defined as the ratio of the transverse field components. If u and v are mutually perpendicular directions in the transverse plane then the wave impedance is given by

$$K_z = \frac{E_u}{H_v} = -\frac{E_v}{H_u}. \tag{51}$$

The senses of u and v are chosen to make K_z positive for a positive-going wave. In a rectangular waveguide carrying fields of the type discussed above, we have from equations (13) and (16)

$$K_z = -\frac{E_y}{H_x} = \frac{j\omega\mu}{\Gamma} = \frac{\eta}{\left[1 - \left(\frac{\lambda}{\lambda_c}\right)^2\right]^{\frac{1}{2}}}. \tag{52}$$

From equation (29) the flow of power P down a rectangular waveguide is

$$P = \frac{a}{4b} \frac{VV^*}{K_z} \tag{53}$$

where $V = Eb \exp{(j\omega t - \Gamma z)}$ is the integral of the electric field in the medium plane from one wall of the waveguide to the other. By comparing this equation with

$$P = \frac{VV^*}{2Z_0} \tag{54}$$

for a conventional transmission line one is tempted to define the characteristic impedance of a waveguide by the formula

$$Z_0 = \frac{2b}{a} K_z. \tag{55}$$

Such a course cannot lead to error but if subsequently an attempt is made to find an analogue for I, the current in a conventional line, care will be needed. Suppose as is commonly done, that the total longitudinal current

$$\int_0^a H_x dx = -\frac{2\Gamma a}{j\omega\mu\pi} E \exp{(j\omega t - \Gamma z)}$$

crossing a transverse line in the wide wall is called I. Then the two relations between P and I and between V and I will define two more characteristic impedances equal neither to the one already obtained nor to one another. One possibility is to have three different impedances but as they differ only by a numerical factor the course suggested here is to retain one only, the first mentioned, and modify the usual equations containing I.

Thus

$$P = \frac{\pi^2}{16} \cdot \frac{Z_0 II^*}{2} \tag{56}$$

$$V = \frac{\pi}{4} Z_0 I. \tag{57}$$

It is reasonable to make the choice in favour of the equation containing P and V for the quantity V has important significance relating to dielectric failure whereas I might with equal or more utility have been defined in terms of the maximum longitudinal current density as $a(H_x)_{max}$. The advantage of the above procedure lies in having a definite though arbitrary characteristic impedance instead of three or none. It should be stressed that the characteristic impedance has been agreed upon only for the rectangular waveguide carrying the TE_{10} mode but that in general a unique wave impedance is always available. The characteristic impedance has two uses : (a) for simply relating potential difference and power and (b) for calculating reflections at a junction between two TE_{10} waveguides of different dimensions.

A 3-inch × 1-inch waveguide has a wave impedance of 506 ohms and a characteristic impedance of 337 ohms at a wavelength of 10 centimetres. If one megawatt is flowing the maximum voltage is $2 \cdot 6 \times 10^4$ volts and the gradient $1 \cdot 02 \times 10^6$ volts per metre. The maximum longitudinal current is 98 amperes and the maximum current density $1 \cdot 69 \times 10^9$ amperes per square metre for a copper waveguide.

7. Cavity Resonators

Solution of Maxwell's Equations

A dielectric region completely enclosed by a conducting surface constitutes a cavity resonator. Such a resonator possesses an infinite number of natural frequencies and corresponding field patterns. We shall first consider the case of perfectly conducting walls and non-conducting dielectric and then take account of damping.

The differential equation which the electric field in a resonator (or waveguide) must satisfy is obtained from Maxwell's equations as follows :

$$\text{curl } \mathbf{E} = -j\omega\mu\mathbf{H}$$
$$\text{curl } \mathbf{H} = j\omega\epsilon\mathbf{E}. \tag{58}$$

Taking the curl of the first and substituting the second,

$$\text{curl curl } \mathbf{E} = \omega^2\epsilon\mu\mathbf{E} \tag{59}$$

since curl curl $=$ grad div $- \nabla^2$ and since div $\mathbf{E} = 0$ we have

$$\nabla^2\mathbf{E} + \omega^2\epsilon\mu\mathbf{E} = 0. \tag{60}$$

This is the vector wave equation. Solutions to this equation are obtained by separation of variables. In a limited number of coordinate systems (about five all told) the vector wave equation splits into three total second-order differential equations and the field may be expressed as the product of the three functions satisfying these equations. Although direct solution by this method can be carried out for comparatively few shapes of cavity its value lies in the possibility of constructing further solutions by linear superposition. The problem is similar to that of acoustical resonance though more difficult.

Once a solution for **E** has been found **H** can be obtained from

$$\mathbf{H} = \frac{j \operatorname{curl} \mathbf{E}}{\omega \mu} \tag{61}$$

and will be found to satisfy the necessary boundary conditions, viz. that there should be no normal component of magnetic field and no tangential component of electric field on a perfectly conducting surface. For over a closed path in the surface, $\int \mathbf{E} \cdot \mathbf{ds} = 0$ since there is no tangential component of **E**. Therefore $\iint \operatorname{curl} \mathbf{E} \cdot \mathbf{ds} = 0$. Hence the normal component of curl **E** vanishes on the surface and **H** is everywhere tangential.

Rectangular Prism

Consider a rectangular prism whose faces are at $x = 0,a$ $y = 0,b$, $z = 0,c$. By writing the vector wave equation in Cartesian coordinates, separating variables and determining the constants of integration from the boundary conditions viz. $E_x = 0$ when $y = 0,a$ and $z = 0,c$, etc., we have for the various components

$$E_x = -\frac{k_1 k_3}{k_1{}^2 + k_2{}^2} E \cos k_1 x \sin k_2 y \sin k_3 z$$

$$E_y = -\frac{k_2 k_3}{k_1{}^2 + k_2{}^2} E \sin k_1 x \cos k_2 y \sin k_3 z$$

$$E_z = E \sin k_1 x \sin k_2 y \cos k_3 z \tag{62}$$

$$H_x = j\omega\epsilon \frac{k_2}{k_1{}^2 + k_2{}^2} E \sin k_1 x \cos k_2 y \cos k_3 z$$

$$H_y = -j\omega\epsilon \frac{k_1}{k_1{}^2 + k_2{}^2} E \cos k_1 x \sin k_2 y \cos k_3 z$$

$$H_z = 0.$$

where

$$k_1 = l\pi/a, \quad k_2 = m\pi/b, \quad k_3 = n\pi/c, \tag{63}$$

$$(2\pi/\lambda)^2 = k_1{}^2 + k_2{}^2 + k_3{}^2, \tag{64}$$

and l, m and n are integers.

It will be noticed that H_z is zero. One further set of linearly independent solutions exists for which $E_z = 0$. The two types of solution are called Transverse Magnetic (TM) and Transverse Electric (TE) modes. The three integers (l, m, n) generate a triple infinity of each kind of mode. The resonant frequency of each mode is determined from the formula

$$\lambda = \frac{2}{\left[\left(\dfrac{l}{a}\right)^2 + \left(\dfrac{m}{b}\right)^2 + \left(\dfrac{n}{c}\right)^2\right]^{\frac{1}{2}}} \tag{65}$$

for the modes TE_{lmn} and TM_{lmn}. The longest wavelength is obtained when one of the integers is zero and the two associated with the largest dimensions are unity. In this case the electric field is parallel to the least dimension which then plays no part in determining the resonant frequency, i.e. the field is two-dimensional. Only in rectangular prisms is there complete degeneracy between TE and TM modes. In practice slight departures of the shape from squareness will cause the modes to separate. Another type of degeneracy occurs if two dimensions are equal e.g. if $a = b$. Then the modes TE_{pqr} will have the same frequencies as TE_{qpr}. Further degeneracies occurring for particular shapes are called accidental. As smaller and smaller wavelengths are considered the number of resonances which can occur in a resonator becomes very large. For a rectangular prism the number of modes resonating at a wavelength greater tham λ_0 is about 8 × volume/λ_o^3. Thus a three-foot cube contains about 6000 modes with wavelengths greater than 10 centimetres. A similar qualitative

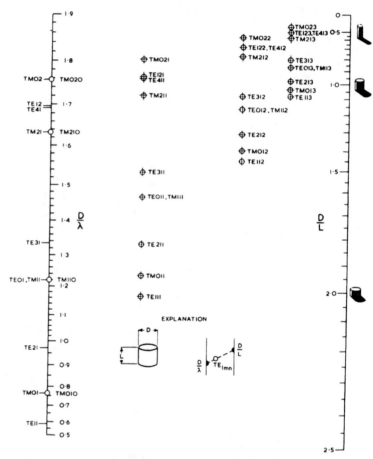

Figure 82.—Mode lattice for cylinder resonators.

situation holds for resonators of any shape. The distribution of modes in a circular cylinder can be obtained from Fig. 82 which is a mode lattice[4] suitable for numerical calculations.

[4] R. N. Bracewell, "Charts for resonant frequencies of cavities," *Proc. I.R.E.*, Vol. 35, 1947, pp. 830-41.

The field components of transverse electric waves in a rectangular prism are :

$$H_x = -\frac{k_1 k_3}{k_1{}^2 + k_2{}^2} H \sin k_1 x \cos k_2 y \cos k_3 z$$

$$H_y = -\frac{k_2 k_3}{k_1{}^2 + k_2{}^2} H \cos k_1 x \sin k_2 y \cos k_3 z$$

$$H_z = H \cos k_1 x \cos k_2 y \sin k_3 z$$

$$E_x = j\omega\mu \frac{k_2}{k_1{}^2 + k_2{}^2} H \cos k_1 x \sin k_2 y \sin k_3 z \qquad (66)$$

$$E_y = -j\omega\mu \frac{k_1}{k_1{}^2 + k_2{}^2} H \sin k_1 x \cos k_2 y \sin k_3 z$$

$$E_z = 0.$$

Some combinations of the integers, l, m, n, lead to trivial solutions or to fields incompatible with the boundary conditions. There are no resonances unless the following conditions are fulfilled : $l + m > 0$ for TM modes and $(l + m)n > 0$ for TE modes.

Circular Cylinder

The field components for the circular cylinder are given for the TM_{lmn} mode by

$$E_r = -\frac{n\pi}{kh} E J_l'(kr) \cos l\theta \sin\frac{n\pi z}{h}$$

$$E_\theta = \frac{n\pi l}{k^2 rh} E J_l(kr) \sin l\theta \sin\frac{n\pi z}{h}$$

$$E_z = E J_l(kr) \cos l\theta \cos\frac{n\pi z}{h}$$

$$\qquad (67)$$

$$H_r = -\frac{j\omega\epsilon l}{k^2 r} E J_l(kr) \sin l\theta \cos\frac{n\pi z}{h}$$

$$H_\theta = -\frac{j\omega\epsilon}{k} E J_l'(kr) \cos l\theta \cos\frac{n\pi z}{h}$$

$$H_z = 0.$$

where (r, θ, z) are cylindrical polar coordinates, l, m and n

are integers, h is the height and a the radius of the cylinder and $x_{lm} = ka$ is the mth root of $J_l(x) = 0$. The resonant wavelength is given by

$$\left(\frac{a}{\lambda}\right)^2 = \left(\frac{x_{lm}}{2\pi}\right)^2 + \frac{n^2}{4}\left(\frac{a}{h}\right)^2. \tag{68}$$

For the TE_{lmn} mode we have the following field components :

$$H_r = \frac{n\pi}{kh} H J_l'(kr) \cos l\theta \cos \frac{n\pi z}{h}$$

$$H_\theta = -\frac{n\pi l}{k^2 rh} H J_l(kr) \sin l\theta \cos \frac{n\pi z}{h}$$

$$H_z = H J_l(kr) \cos l\theta \sin \frac{n\pi z}{h}$$

$$\tag{69}$$

$$E_r = \frac{j\omega\mu l}{k^2 r} H J_l(kr) \sin l\theta \sin \frac{n\pi z}{h}$$

$$E_\theta = \frac{j\omega\mu}{k} H J_l'(kr) \cos l\theta \sin \frac{n\pi z}{h}$$

$$E_z = 0.$$

In this case $x_{lm} = ka$ is the mth root of $J_l'(x) = 0$. Table 5 gives a few values of these roots.

TABLE 5—ROOTS OF BESSEL FUNCTIONS

Mode	x_{lm}
TE_{11}	1·8412
TE_{21}	3·0543
TE_{01}	3·8317
TE_{31}	4·2014
TE_{41}	5·3175
TE_{12}	5·3315
TM_{01}	2·4048
TM_{11}	3·8317
TM_{21}	5·1356
TM_{02}	5·5201

The integers l, m, n, specifying the mode may be described as follows :

l = the number of full period variations undergone by any non-zero field component as θ varies from 0 to 2π.

m = the number of zeros of angular component of electric field in the case of TE waves or of axial component of electric

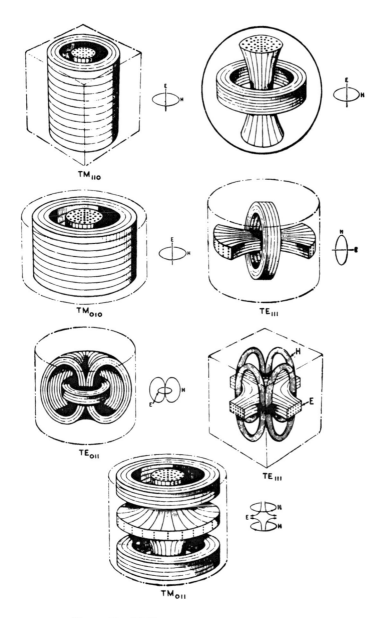

Figure 83.—Field patterns in cavity resonators.

field in the case of TM waves, lying on a radius, counting that at the wall but not that which may occur at the centre.

n = the number of half-wave variations through which the standing wave pattern extends in the z-direction.

Not all integral values may be assumed by l, m and n. Thus m cannot be zero. In the case of TE modes n cannot be zero, for this would involve a magnetic field parallel to the axis, which is not consistent with the boundary condition for magnetic fields on the conducting discs closing the cavity.

To calculate resonant frequencies for the circular cylinder the quickest procedure is to use the mode lattice shown in Fig. 82.

Field patterns of simple modes in various cavities are illustrated in Fig. 83. The drawings give three-dimensional views of the electric and magnetic fields which bring out some general properties of standing wave patterns in resonators, viz. that where the magnetic field is strong, the electric field is weak and that the magnetic lines form loops enclosing the electric lines. It can be seen that the magnetic lines run tangentially to the surface whilst the electric lines terminate normally. In one case, the TE_{011} mode in the circular cylinder, electric lines are shown which do not terminate at all but are closed on themselves. The phase relation of the two fields is such

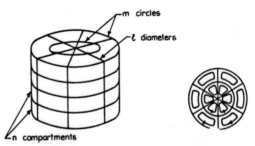

Figure 84.—The subscript notation for circular cylinders. The TM_{324} mode is used as a model and H-lines in the transverse plane are shown.

that the stored energy passes from one to the other each quarter-cycle or one is zero when the other is at a maximum.

It is of help in picturing a given mode in a circular cylinder to know that metallic partitions can be inserted in the resonator without disturbing the field and that the cells into which the resonator is thus divided contain fields resembling the TE_{011}, TE_{111}, TM_{010} or TM_{011} modes all of which are shown in Fig. 83.   The partitions are planes containing the axis, planes normal to the axis and coaxial cylinders.   Altogether there are l nodal diametral planes, m nodal circles and n axial com-

Figure 85.—The TE_{023} mode in a cylindrical cavity.

partments.   This is illustrated in Fig. 84 which shows the partitions for the TM_{324} mode and the H-lines in the transverse plane.   Each cell contains a field resembling the TM_{011} mode. Fig. 85 shows a cylinder containing the TE_{023} mode.

Travelling waves in circular pipes are named just as in circular resonators save that only the first two subscripts are necessary. Travelling wave patterns, such as that of Fig. 69, may be obtained from resonator patterns by displacing the electric field one quarter-wavelength axially. Comparison of the TE_{10} travelling wave Fig. 69 and the TE_{101} standing wave in a rectangular pipe, Fig. 83, shows the space relation between electric and magnetic fields in the two cases. In the first, one field type is at a maximum where the other is zero ; in the second, the transverse components of the two field types are everywhere in the same proportion.

Field equations for the circular waveguide may be obtained from equations (67) and (69) by replacing cos $n\pi z/h$ and sin $n\pi z/h$ by exp $(-j2\pi z/\lambda_g)$ and j exp $(-j2\pi z/\lambda_g)$ respectively and multiplying by j the expressions for electric field, where $(\lambda/\lambda_g)^2 + (\lambda/\lambda_c)^2 = 1$ and the critical wavelength λ_c is related to the diameter of the pipe by

$$\frac{D}{\lambda_c} = \frac{x_{lm}}{\pi} . \tag{70}$$

Table 5 allows D/λ_c to be evaluated or it may be taken directly from the D/λ-scale of Fig. 82. The least value of D/λ_c is 0·586. Hence the longest wavelength which can be propagated in a circular pipe is 1·70 times the diameter and the mode is the TE_{11}.

Instead of deriving the standing waves in a resonator directly it is possible (in cavities of constant cross-section) to build up the fields by superposition of equal waves travelling in opposite directions. Thus the circular cylindrical resonator may be regarded as a circular waveguide closed by plane ends an integral number of guide-wavelengths apart and the field as due to successive reflections of a travelling wave.

Modes in resonators formed of coaxial cylinders closed by annular ends are named after the corresponding cylindrical modes into which the patterns pass continuously as the inner cylinder shrinks to zero radius. The TEM modes however, cannot exist in the absence of the inner cylinder.

Coupling[5]

The cavities so far considered have been completely enclosed in a perfectly conducting envelope. We have shown that if any electromagnetic energy resides within certain cylindrical and prismatic cavities then it must be stored in one or more of the normal modes at the discrete natural frequencies of the cavity. If the slight spread due to finite conductivity of the walls is neglected, there can be no fields within the cavity, at frequencies other than the natural frequencies, in the absence of a generator. To sustain the field in a resonator a source of power must be introduced. Usually one desires to retain the properties of an undisturbed cavity so a small orifice in the wall is made and a field maintained across it resembling the field which would be there in the presence of the mode which it is desired to excite. Let the impressed fields at the point of coupling be $\mathbf{E}_0 \exp j\omega t$ and $\mathbf{H}_0 \exp j\omega t$ and let the fields at this point due to unit energy stored in the nth natural mode be $\mathbf{E}_n \exp j\omega_n t$ and $\mathbf{H}_n \exp j\omega_n t$. Then every mode will be excited for which the scalar products $\mathbf{E}_0 \cdot \mathbf{E}_n$ and $\mathbf{H}_0 \cdot \mathbf{H}_n$ do not vanish. This does not mean that oscillations are set up at the various natural frequencies, but that the various natural patterns execute forced vibrations at the angular frequency ω. The amplitude of excitation is proportional to the scalar products above but if $\omega \neq \omega_n$, it is small. Often only the mode whose natural frequency is nearest to the impressed frequency need be considered. When the impressed frequency is very close to a natural frequency, the amplitude of excitation of that mode is enormous in comparison with all others and limited by the damping due to finite conductivity of the walls.

It is worth while stressing the difference between free and

[5] Coupling is discussed in the following: E. V. Condon, "Forced oscillations in cavity resonators," *J. Appl. Phys.*, Vol. 12, 1941, pp. 129-32; S. A. Schelkunoff, "Representation of impedance functions in terms of resonant frequencies," *Proc. I.R.E.*, Vol. 32, 1944, pp. 83-90; C. G. Montgomery, R. H. Dicke and E. M. Purcell (Eds.), *Principles of Microwave Circuits* (Massachusetts Institute of Technology, Radiation Laboratory Series, Vol. 8), McGraw-Hill, New York, 1948.

forced vibrations in a cavity. A cavity with no source of power can only contain the natural field patterns at their respective natural frequencies whereas a cavity continuously excited by a simple harmonic oscillator contains, in addition to possible natural oscillations, any natural field pattern varying at the oscillator frequency. Figs. 67 and 68 show fields of the TM_{01} mode in circular waveguides of different sizes. In one case the fields extend with unchanged amplitude along the guide and in the other they are attenuated rapidly with distance. Forced patterns in a resonator behave in either of these ways and may therefore differ from the corresponding natural pattern to the extent that dependence on one coordinate may resemble a damped wave rather than a standing wave with nodes and antinodes. Some will make their presence felt throughout the cavity whereas others will be localised in the vicinity of the coupling point.

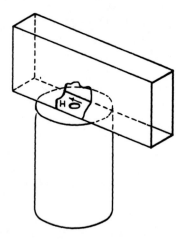

Figure 86.—Window-coupling to a cylindrical cavity.

Practical means of exciting cavities are divided into electric and magnetic according to the predominating component of the desired mode at the point of coupling. Both electric and magnetic coupling may be effected by feeding the cavity through a hole in a waveguide. Fig. 86 shows a cylindrical cavity excited from a rectangular waveguide.   At the coupling orifice there is normally no electric field so the coupling is magnetic.   The magnetic field in the guide is tangential to the wall in the direction shown and any natural mode of the cavity, which calls for such a magnetic field at the orifice, will tend to be excited. A TE_{10} wave in the waveguide of Fig. 86 would be suitable for exciting the TE_{111} mode in the circular cylinder.   Fig. 83 illustrates the latter mode.

Fig. 87 illustrates two methods which are commonly used

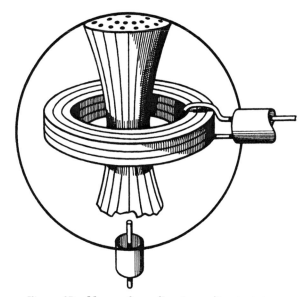

Figure 87.—Means of coupling to cavity resonators.

for coupling a coaxial transmission line to a cavity.   The magnetic form of coupling comprises a loop which links with magnetic lines of force.   The electric coupling may be regarded as a small aerial exciting modes which have a normal component of electric field.   In the last case the degree of coupling may be controlled by withdrawing the probe and in the first, one may simply rotate the loop.   If only one mode is excited two positions of zero coupling will exist, but unless the loop is situated at a point free from electric field, the

positions of zero coupling are not in general 180 degrees apart
nor are the amplitudes of excitation equal at the two positions
of maximum coupling.

Tuning

Adjustments to the resonant frequency of a cavity resonator
are conveniently made by making changes in the shape. A
common way of doing this is to insert plugs through the walls.
Often the effect on resonant frequency cannot be simply
calculated but it is possible to say whether the frequency will
be increased or decreased from a knowledge of the undisturbed fields. If a conducting object is introduced into the
resonator at a place where the electric field predominates
then the resonant frequency will fall (Fig. 88). The frequency

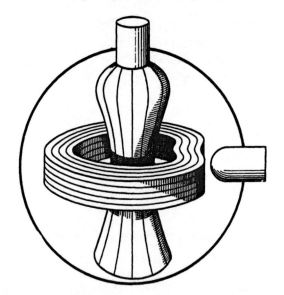

Figure 88.—Effect of tuning slugs on resonant frequency.

will rise if the object is placed in a strong magnetic field. A
resonant coil and condenser show the same effect, for if metal
is placed between the condenser plates the capacity increases
but metal in the coil lowers its inductance. To estimate the

order of magnitude of the change we may assume that the proportional change in resonant frequency is equal to the average proportion of the total energy of the resonator which is stored in the electric or the magnetic form within the volume to be occupied by the plug.   If the plug is to be inserted where there is both electric and magnetic energy, then the difference between the two average values must be taken.   It follows that there are places in the walls of most resonators where plugs may be inserted without much change of frequency.

For small plugs and small insertions the effect is roughly proportional to the volume of the plug and if the point is one of pure magnetic or electric field the proportional change in wavelength is approximately equal to the ratio of the volume of the plug to the volume of the cavity.   A perturbation theory for obstacles in cavities is given by Slater in the paper mentioned in the previous section.   The approximate results given above represent the first term of a series developed by Slater.

Damping

The damping of fields in resonators results from energy loss due to the conductivity of the walls and dielectric.   Where low damping is required dielectric loss can be rendered negligible in comparison with conductor loss by using air dielectric. Where both sorts of loss are present the resultant damping is the sum of the separate dampings.   If the conductivity of a uniform dielectric completely filling the resonator is g and its permittivity ϵ, then the damping is simply $g/\omega\epsilon$ for any shape or mode.   In the more important case of conductor loss however the shape and mode are very important.   It is usual to work in terms of the reciprocal of the damping viz. Q which may be defined as 2π times the ratio of the mean energy stored in the resonator to the energy lost per cycle.   Thus

$$Q = \omega U/W \qquad (71)$$

where U is the stored energy and W the power dissipated.   To calculate the Q of the resonator we must therefore first deter-

mine the fields inside it and then find by integration the energy
of the field and the flow into the walls. Taking the scalar
product of the first of Maxwell's equations with $\mathbf{H^*}$ and the
conjugate of the second with $\mathbf{E}$ and subtracting we have

$$\mathbf{H^*}\cdot \operatorname{curl} \mathbf{E} - \mathbf{E}\cdot \operatorname{curl} \mathbf{H^*} = -g\mathbf{E}\cdot \mathbf{E^*} + j\omega \epsilon \mathbf{E}\cdot \mathbf{E^*} - j\omega \mu \mathbf{H}\cdot \mathbf{H^*}$$

$$= \operatorname{div}(\mathbf{E}\times \mathbf{H^*}). \tag{72}$$

Integrating over a volume bounded by a surface S,

$$\tfrac{1}{2}\iint (\mathbf{E}\times \mathbf{H^*})\cdot d\mathbf{S} + \tfrac{1}{2}j\omega \iiint \mu \mathbf{H}\cdot \mathbf{H^*}dv - \tfrac{1}{2}j\omega \iiint \epsilon \mathbf{E}\cdot \mathbf{E^*}dv$$

$$+ \tfrac{1}{2}\iiint g\mathbf{E}\cdot \mathbf{E^*}dv = 0. \tag{73}$$

The real part of the complex Poynting vector $\tfrac{1}{2}\mathbf{E}\times \mathbf{H^*}$ gives the
mean power crossing unit area. In a uniform static field of
strength $\mathbf{E}_0$, the electric energy density is $\tfrac{1}{2}\epsilon E_0^2$; in an harmonic field $\mathbf{E}$, $\mathbf{H}$ $(x,\ y,\ z)\ \exp j\ (\omega t + \alpha)$, the mean electric
and magnetic energy densities are $\tfrac{1}{4}\epsilon \mathbf{E}\cdot \mathbf{E^*}$ and $\tfrac{1}{4}\mu \mathbf{H}\cdot \mathbf{H^*}$.
If the conductivity of the dielectric is zero, the last term in
equation (73) representing the energy lost in the dielectric,
is zero. The first term is equal to the complex flow of power
across the surface S. If the surface is perfectly conducting,
the vector product of $\mathbf{E}$ and $\mathbf{H}$ has no component normal to
the surface.

Consequently

$$\tfrac{1}{2}j\omega \iiint u\mathbf{H}\cdot \mathbf{H^*}\ dv - \tfrac{1}{2}j\omega \iiint \epsilon \mathbf{E}\cdot \mathbf{E^*}\ dv = 0 \tag{74}$$

and it follows that the mean values of the oscillating electric
and magnetic energies are equal. The total energy stored in a

resonator is therefore twice the mean magnetic energy and is given by

$$U = \tfrac{1}{2} \iiint \mu \mathbf{H} \cdot \mathbf{H}^* \, dv. \tag{75}$$

The mean flow of energy W into the walls is the real part of $\tfrac{1}{2} \iint E_t H_t^* dS$ where E_t is the small tangential component of electric field and dS is an element of area of the wall. Now

$$E_t = \mathsf{R}(1 + j) \, H_t, \tag{76}$$

therefore

$$W = \frac{\mathsf{R}}{2} \iint H_t H_t^* dS \tag{77}$$

$$= \frac{\mathsf{R}}{2} \iint \mathbf{H} \cdot \mathbf{H}^* dS.$$

Hence we have for the Q, assuming that the loss of energy is too small to appreciably affect the field distribution,

$$Q = \frac{\omega U}{W} = \frac{\mu \omega \iiint \mathbf{H} \cdot \mathbf{H}^* dv}{\mathsf{R} \iint \mathbf{H} \cdot \mathbf{H}^* dS} \tag{78}$$

where μ is the permeability of the dielectric and R the intrinsic resistance of the conductor. If we assume that dielectric and conductor have equal permeabilities, the formula becomes

$$Q = \frac{2 \iiint \mathbf{H} \cdot \mathbf{H}^* dv}{\delta \iiint \mathbf{H} \cdot \mathbf{H}^* dS} \tag{79}$$

where δ is the skin depth.

An idea can be obtained of the order of magnitude of Q's by assuming a uniform distribution of magnetic energy. A better approximation should result from halving the value obtained as magnetic fields tend to be strongest near the conducting surface.

Then

$$Q \doteq \frac{V}{\delta S} \tag{80}$$

where V is the volume and S the surface area. If the highest Q is to be obtained, the largest ratio of volume to surface area should be aimed at for the losses occur at the surface whereas the energy is stored throughout the volume. As the simple formula is approximate one cannot conclude that the resonator of highest Q for a given wavelength is spherical in shape. However the sphere is much better than either the cylinder or cube.

The Q of cylindrical resonators may be calculated from the following dimensionless formulae derived from equations (67). (69) and (79). The quantity $Q\delta/\lambda$ is a function of mode and shape only and is of the order of magnitude unity. In *TM* modes,

$$\frac{Q\delta}{\lambda} = \frac{\left\{ \left(\frac{x_{lm}}{2\pi}\right)^2 + \frac{n^2}{4}\left(\frac{a}{h}\right)^2 \right\}^{\frac{1}{2}}}{1 + \frac{2a}{h}} \qquad n \neq 0$$

$$\tag{81}$$

$$= \frac{\left\{ \left(\frac{x_{lm}}{2\pi}\right)^2 + \frac{n^2}{4}\left(\frac{a}{h}\right)^2 \right\}^{\frac{1}{2}}}{1 + \frac{a}{h}} \qquad n = 0$$

and for *TE* modes,

$$\frac{Q\delta}{\lambda} = \frac{4\pi^2 \left\{ \left(\frac{x_{lm}}{2\pi}\right)^2 + \frac{n^2}{4}\left(\frac{a}{h}\right)^2 \right\}^{3/2} \left\{ 1 - \left(\frac{l}{x_{lm}}\right)^2 \right\}}{x^2_{lm} + 2n^2\pi^2\left(\frac{a}{h}\right)^3 + \frac{l^2 n^2 \pi^2}{x^2_{lm}}\left(\frac{a}{h}\right)^3\left(\frac{h}{a} - 2\right)}. \tag{82}$$

These formulae have been plotted for a number of modes in Figs. 89 and 90.

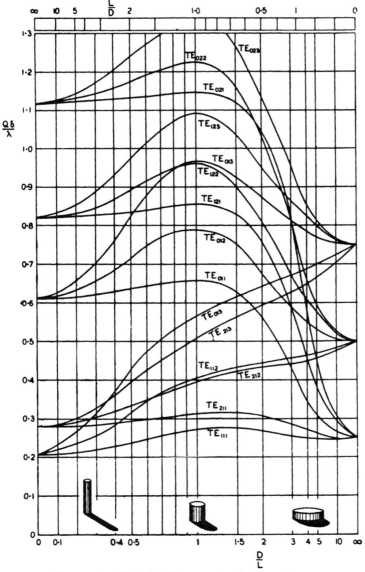

Figure 89.—$Q\delta/\lambda$ for TE modes in circular cylinders.

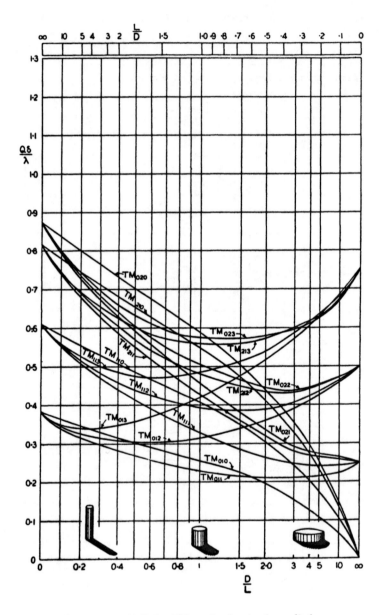

Figure 90.—$Q\delta/\lambda$ for TM modes in circular cylinders.

Equivalent Circuit

The behaviour of a resonant cavity in the neighbourhood of a resonance is reminiscent of the parallel resonant circuit with small losses and suggests the existence of a parallel combination of L, C and R which would accurately represent the resonator. To specify such a parallel circuit three quantities are necessary. These might be chosen as $(LC)^{-\frac{1}{2}}$, $R(C/L)^{\frac{1}{2}}$ and $(L/C)^{\frac{1}{2}}$ which are respectively the resonant frequency ω_0, Q and characteristic impedance. Now only two quantities are intrinsic in a resonator—resonant frequency and Q. Since there is nothing corresponding to characteristic impedance, none of the quantities L, C and R is defined. However by making one arbitrary assumption one may obtain an equivalent circuit which is then not unique. For example one may assume a value of L, C, R or $(L/C)^{\frac{1}{2}}$ or arbitrarily define the voltage or current associated with the circuit.

Cavity resonators, which are used for accelerating electron streams Fig 91, have a gap across which a voltage $V = V_0 \exp j(\omega t + \alpha)$ appears when electromagnetic power W

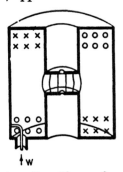

Figure 91.—Resonant cavity with gap for accelerating electrons.

is fed into the cavity. If the beam loads the cavity lightly, this power is absorbed in the walls. The suitability of the cavity for producing a high gap voltage from a given input power is measured by the shunt resistance R_{sh} which is defined so that

$$\frac{VV^*}{2R_{sh}} = W, \tag{83}$$

i.e. if the gap voltage were applied across the shunt resistance the dissipation would equal that in the cavity. Shunt resistance is a useful quantity in cavity resonators used for this purpose and owes its existence to the focusing of attention upon a particular line integral of electric force.

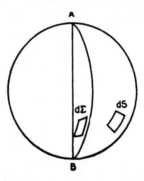

Figure 92.—Shunt resistance of a cavity. The power dissipated in the cavity is equal to that which the voltage along the line of force AB would deliver to a resistance R_{sh}.

In general, if A and B are two points on the wall of a cavity at opposite ends of a line of electric force (Fig. 92), then since

$$W = \tfrac{1}{2}\mathsf{R} \iiint \mathbf{H}\cdot\mathbf{H}^* dS \qquad (84)$$

and

$$V = \int_A^B \mathbf{E}\cdot\mathbf{ds}, \qquad (85)$$

we may define the shunt resistance for the path AB as

$$R_{sh} = \frac{\left| \int_A^B \mathbf{E}\cdot\mathbf{ds} \right|^2}{\mathsf{R}\iint \mathbf{H}\cdot\mathbf{H}^* dS} \qquad (86)$$

where dS is an element of area of the wall and the integral in the numerator is taken along the line of force AB. Substituting for the line integral of $\mathbf{E}$ a surface integral of curl $\mathbf{E}$

and recalling that curl $\mathbf{E} = -j\omega\mu\mathbf{H}$ we have for R_{sh} a formula involving magnetic field only,

$$R_{sh} = \frac{\omega^2\mu^2 \left| \iint \mathbf{H}\cdot d\mathbf{\Sigma} \right|^2}{\mathsf{R} \iint \mathbf{H}\cdot\mathbf{H}^* dS} \tag{87}$$

where Σ is any surface of which AB is part of the rim and the rest of the rim lies in the wall. If the relative permeabilities of dielectric and conductor are unity, the factor $\omega^2\mu^2/\mathsf{R}$ equals $2\omega\mu_0/\delta$.

Values of L and C may be calculated from the Q and resonant frequency once R_{sh} is known. However they may be calculated by direct integration, when the path AB is specified, by using the relations

$$\tfrac{1}{2}CVV^* = U \tag{88}$$

$$LC\omega_0^2 = 1 \tag{89}$$

where U is the energy stored in the resonator.

Thus

$$C = \frac{\iiint \mathbf{H}\cdot\mathbf{H}^* dv}{\omega^2\mu \left| \iint \mathbf{H}\cdot d\mathbf{\Sigma} \right|^2} \tag{90}$$

$$L = \frac{\mu \left| \iint \mathbf{H}\cdot d\mathbf{\Sigma} \right|^2}{\iiint \mathbf{H}\cdot\mathbf{H}^* dv} \tag{91}$$

Equivalent circuits which may be used for calculations on cavities are given in Fig. 93. Only one resonant mode is allowed for and this is usually satisfactory as the behaviour of a resonator is dominated by the mode whose resonant frequency is close to the exciting frequency.

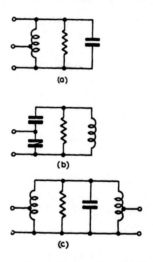

Figure 93.—Equivalent circuits for cavity resonators. (a) Cavity with magnetic coupling. (b) Cavity with electric coupling. (c) Cavity with two coupling loops.

BIBLIOGRAPHY

Proceedings at the Radiolocation Convention March-May 1946, *J. Instn. Elect. Engrs.*, Vol. 93, Part IIIA, 1946.

A. B. Bronwell and R. E. Beam, *Theory and Application of Microwaves*, McGraw-Hill, New York, 1947.

L. G. H. Huxley, *Survey of the Principles and Practice of Wave Guides*, Cambridge University Press, 1947.

W. Jackson, *High Frequency Transmission Lines*. Methuen, London, 1945.

R. L. Lamont, *Waveguides*. Methuen, London, 1942.

S. Ramo and J. R. Whinnery, *Fields and Waves in Modern Radio*. Wiley, New York, 1944.

R. I. Sarbacher and W. A. Edson, *Hyper and Ultrahigh Frequency Engineering*. Wiley, New York, 1943.

S. A. Schelkunoff, *Electromagnetic Waves*. Van Nostrand, New York, 1943.

J. C. Slater, *Microwave Transmission*. McGraw-Hill, New York, 1942.

CHAPTER VII

TRANSMISSION LINE AND RESONATOR TECHNIQUES

IN the previous chapter the theory of waveguides and cavity resonators has been discussed. It is now necessary to relate that theory to practical applications, and to discuss in some detail the more significant of these applications. In actual fact, more than the direct transfer of theory to practical equipment (i.e. working models) is involved ; such aspects as economic considerations, ease of operation, wide utility etc., must also be considered. It is with these aspects as well as the application of theory that the following discussion will be concerned.

When a particular system is to be described, it will be designated as either coaxial line or waveguide, these being the only two which are of concern in this chapter.

1. Feeder Systems in General

In a radar set, the feeder system connects the transmitter to the aerial and the aerial to the receiver and includes the auxiliary parts such as TR and RT switches.

The main requirements of such a system are that the RF power be transmitted with negligible or small losses, and with freedom from electrical breakdown. In addition, as the radar set may be required to work on any one of a number of frequencies in a given frequency band, the feeder system and its associated parts must function equally well over the specified band. This eliminates the necessity for readjustment and tuning of the apparatus every time a change of frequency is made. In practice however, it may not be possible to obtain uniformly good performance over the entire band, due to the frequency-selective elements employed, and so a limit is set for the worst performance that can be tolerated.

This limit varies according to the equipment used, but in some cases a voltage standing wave ratio as low as 0·5 is permissible, which corresponds to 10 per cent reflected power. Naturally it is preferable to obtain higher values of standing wave ratio.

As the feeder systems usually employed are several wavelengths long, the characteristics of the guide selected must be carefully considered. If, as is permissible, the losses in the guiding system are neglected as a first approximation, it is found that in the case of either the coaxial line or the waveguide, the " characteristic impedance of the line " is purely resistive. It can be shown that, for a transmission system with negligible losses and on the assumption of a linear power source, maximum power transfer occurs when at any point in the system the impedance looking in one direction is the complex conjugate of that in the other direction.

In practice it is not possible to obtain a linear power source. With most types of power source, there is a certain load impedance into which maximum power will be delivered. Unfortunately this load value usually corresponds to a point of frequency instability and so a compromise has to be made to determine the operating conditions. If the feeder has a certain characteristic impedance, it is then necessary that a device be used to transform that value to the impedance required at the source.

If the feeder be terminated by an impedance, reflection of power will occur except in the case where the load impedance is equal to the characteristic impedance of the feeder. Thus maximum power is only transferred to the load when its impedance is equal or " matched " to the feeder. The presence of reflected power, which produces standing wave patterns of voltage and current, causes increased conductor losses and increased electric stresses at voltage anti-node points in the feeder. In addition the feeder becomes sensitive to small frequency changes, an effect which is more important when the feeder is many wavelengths long. The presence of standing waves may also react on the power source to produce a change in the oscillating frequency.

Once the characteristic impedance or admittance has been determined by the physical dimensions of the transmission line and the mode employed, it is usual to standardise on that value of impedance or admittance for the rest of the elements involved in the system. It then becomes the reference impedance in terms of which other impedances are quoted. (See section 2.)

For a coaxial line, $Z_o = 138 \, k^{-\frac{1}{2}} \log_{10} (r_2/r_1)$ ohms where r_1 and r_2 are the radii of the inner and outer conductors respectively, and k is the dielectric constant of the medium.

The wave impedance and characteristic impedance of waveguides have been discussed in section 6 of the previous chapter. For waveguides in general, the wave impedance is given by $K_z = \eta \lambda_g / \lambda$ (TE modes) and $K_z = \eta \lambda / \lambda_g$ (TM modes) where η is the intrinsic impedance of the medium, λ_g is the wavelength in the guide and λ the wavelength in air. Now for the TE_{10} mode, $\lambda_g / \lambda = [1 - (\lambda/2a)^2]^{-\frac{1}{2}}$, and therefore $K_z = 377 \, k^{-\frac{1}{2}} [1 - (\lambda/2a)^2]^{-\frac{1}{2}}$, where k is the dielectric constant of the medium and a is the broad dimension of the guide. The characteristic impedance is defined only for a rectangular waveguide carrying the TE_{10} mode as :—

$$Z_o = \frac{2b}{a} K_z.$$

2. Matching

Matching is here defined to mean the electrical transformation of some arbitrary impedance to a specified value. When the reflections caused by a load on a transmission line have been cancelled out by some means or other, the load is said to be matched to the line. Thus, matching in radar practice usually means designing a line network such that an arbitrary load will entirely absorb the power in a wave incident upon it from a transmission line.

Consider the two cases illustrated in Fig. 94. In case (a) with $Z_o = Z$, at any point A the impedance looking towards the load is equal to Z_o, and so there is no reflected power. The load is then said to be matched to the line.

In case (b) where $Z_o \neq Z$, a matching device whose nature is of no concern here must be inserted as shown. At A the impedance looking towards the load, assuming of course that the matching device has been correctly adjusted, is equal to Z_o. Then, if the matching device is dissipationless, there is a maximum transfer of energy to the load. The matching

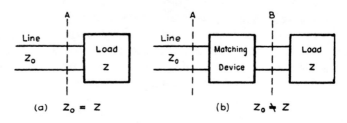

Figure 94.—Various line terminations showing impedance and matching relationships.

device thus transforms the load impedance Z to the line impedance Z_o, and hence is often termed a transformer.

In matching, it is first necessary to know with some precision the impedance of the load. The most convenient method at very high frequencies is to make this measurement by determining the standing waves produced when this load is used to terminate a line of known impedance. The relations between impedance, voltage standing wave ratio S, and the physical displacement of the voltage node from the load have been discussed in the previous chapter. Although the equipment used will be discussed fairly fully in the section on measurements, it is advisable to give at this stage at least an idea of its nature. In the case of coaxial transmission lines, the necessary information may be obtained by a pick-up probe which projects a small distance into a narrow longitudinal slot in the line, and which may be moved along it. Fig. 95 shows the complete " set-up " with detecting crystal and meter.

The arrangement of the slot does not affect the field distribution inside the line, because for the *TEM* mode, which has only axial current flow, it intercepts no current, assuming that the slot width is small in comparison with the main dimensions. Plate X (a) shows a photograph of this type of equipment.

Once the load impedance has been determined it is convenient to work in terms of normalised impedance. This is

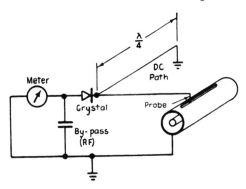

Figure 95.—Impedance measuring equipment. Electrical circuit details.

obtained by dividing the impedance by the characteristic impedance. It is equally suitable to work in terms of admittances rather than impedances, the actual case usually dictating the choice. When the normalised impedance or admittance is known it is necessary to decide how the transformation to the required value is to be made. This is easily done by first plotting the load impedance on either the Smith or Kennelly charts.

Consider for example the point P in Fig. 96 representing an impedance $0.6 - j0.7$ (normalised values) plotted on the Smith chart. To transform this value to that of $1 + j0$, point O, there are several possible devices that may be used :

(a) A reactance of magnitude $+ j0.7$ may be added at the load. This transforms point P to S along the path shown, giving a resistive value 0.6. Then, by means of a quarterwave section, this value is transformed to $1.0 + j0$. [If Z is the normalised characteristic impedance of the $\lambda/4$ section, then $Z^2 = (0.6 \times 1.0)$. Generally $Z^2 = (Z_1 Z_2)$. For further explanation see standard reference books.]

(*b*) At certain distances from the load the impedance corresponds to either points *Q* or *U*, depending on the distance traversed. This distance corresponds to so many electrical

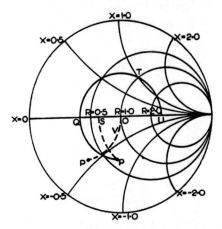

Figure 96.—Smith Chart. Methods of matching a load to a line are shown graphically on this chart.

degrees θ in terms of the wavelength. At either point it is possible, by means of a $\lambda/4$ section of the correct value, to transform to the value $1 + j0$. [See Fig. 97 (*a*).] It should be observed that when dealing with waveguides, the λ referred to is the guide wavelength λ_g of the previous discussions.

(*c*) At a distance θ degrees from the load, and corresponding to point *T*, the impedance is $1 + j1$. By adding a reactance in series at this point, and of magnitude $-j1$ the impedance is transformed along the constant resistance contour to the point $1 + j0$. This particular solution is actually a very neat one, because it requires only one matching element. The disadvantage is that it is often inconvenient to place the reactance at the required point because of mechanical considerations.

(*d*) It is sometimes possible by means of a section of line or guide of the correct impedance and correct length, to transform directly from a point P' to $1 + j0$. This is in effect a combination of the two transforming devices of method (*a*).

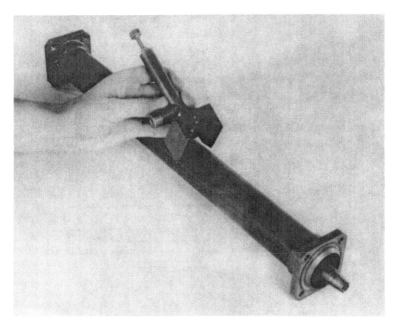

(a) Slotted line impedance gear.

Equipment for use in the wavelength range 15-40 centimetres.

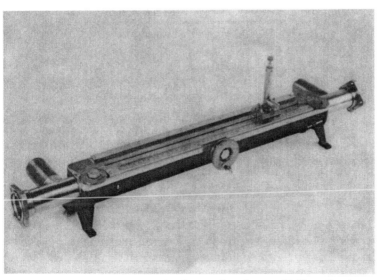

(b) Impedance measuring gear for the wavelength range 20-30 centimetres (see p. 214).

PLATE X.

The determination of the actual dimensions is a little more difficult in this case however, as the transformation follows the curve $P'V$ in Fig. 96. [See Fig. 97 (*b*).]

So far, the discussion has been limited to the matching of

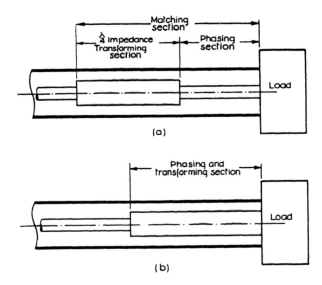

(a)

(b)

Figure 97.—Matching methods. Physical arrangement of matching devices.

particular elements and the methods that may be employed. It is also desirable to have a unit which may be used to match any arbitrary element that may be chosen. This immediately presupposes variable elements, thus eliminating the possibility of the normal $\lambda/4$ sections. Such units have provision for adding shunt reactances at various points of the line.

A convenient unit has two reactance elements spaced $\lambda/8$ apart and placed so that they are effectively in parallel with the transmission line. Fig. 98 shows suitable elements for coaxial lines. The actual function of this transformer is best illustrated by reference to the Smith chart (see Fig. 99), which may be used for admittances by reading off values as G and B, instead of R and X, as in the case of impedances. The admittance is the reciprocal of the complex quantity $R + jX$, and this transformation may be carried out on the Smith chart by rotating the impedance point 180 geometrical degrees about the centre and reading the real and imaginary components obtained as conductance and susceptance respectively.

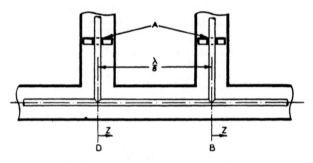

Figure 98.—Two stub matching transformer. Physical arrangement showing stubs and their movable shorting plungers *A*.

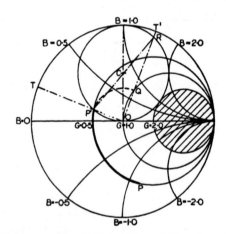

Figure 99.—Tuning range of two stub matching transformer. The transformer may be used to match any impedances except those which lie within the shaded circle.

Assume that point *P* corresponds to the load admittance which is to be matched to the line admittance (point *O*). The further assumption is made that the reactance elements may be adjusted to give any reactance between $+ \infty$ and $- \infty$.

At point B the admittance looking towards the load is that corresponding to P on the Smith chart. At B a susceptance is added which transforms the admittance along a constant G circle to point P' on the chart. The admittance at D is obtained by rotating P' about O through $\pi/4$ electrical degrees to the point Q. This is done by rotating point P' about O for a quarter of a revolution. At this latter point a susceptance is added to transform Q along the $G = 1$ circle to point O.

Consider now the circle $G = 1$ on the diagram. An admittance corresponding to any point on this circle may be matched by a single reactance element. At a distance $\lambda/8$ wavelengths away (toward the generator) any point on this circle is given by a corresponding point on circle $OP'R$. That is, the circle $G = 1$ is transformed through 90 geometrical degrees (45 electrical degrees) as shown, with its centre at C. If now the load admittance is adjusted by means of the reactance element nearest the load to any point on the circle $OP'R$, then the admittance presented at D will be on the $G = 1$ circle as shown by. the previous example and may be matched by a single reactance element. The location of any point corresponding to Q may be determined by drawing $P'CR$ and then joining OR. The intersection of OR with $G = 1$ gives the point Q. Angle $P'OQ$ or TOT' is then a right angle. Alternatively Q may be obtained by the intersection of the circle with OP' as radius and centre O, with the circle $G = 1$.

It will be seen that it is impossible to match any admittance which lies within the circle $G = 2$, although in some cases it is possible to obtain a fairly near match.

The principal use of this example is to show how the Smith chart may give a pictorial presentation of possible transformations with a two stub transformer. In addition, it is also a good example of the use of such a chart for quantitative thinking. The Kennelly chart may be used in exactly the same way for such problems. Other uses will naturally suggest themselves as familiarity with the charts is gained, such as the determination of the input impedance of a line for any loading, etc.

Another matching arrangement has three such elements spaced $\lambda/4$ from each other and allows, in theory, the matching of any given impedance to any required value. The adjustment of these elements offers an infinitude of solutions, so that some procedure must be adopted for tuning. It is feasible to gang the two outer elements and thus have two adjustments. In practice it is often difficult to match even small differences of impedances.

In place of several elements fixed in position, a single variable element whose position may be varied, can be used to match any impedance. This is readily appreciated by reference to the Smith chart.

Another type of variable transformer has two $\lambda/4$ dielectric sections separated by a distance d. The distance d and the position of both slugs in the line may be altered, and it is possible to match any impedance which produces a voltage standing wave ratio of better than $1/k^2$ where k is the dielectric constant. This can be proved mathematically by considering

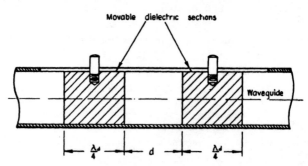

Figure 100.—Dielectric matching transformer, showing the movable dielectric blocks inserted in a waveguide.

the change in impedance of the line due to the dielectric sections. The arrangement is shown diagrammatically in Fig. 100.

Further information as to the matching limits of the above-mentioned methods is given in "Microwave Transmission

Design Data."[1]   Practical construction of these elements and other considerations will be discussed in section on design aspects. This does not exhaust the possible methods of matching, but it gives in essence the principal arrangements.

3. Broadbanding

In section 1 reference was made to the necessity for having good response over a specified band. It is now necessary to discuss this in more detail. The allowable performance is usually specified by a limit of voltage standing wave ratio that is permissible (usually greater than 0·5) in the feeder system.

To discuss broadbanding devices it is convenient to define a term "frequency tolerance" as $(B/f_o) \times 100$ per cent, where B is the bandwidth within which the voltage standing wave ratio is greater than the specified value, and f_o is the mid-frequency. It should be understood that this definition is entirely arbitrary, and is only selected because of its convenience. Speaking generally the tolerance decreases with the number of reactive elements and transforming sections. Hence, in any system with many elbows, stub supports, junctions etc., it is much more difficult to satisfy the bandwidth requirements. It may also be stated as a rough maxim that frequency tolerance is a function of the magnitude of the matching transformation Z_1/Z_2. If the change is small, or several small changes are used to give the required overall change, the bandwidth is greater.

For example, consider the use of a $\lambda/4$ transforming section to change an impedance of say $10 + j0$ ohms to $50 + j0$ ohms.

Then
$$Z = (10 \times 50)^{\frac{1}{2}}$$
$$= 22·4 \text{ ohms.}$$

This transformation is plotted on a Kennelly diagram in Fig. 101, where Z is the transformer impedance. At fre-

[1] T. Moreno, *Microwave Transmission Design Data*, McGraw-Hill, New York, 1948.

quencies above and below the design frequency the transformer is not $\lambda/4$, and so introduces a reactance which in turn means a reflection of power. This is shown in Fig. 101 by points Z_H and Z_S.

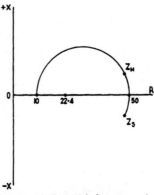

Figure 101.—Impedance variation with frequency for a high ratio quarter-wave transformer.

Consider now the transformation from say $40 + j0$ to $50 + j0$

$$Z = (40 \times 50)^{\frac{1}{2}}$$
$$= 44 \cdot 6 \text{ ohms.}$$

This transformation is plotted to the same scale in Fig. 102.

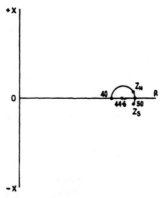

Figure 102.—Impedance variation with frequency for a low ratio quarter-wave transformer.

For the same percentage change of frequency as in Fig. 101, the corresponding points Z_H and Z_S are closer to the required value of 50 ohms, giving much smaller reflections. The same results may be deduced analytically without much difficulty.

If it is desired to transform from 10 to 50 ohms (resistive) it will be seen that it is preferable for this to be done in small steps. The change may be made in as many steps as desirable, and the selection of the ratios of each change may be made in a number of ways. A logarithmic ratio may be used, or alternatively the ratios may be made to conform to the binomial coefficients, 1, 2, 1. (See Slater[2] pp. 55 *et seq.* for a discussion of these ratios).

In the case of reactance (or susceptance) cancellation by the use of correctly located matching elements, there is some scope for increasing the frequency tolerance.

Fig. 103 shows a plot of input susceptance of a short circuited

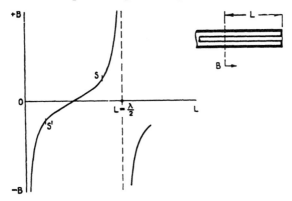

Figure 103.—Susceptance of short-circuited line as a function of its length.

length of transmission line as a function of the length L. When using this device as a shunt matching element it is preferable to make use of the region between the points S and S', both from the point of view of linearity, as well as smaller percentage changes in B when the length is altered slightly. The latter corresponds to small frequency changes with a constant length. By using varying values for the characteristic admittance of the element it is possible to control the slope of the region $S - S'$ or the dependence of the shunt susceptance on frequency. In general, the least slope compatible with the matching requirements should be used. If a shunt element is required at a particular point it may equally well be placed one half-wavelength away. By making the proper choice the net frequency tolerance may be increased.

After the frequency tolerance has been increased by the above methods, there is yet another device which may be used to increase the tolerance of the whole system. Consider a half wave section of line ; from transmission theory the impedance is identical at either end, and is not dependent on the actual characteristic impedance of the section. (Unduly large changes of section will however have some effect, due to the distortion of the fields by fringing effects.) If now a $\lambda/2$ length of line of different characteristic impedance is introduced into the main line, it will have no effect at the correct frequency. At either higher or lower wavelengths its length will not be $\lambda/2$, and so some reactance will be introduced, which may be used to cancel out the existing reactance

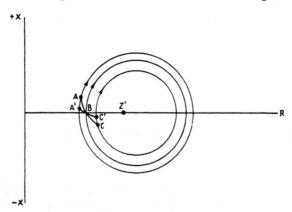

Figure 104.—Diagram showing the impedance transformation ABC to $A'BC'$ by means of a half-wavelength section, and the resulting increase in frequency tolerance.

Fig. 104, shows points A, B and C representing the impedance of a load at low, centre and high frequency points of the required band. If a $\lambda/2$ section of line of characteristic impedance Z' is inserted, the points are transformed as shown, to A',B,C', which are closer to the required impedance at B. By adjusting the magnitude of Z', considerable control of the points A' and C' may be obtained. If however, the points A and C had been reversed, it would have been necessary to use a value of Z' smaller than B.

When using the $\lambda/2$ section it should be located so that the locus of the impedance is vertical at the load side of the section.

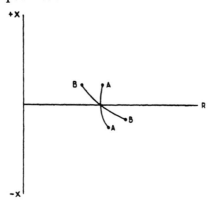

Figure 105.—Diagram showing the optimum frequency-impedance curve A-A to obtain the best results from a half-wave broadbanding section.

This corresponds to locus AA of the diagram (Fig. 105) and gives the optimum matching conditions. When the locus corresponds to BB, only very slight improvements may be obtained by using a $\lambda/2$ section. As only reactances are being introduced by the section, the resistive components are left unchanged.

4. Losses and Power Capacities of Transmission Line

Before considering practical design aspects it is advantageous to discuss briefly the properties of various materials in common use at microwavelengths. So far, all the systems have been treated as perfect, lossless systems ; but it is now necessary to see where losses do occur, and to gain some idea of their magnitude. It will be found that actually these are so small as to have no appreciable effect on the previous calculations.

In practice, most equipment is designed assuming perfect conductors, and then with the magnitudes of the current flow thus obtained and a knowledge of the resistive properties of the materials involved, the actual losses are determined.

At microwave frequencies, current flow in a conductor is determined principally by " skin effect," a discussion of which has already been given in the previous chapter. If δ, the depth of penetration, is defined as the depth below the surface of a conductor at which the current density drops to $1/e$ of its surface value,

$$\delta = 0.029\left(\frac{\lambda}{\mu_r g}\right)^{\frac{1}{2}}$$

where g is the conductivity of the conductor in mhos per centimetre, μ_r is the relative permeability (μ/μ_o), and λ is the wavelength in free space in centimetres.

Fig. 106 gives values of δ for different materials for wavelengths up to 30 centimetres. As δ becomes smaller, the current flow becomes greater at the surface of the conductor,

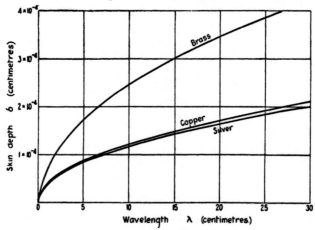

Figure 106.—The depth of penetration (δ) of currents in various conductors at very short wavelengths.

and it is because of the excessively high surface current densities that there are considerable losses.

The actual surface of the material is of some importance, depending particularly on the application. Very little quantitative information of much significance is available however, and it is often necessary to make tests on individual materials. More will be said in section 6 regarding the surfaces of high Q resonators.

The dielectric in most feeder systems is air at atmospheric pressure. It is usual to accept its dielectric constant as unity and to assume that its losses are entirely negligible. In actual fact, dry air has a dielectric constant of 1·006, and its losses are zero until a wavelength of 0·5 centimetres is reached, when there is a pronounced resonance absorption due to the magnetic moment of the oxygen.[3] Atmospheric air differs considerably from the ideal, particularly in its losses, which vary with both temperature and moisture content as described in Chapter II, section 6. There is also a resonance absorption band in the neighbourhood of 1·3 centimetres due to the electric moment of the water vapour molecule.

The properties of most dielectric materials in common use are available in various recent publications.[4]

Any loss in a dielectric can be characterised by the complex dielectric constant

$$k' = k - jk''$$

where $k'' = g/\omega\epsilon_0$. The loss tangent of the dielectric is defined by

$$\tan \zeta = \frac{k''}{k},$$

and is equal to $g/\omega\epsilon$.

The attenuation in a coaxial cable, resulting from dielectric losses is given by

$$\alpha_d = 27\cdot3\,\frac{k^{\frac{1}{2}}}{\lambda}\tan \zeta \text{ decibels per metre.}$$

The attenuation due to conductor losses in a coaxial line

[3] J. H. Van Vleck, "Absorption of microwaves by oxygen," *Phys. Rev.*, Vol. 71, 1947, pp. 413-24.
[4] "Tables of dielectric materials," Laboratory for Insulation Materials, Massachusetts Institute of Technology, National Defence Research Committee Reports 14-237 and 14-425, 1945.

with the same material for both conductors is given by

$$\alpha_c = 819 \frac{\delta}{Z_0 \lambda}\left(\frac{1}{r_1} + \frac{1}{r_2}\right) \text{ decibels per metre}$$

where r_2 is the internal radius of the outer conductor, and r_1 is the radius of the inner conductor.

The losses in the conductor of a waveguide are dependent on the mode that is set up. For the TE_{10} mode (the reason for this choice will be seen later) the attenuation due to conductor loss is[5]

$$\alpha_c = 27 \cdot 3 \frac{\delta}{a\lambda} \frac{\left[\dfrac{a}{b} + 2\left(\dfrac{\lambda}{2a}\right)^2\right]}{\sqrt{1 - \left(\dfrac{\lambda}{2a}\right)^2}} \text{ decibels per metre}$$

where a is the broad dimension of the guide, and b is the narrow dimension of the guide. When there is attenuation due to both conductor and dielectric losses, the total attenuation is the sum of the two :

$$\alpha = \alpha_c + \alpha_d.$$

A good visual comparison of the attenuation in coaxial lines

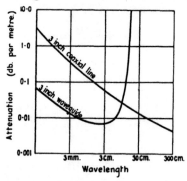

Figure 107.—Comparison of attenuation in a coaxial line and waveguide of identical outside dimensions.

and waveguides is given in Fig. 107 which shows the attenuation for a 3 inch coaxial line of optimum impedance and a 3 inch circular guide with a TM_{01} mode.

[5] See Chapter VI, equation (45).

Table 6 shows attenuation constants of some transmission lines.

TABLE 6—ATTENUATION IN TRANSMISSION LINES

Transmission line dimensions (inches)	*Wavelength (centimetres)*	*Attenuation constant (decibels per metre)*	
Waveguide			
$6\frac{1}{4} \times 2\frac{1}{2}$	25	0·0060	
$3 \times 1\frac{1}{2}$	10	0·0130	
$1 \times \frac{1}{2}$	3	0·067	
Air-filled coaxial			
$1\frac{5}{8}$	25	0·020	
$1\frac{5}{8}$	10	0·0315	
$\frac{7}{8}$	10	0·059	
Polythene cable		Conductor attenuation	Dielectric attenuation
$\frac{3}{8}$	25	0·165	0·097
$\frac{3}{8}$	10	0·260	0·24
$\frac{3}{8}$	3	0·480	0·8

The waveguides are specified by the internal dimensions, and coaxial lines by the internal diameter of the outer conductor. In all cases the coaxial lines have been assumed to have a characteristic impedance of 50 ohms. With the polythene cable, a dielectric constant of 2·2 and a loss factor of 0·0006 have been used. The attenuation in all cases has been calculated for copper conductors.

Before discussing the practical design aspects of feeder systems, it will be profitable to discuss very briefly the safety factor of such systems. Electrical breakdown occurs for two reasons, arc-over because of high electric stresses in the feeder, and localised heating and oxidation at points of high RF resistance.

The former is minimised by careful choice of dimensions, and by ensuring that no sharp edges occur in either the coaxial line or the guide where there is increased dielectric stress.

Breakdown of dry air at atmospheric pressure occurs with an electric stress of about 30,000 volts per centimetre, but this value decreases considerably with moisture content and lower pressures. It is therefore necessary that this value be decreased by a factor of safety to give a suitable working potential difference. This will make allowance for resonant voltage step-ups which occur in matching devices, chokes, etc., as well as preclude the possibility of breakdown due to varying atmospheric conditions. The breakdown stress is proportional to the square root of the pressure, a fact of considerable importance in equipment designed for use at high altitude.

The power that can be transmitted in a feeder is given by

$$P = 6 \cdot 63 \, E^2. \, a.b. \, \frac{\lambda}{\lambda_g} \, 10^{-4} \text{ watts}$$

for a waveguide[6] (TE_{10} mode) and

$$P = \frac{E^2 r_1^2 Z_0}{7200} \text{ watts}$$

for a coaxial line where E is the peak electric field strength.

At $\lambda = 10$ centimetres and taking $1 \cdot 5 \times 10^4$ volts per centimetre as the maximum-permissible electric stress, a waveguide 3 inches $\times$ $1\frac{1}{4}$ inches will carry $2 \cdot 4$ megawatts of power, whereas a 50 ohm coaxial line, $1\frac{3}{8}$ inches outer conductor, will only carry $1 \cdot 0$ megawatts.

In practice, it is usual to determine the factor of safety on a power basis and a value of $0 \cdot 2$ to $0 \cdot 25$ of the maximum value is commonly used.

5. Design Aspects

The first point to be considered in designing feeders and associated equipment is the type of feeder to be used. Twin wire lines are automatically excluded because of their high radiation at microwave frequencies. Coaxial lines may be used at frequencies as high as 3,000 megacycles per second but attenuation is fairly severe at the higher limit. In addition, the physical dimensions must be kept sufficiently small to prevent a higher transmission mode existing in the line. It should be emphasised that in all transmission systems there should be only one mode of energy propagation. If two or more exist, the field patterns become complicated and considerable difficulty would exist in designing equipment to

[6] See Chapter VI, equation (29).

perform desired functions. Therefore in the following remarks, a single propagation mode is assumed.

Waveguides are most conveniently used for frequencies
above about 1,000 megacycles per second ; at lower frequencies
they become bulky. Thus there is considerable overlap in
the useful frequency range of the two types of feeder (see
Fig. 65, Chapter VI) and a selection is usually made by taking
into consideration the power to be handled, attenuation and
ease of fabrication.

Up to date, in discussing the theoretical aspects of feeders
no distinction has been made between coaxial lines and waveguides. However, in treating the actual design of such elements, it is necessary to consider them separately because of
their physical difference.

(a) Coaxial Lines

For most low power work and on frequencies below about
3,000 megacycles per second flexible coaxial conductors are
preferred for inter-connecting apparatus. At 10,000 megacycles per second losses become too high to allow them to be
much used. They are usually made with a solid or stranded
copper inner conductor, a waxy type solid flexible dielectric and a copper braid outer conductor. The dielectric is as a rule one of the hydrocarbon polymers
(polystyrene, polyethylene etc.) which have a complex dielectric constant of the order of $k' = 2 \cdot 4 - j0 \cdot 0014$.

Where the dielectric losses of a solid dielectric are too high,
spaced dielectric supporting beads may be used. These
unfortunately introduce a reflection of power at every bead,
due to the change in the characteristic impedance at that point.
This difficulty may be overcome by the correct spacing of the
beads so that the reflections cancel each other. There are
numerous combinations that will allow this—firstly a bead
distance of slightly less than $\lambda/4$ will give good results for
short lines, whereas the spacing for long lines must be increased slightly above $\lambda/4$ for optimum results. Another
method starts with two beads correctly spaced for minimum
reflection, and places similar units with their equivalent

electrical centres at distances of $(\tfrac{1}{2}n - \tfrac{1}{4})\lambda$ (where n is any positive integer). A further method is to make the beads $\lambda/2$ long, which unfortunately makes them fairly frequency selective. An additional, and in many ways a preferable method,

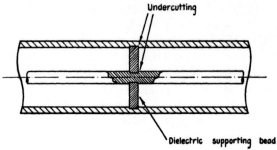

Figure 108.—Undercut bead support for coaxial line to reduce reflection of power.

is to undercut the inner and/or outer conductors in order to make the equivalent impedance at the bead identical with the rest of the line. Negligible reflections are set up by this device which is shown in Fig. 108.

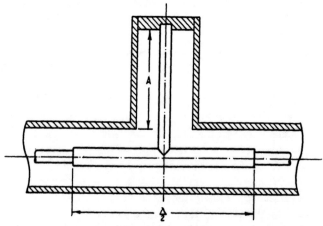

Figure 109.—Broadband stub support, showing the combination of a half-wave broadbanding section with the quarter-wave supporting stub (dimension *A*).

For high power work, any such beads in the line cause distortion of the field in their immediate vicinity. This may be sufficient to raise the electric stress above the breakdown point. Hence an alternative scheme is necessary to support the inner conductor. A quarter-wave shorted section provides a very convenient means of doing this, and the arrangement is shown diagrammatically in Fig. 109. The dimension A which is approximately one quarter wavelength is determined experimentally so that no reflection is produced at the mid-frequency. As the short circuited section is fairly frequency selective, it is combined with the half-wave section as shown to provide broadband response. Reference to an impedance chart shows that variation of the electrical length of the $\lambda/2$ section with frequency tends to cancel the charging stub reactance. The final result is shown in Fig. 110 which also shows the response of a plain quarter wave section supporting stub. In addition to good electrical performance this broadband stub gives excellent mechanical support for the inner conductor.

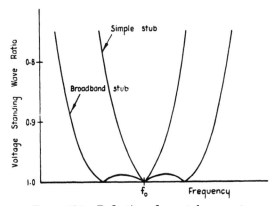

Figure 110.—Reflections from stub supports.

The selection of the characteristic impedance of the coaxial line depends on many considerations, loss, factor of safety, etc. In Fig. 111 the essential characteristics of the coaxial line are graphed against characteristic impedance Z_o.

The final choice of Z_o will usually be a compromise between the various factors, and values of 50 and 75 ohms have been extensively used. After the ratio of the diameters (or Z_o)

has been fixed, it is necessary to limit the actual dimensions, so that the mean circumference $\pi(r_1 + r_2)$ is less than λ. If this is not done the TE_{11} mode is capable of being propagated.

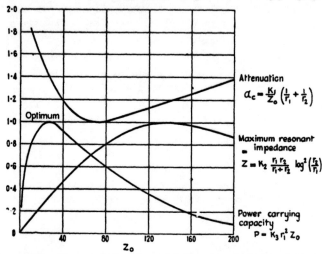

Attenuation

$$\alpha_c = \frac{K_1}{Z_0}\left(\frac{1}{r_1} + \frac{1}{r_2}\right)$$

Maximum resonant impedance

$$Z = K_2 \frac{r_1 r_2}{r_1 + r_2}\log^2\left(\frac{r_2}{r_1}\right)$$

Power carrying capacity

$$P = K_3 r_1^2 Z_0$$

Figure 111.—Characteristics of coaxial lines as a function of Z_o for a fixed value of r_2. Curves are referred to an optimum value of unity.

Leaving aside connection by cables it is apparent that for high power work, rigid elements must be considered, which for convenience do not normally exceed about three feet in length.

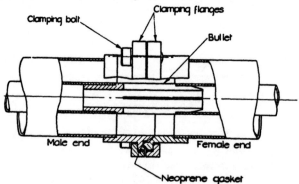

Figure 112.—Coaxial line joint for a $1\frac{1}{8}$ in. O.D. line.

The problem is then how to make joints, bends, allow rotating motion, etc.

The method of joining straight sections of line is best seen by reference to Fig. 112 which shows the mechanical details. The result is a satisfactory rigid joint which may be water-proofed by a " Neoprene " or rubber gasket at A. The original design of this type of joint is due to the Radiation Laboratory of Massachusetts Institute of Technology.

Normal manufacturing tolerances allow 1/64 inch, thus giving the possibility of a gap 1/32 inch at the joint of the inner conductor. It will be apparent that the inner conductor must always have a negative tolerance in order to avoid compression and buckling which would set up reflections.

For bends, a right angle stub support device is used (see Fig. 113).

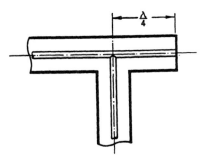

Figure 113.—Right-angle bend for coaxial line with a quarter-wave supporting stub.

For changing from one size of coaxial line to another, taper sections are used in which the change of dimension occurs over an integral number of half-wavelengths. The reflection from one of these sections is fairly small and decreases with increase in taper length.

An important device in most radar systems is a " rotating joint " in the feeder line, to allow rotation of the aerial and feed system. There are a number of methods by which this may be done, but they are in general derivatives of the type shown in Fig. 114.

In the construction shown there are two open circuited $\lambda/4$ lengths of line which produce an effective short circuit at point *A*. To the RF energy there is then no discontinuity and the device behaves like a continuous line. (The concept of a plain capacity by-pass is not directly applicable to this device because the distance involved is comparable with the wavelength). As the frequency varies, the open circuited

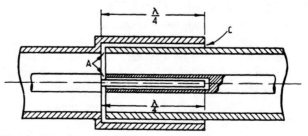

Figure 114.—Rotating joint for an air dielectric coaxial line.

lengths are no longer quarter wavelengths and so there is a small reactance at *A*. Actually there is always some radiation from point *C*, but it is so small as to be unimportant. It would of course be possible to make spring contacts in place of the quarter wave sections shown, to allow relative motion of the two sections, but experience has shown that this is a most unsatisfactory method. Excessive wear, dirt and grease cause high losses and necessitate increased maintenance. Although this design is quite a suitable one, in practice more elaborate ones

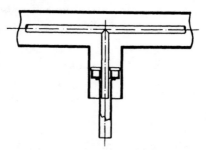

Figure 115.—Variable reactance element showing a movable shorting plunger in the side coaxial line.

are used to obtain better performance, lower radiation, better power handling etc.

Reactance elements referred to in the section on matching

generally take the form of a movable shorting plunger in a stub line. Fig. 115 shows the arrangement. A little thought will show that the reactance produced is effectively in shunt with the main line—hence it is usually more convenient to discuss the susceptance introduced.

(b) Waveguides

The design of elements using waveguides is simplified by virtue of the absence of an inner conductor which would require careful supporting ; however, more complex transmission modes are used making calculation somewhat more difficult. As described in the previous chapter there is an infinite number of possible transmission modes, each with its own field pattern. In practice only a few of the modes of lowest order are of importance for reasons of smaller physical size, lower attenuation and greater separation of the modes. Whilst any cross-section of guide may be employed, two only are of common use, the circular and rectangular types.

The circular waveguide is used in limited applications only,

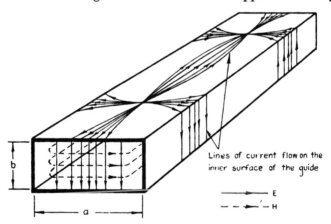

Lines of current flow on the inner surface of the guide

———— E
— — ▸ — H

Figure 116.—Field and current distributions in a rectangular waveguide for the TE_{10} mode.

because of the difficulty in maintaining the polarisation of the TE_{11} mode in the correct plane. In rectangular guides this difficulty is absent, hence their more general application.

The determination of the fields in a waveguide is treated in Chapter VI.

The selection of the dimensions of a waveguide for a particular wavelength is governed mainly by the cut-off or critical frequency, the power capacity, and the attenuation desired. With rectangular guides, the most commonly used mode, which gives the smallest physical dimension is the TE_{10}. This has a field distribution as shown in Fig. 116. The current

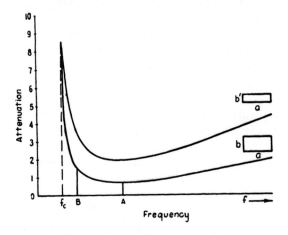

Figure 117.—Variation of attenuation with frequency for waveguides having different dimensions b. ($T\ddot{E}_{10}$ mode.)

flow is shown on the surface of the guide and the field distribution across the cross section.

Fig. 117 shows the variation of attenuation in a waveguide with frequency and the ratio a/b for a TE_{10} mode. Although point A corresponds to minimum attenuation the dimension a is usually chosen to give an operating frequency B approximately 1·5 times f_c. Greater values of B allow the possibility of higher order modes being propagated and must be avoided even though the attenuation is slightly increased.

In the case of the TE_{10} mode where the critical frequency is independent of the dimension b, this dimension must be chosen from a consideration of the permissible attenuation and the power to be handled by the waveguide. The question of power transfer has already been dealt with in section 4. There is however a limit in this regard, namely that it must be sufficiently small to prevent the propagation of any other undesired modes.

In circular waveguides the dominant mode is the TE_{11}. The field pattern for this mode, which is polarised, may be distorted when bends or corners are introduced in the guide. This in turn means that the plane of polarisation may be rotated, and in some cases elliptical polarisation may be produced causing difficulty in receiving energy from the guide. For rotating joints, the TM_{01} mode is preferred because of its circular symmetry, but filter devices are then necessary to prevent the lower TE_{11} mode from propagating. These filters will be described later in this section.

The jointing of waveguides may be effected in two distinct ways—plain butt joints bolted together, or choke type joints. The former suffers from the disadvantage of requiring careful machining in order to produce small losses. The choke type of coupling is shown in Fig. 118.

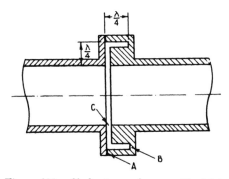

Figure 118.—Choke type of waveguide joint.

In effect, there is a short-circuited half wavelength of radial line which gives a short circuit condition at c. The joint then becomes equivalent to a continuous length of line. This type of joint may be made equally well in either round or rectangular guides. A-B is a short-circuited quarter wavelength of line giving a current node at A,—hence the metallic

connection need not be first class, thus allowing coarser tolerances in the machining. The loss in this type of joint at
10 centimetres is of the order of 0·01 to 0·03 decibels whereas
the average butt type gives a loss of about 0·05 decibels or
higher.

Changes of direction of waveguide runs, and right angle
bends are easily accomplished. A circular guide may be bent
with a radius of curvature large compared with λ, but unfortunately this introduces elliptical polarisation of the wave

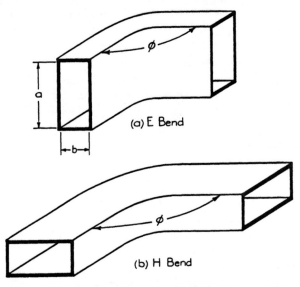

Figure 119.—Types of bends for rectangular waveguides.

which in general is somewhat troublesome. Although it is
theoretically possible to correct this, it presents considerable
difficulty in practice.

With rectangular guides, a bend may be made in two ways—
either on the broad or narrow side. Fig. 119 shows the two
types of bend and, as before, the radius of curvature must be
large compared with λ if a negligible reflection is to be introduced. The H-bend is so named through being bent in the
plane of the magnetic field. Another possible type of bend

is shown in Fig. 120.   By inserting a plane at the corner as
shown, conditions for the continuity of fields around the corner
may be established.   There is as yet very little satisfactory
theoretical treatment of this problem, most results having

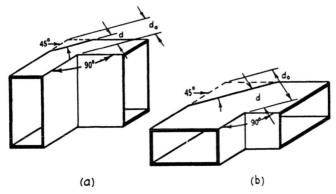

(a)   (b)

Figure 120.—Corners in waveguides, showing the reflecting plates.

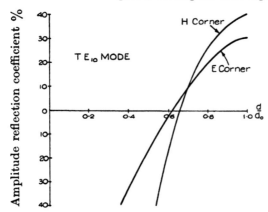

Figure 121.—Reflections from waveguide corners, as a function of the
ratio d/d_0. (See Fig. 120).

been obtained experimentally.   Fig. 121 shows curves of
reflection coefficient versus the ratio d/d_o for right angle
bends, due to English investigators.   (Negative values have
been plotted in this graph to avoid cusps).   It is evident
that the E-bends are much less critical of adjustment and
thus more satisfactory in practice.

Twists may be made in waveguides provided that the guide is not deformed unduly in cross section. For small reflection, the length of the twist should be of the order of one wavelength, or more. Flexible guide may also be used in certain applications ; however the attenuation is somewhat greater than for solid guide, and some reflection of power is present. Flexible guide may be made in a number of ways—in bellows form with each fold a half wavelength long or wound from strip as in the construction for flexible metallic piping.

A taper may be used in a waveguide to transform from one mode to another ; for example, from a TE_{11} mode in a round guide to a TE_{10} in a rectangular guide. This will be seen to be a transformation from the lowest mode in one guide to the lowest in another. Transformation between higher modes usually introduces unwanted modes which then necessitate filter devices. The reflections set up at such a taper are a function of the length of the taper.

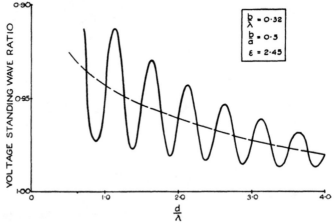

Figure 122.—Reflections from a tapered dielectric section of length d inserted in a waveguide ; taper in E-plane.

A change in dielectric in a guide may be conveniently made by taper sections, starting from zero thickness and increasing to full guide section in a distance d. Fig. 122 shows the theoretical reflection set up from such a taper. The transition could of course have been made by means of quarter-wave transforming sections, but the taper method obviates the need for an intermediate dielectric and is less frequency selective.

If an abrupt change of section occurs in a guide, or an obstacle is inserted into the guide, field distortions occur which are equivalent to the addition of a susceptance at that part of the guide. Because of this distortion, which requires several wave functions to satisfy the boundary conditions, higher order modes are excited in the guide, but these usually suffer high reactive attenuation and die away rapidly in either direction from the discontinuity. These higher order waves store energy in a similar manner to an inductance or capacitor, and therefore give the same effect as a shunt susceptance at the point of discontinuity. For this reason, admittance or impedance measurements should be made at a distance of the order of the greater transverse dimension from such a point. The wave admittance may be written $1 + j0$ and the element or section under consideration may be regarded as adding a normalised susceptance $\overline{B}$. The admittance at any point may then be written $Y = 1 + j\overline{B}$, and used in calculations in the normal transmission theory form.

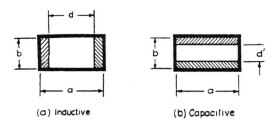

(a) Inductive   (b) Capacitive

Figure 123.—Waveguide irises. Reactive elements which may be inserted in waveguides.

As mentioned in the previous chapter irises in waveguides may produce either positive or negative susceptances depending on the way in which they are formed. The two main forms, inductive and capacitive, are as shown in Fig. 123.

$$\bar{B} \text{ inductive} = -\frac{\lambda_g}{a} \cot^2 \frac{\pi d}{2a}$$

$$\bar{B} \text{ capacitative} = \frac{4b}{\lambda_g} \log_e \left(\operatorname{cosec} \frac{\pi d}{2b} \right).$$

Asymmetrical types may be used but an allowance must then be made for the asymmetry.

Screws or metal cylinders inserted into the broad side of waveguides add a capacitative susceptance, which changes abruptly to an inductive value as the insertion is increased. The changeover occurs when the insertion is approximately one quarter of the free space wavelength. In most cases, only the capacitative values are used as the adjustment in the neighbourhood of the changeover point is too critical. Unfortunately both the screw and the capacitative iris are of little value at high power because of the increased liability of arc-over.

Under certain conditions irises of appropriate dimensions and obstacles such as rings display resonance phenomena in

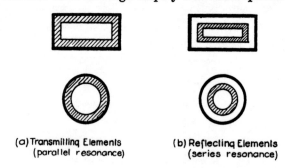

(a) Transmitting Elements   (b) Reflecting Elements
(parallel resonance)   (series resonance)

Figure 124.—Resonant waveguide irises, similar to parallel and series resonant circuits.

that they either transmit all the incident power or completely reflect it. As it is not possible to discuss in any detail the determination of the resonant frequency of such elements, it is sufficient to indicate their general configuration.

Fig. 124 shows the main types of elements, which are usually constructed from thin sheet metal. The application of such devices to filters of all types will be immediately apparent.

T-junctions in waveguides form an important class of elements, both as dividing networks and as matching elements, and they may be treated in much the same way as conventional six terminal networks as far as impedance properties are con-

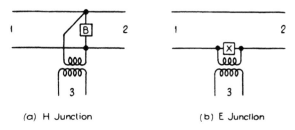

(a) H Junction   (b) E Junction

Figure 125.—Equivalent circuits of T-junctions in waveguides.

cerned. With rectangular guides, junctions may be made in two ways, either in the broad or narrow sides. The impedance relations in either case are obtained by setting up the wave equations at the junction and solving for the correct boundary conditions. This involves higher order modes which, as in the case of irises are rapidly attenuated at a distance from the junction. With a TE_{10} wave, a junction in the broad side of the guide produces a predominantly series loading to the main guide and is called an E-junction. A junction in the broad side produces predominantly shunt loading and is termed an H-junction. This is not necessarily true for other main modes of propagation.

The T-junction can best be represented as shown in Fig. 125, where 1 and 2 are the main arms, and 3 is the side arm. In either the H or E types, there is a common susceptance or reactance at the junction and the side arm is connected through a step down transformer. The effect, besides giving a step down of impedances from the side arm to the main arm, produces a phase shift at the junction which must be considered in the design. To overcome the step down transformer action, an iris of suitable dimensions or some matching section must be inserted in the side arm guide. If this is not done, a correct termination in the side arm will not appear as a correct termination in the main guide when a shorting plunger is placed in the main guide on the opposite side of the junction, and at a correct distance from it.

The impedance or admittance relationships may easily be determined experimentally by placing known impedances at two points, and measuring the third impedance. This simple method will give quite reasonable results.

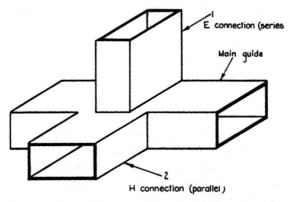

Figure 126.—" Magic tee " junction in a rectangular waveguide, showing the side *E* and *H* connections.

By placing a shorting plunger in the side waveguide of a junction, varying values of reactance or susceptance may be introduced into the main line. Thus, very convenient matching elements may be made.

By adding a second side waveguide at the junction an eight terminal network is produced which has also certain impedance symmetries. One special case of interest is that of the " magic tee," in which there is one side connection to the broad side and one to the narrow side of the main guide. Fig. 126 shows the general arrangement.

Consider now a TE_{10} wave incident at the junction in guide 2. The magnetic field will couple into both arms of the main guide and excite waves in them which are in phase. No wave will be excited in guide 1 because neither the magnetic nor the electric fields present will couple to it to produce a TE_{10} wave. If now a TE_{10} wave in guide 1 is incident at the junction, the electric field couples to the main guide and excites waves in either arm which are 180 electrical degrees out of phase. Again there is no coupling between 1 and 2.

These properties allow the device to be used in the way that a hybrid coil is used at low frequencies. Further uses are as impedance bridges and in special mixer circuits where high attenuation is required between input and local oscillator circuits.

The non-directional line coupler is also an example of an eight terminal network. It consists of one transmission line (coaxial or waveguide) coupled to another line by means of a small hole as shown in Fig. 127. The coupling between the lines is determined by the size of the hole. If a

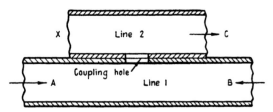

Figure 127.—Non-directional coupler for coupling two waveguides (or coaxial lines).

short circuit is placed in line 2, one quarter wavelength from the hole (position X), power will be transferred from line 1 to line 2 and will flow in line 2 regardless of the direction of flow in line 1. By making the hole small enough a high degree of attenuation may be introduced between the two lines.

This device is considerably affected by standing waves in the line ; in other words, the coupled power is dependent on the location of the hole with respect to voltage maxima. By making use of two such holes a quarter wavelength apart, independence of location may be achieved giving the twin-hole " waveselector " or " directional coupler," the waveguide version of which is shown in Fig. 128.

Power flowing in line 1 in direction A couples some power into line 2. This power can flow only in direction C, as for direction D there is a path difference of 180 electrical degrees for power coupled through the two holes, resulting in cancellation. Power flowing in direction B in line 1 will give a power flow in direction D in line 2. This power is completely

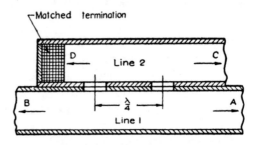

Figure 128.—Directional coupler for coupling two waveguides.

absorbed by a matched termination and hence no power flows in direction C, either because of reflection or direct coupling. Thus it may be seen that the device may be used to determine the ratio of direct and reflected power in a transmission line by merely reversing C and D and measuring the power in line 2 in both cases.

The production of a rotating joint in waveguides usually involves firstly the transformation of the main transmission mode (TE_{10} etc.) to one which has circular symmetry. The most commonly used waves are the TM_{01}, and TE_{01}. As these are both higher modes for a circular guide than the TE_{11}, care must be taken to prevent it from being propagated by employing a filter or resonant ring in the guide. Once circular symmetry has been obtained then relative motion can be easily provided by using a device similar to that for the coaxial line (an open-circuited quarter-wave radial transmission line without an inner conductor assembly). In the most common set-up, a transfer is made from a TE_{10} mode in a rectangular guide to a TM_{01} mode in a circular guide. This may be done in several ways, the criterion of performance being that reflections should be negligible and that only the TM_{01} mode exists in the circular guide. Fig. 129 indicates one way in which it may be done.

From previous remarks the stub will be seen to be effectively in series with both guides.

The resonant ring is $\lambda_{TE11}/4$ from the centre of the junction and presents a high series impedance to the TE_{11} mode. The main stub being $\lambda_{TM01}/2$, (d_2), presents a low series impedance to the TM_{01} mode. Thus the TE_{11} mode in the circular guide

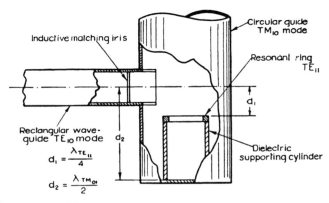

Figure 129.—TE_{10}–TM_{01} transformer for transforming a TE_{10} wave in a rectangular guide to a TM_{01} wave in a circular guide.

is inhibited and the TM_{01} encouraged. In this particular transformation an inductive matching iris is necessary in the rectangular guide to provide correct matching into the transformer. If the device is dissipationless, then by the reciprocity theorem, the transformation from TM_{01} to TE_{10} may be made with this same device by reversing the power flow.

For transforming from coaxial line to waveguide, a probe or loop from the coaxial line must be inserted in the waveguide. For a matched condition to be obtained, the probe or loop must be resonant ; this means the probe must be approximately a quarter of a wavelength, and the loop approximately a half wavelength long. Two typical couplings are illustrated in Fig. 130. It will be noticed that in the high power feed (*b*),

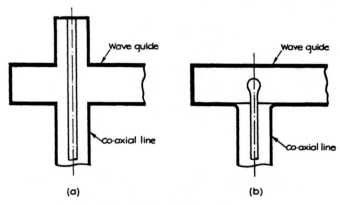

(*a*)   (*b*)

Figure 130.—Coaxial to waveguide couplings.

edges are carefully rounded to prevent high localised electric stresses.

6. Cavity Resonators

The uses of cavity resonators are numerous, klystrons, magnetrons, wavemeters and filters being but a few of the many applications. Although in theory resonators may have any shape, in practice they are usually restricted to about three main forms—circular cylinders, coaxial cylinders and rectangular prisms. Because of ease of fabrication, the circular cylindrical types are the most common. The resonant frequencies of these types of resonator have been given in the previous chapter.

The coupling to a cavity can be considered as a means of forming a flux distribution inside the cavity. As described in the previous chapter, this may take the form of either loops, probes or windows, depending on the feed system employed. Now, the nature and position of the feed influence the type of mode excited, and by careful design it is possible to inhibit spurious modes. These spurious modes are usually unwanted ones which occur at frequencies close to those on which it is desired to operate. As a rule they also have lower Q values. If for example a TE_{011} mode is used in a tunable cylindrical cavity, there is a TE_{311} mode of lower Q very close in frequency and it is possible that in tuning, the two points may be confused. For the type of coupling used (loop) it is not convenient to vary its location except to a restricted degree, and so it is not possible to excite only the desired mode. To avoid this a study is made of the field patterns of both modes, TE_{011} and TE_{311}. This reveals the fact that for the TE_{011} mode it is not necessary to establish electrical continuity between the ends and walls of the cylinder, whereas the TE_{311} mode requires this continuity. The first step then is to make a false bottom for the device. In addition, by placing an absorbing material, wood bakelite, behind the bottom plate, the fields of the unwanted mode which penetrate to this space are absorbed and so cause it to be highly damped. It is possible by this method to reduce the ratio of wanted to unwanted response from say 3 to 1 to about 100 to 1.

Other methods are available for reducing the response to unwanted modes, such devices as wires correctly positioned in the cavity, absorbing materials etc., being fairly common. In each case, a study of the field patterns of the required and spurious modes must be made to determine the method to be employed.

In the case of a hollow cylindrical cavity, the resonant frequency is usually altered by moving one of the end plates in or out. It is essential in such a case that the mode of oscillation used requires no electrical continuity at the junction of wall and end plate. The TE_{011} mode is particularly suited to this application. Coaxial type cavities may be made with a gap in the centre post, Fig. 131. By varying the distance d the resonant frequency may be altered conveniently. At point A the current is small and the contact problem is not unduly important.

The dependence of Q on dimensions etc. has already been derived in the previous chapter and has been given by the approximation

$$Q \doteq \frac{1}{\delta} \frac{V}{S}.$$

The actual value of Q for any particular mode in a resonator is a fairly complicated function of the dimensions and the

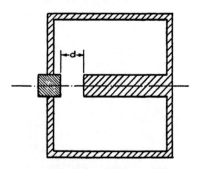

Figure 131.—Semi-coaxial cavity resonator. The resonant frequency is a function of the distance " d."

material of which it is made. Graphs are given in Chapter VI for the determination of Q for common shapes and the more important modes.

As can be seen from the above equation, for a fixed wavelength the Q value increases with size (ratio V/S becomes larger) which means higher order modes. The TE_{01n} modes for. cylindrical resonators have high values of Q and are generally used in applications where high Q is essential.

The production of cavities with high Q values is relatively easy, values up to 100,000 at a wavelength of 3 centimetres being attained without undue bulkiness. The interior surface of the cavity appears from observation to be highly important as regards the achievement of a high Q. A highly polished silver surface gives best results. The presence of water in droplet form, oxidation and dirt appear to lower the Q very substantially.

The effects of the deformation of cavities are important, particularly with prisms and cylinders. With the cube with its 12-fold degeneracy, changes in parallelism or length of one side have the effect of separating the TE and TM modes having the same frequencies. With deformation of the end plates of a cylindrical cavity different modes and frequencies will be obtained. In practice, it is found that with reasonable tolerances in workmanship little or no difficulty is experienced.

Because of the high Q values obtainable with resonant cavities, they may be used as either band pass or band stop filters. They may also be used as impedance transformers, examples of which are the aerial duplexing cavities described in Chapter IX. In many ways a resonant cavity behaves in a similar manner to a lumped element resonant circuit. For that reason, the analysis of the operation of resonators is conveniently carried out by lumped element analysis. At low frequencies a resonant circuit has intrinsic parameters L, C and R_{sh}, which are related to ω_o and Q_o; the resonator possesses only ω_o and Q_o as intrinsic parameters for any particular mode of oscillation.

It is possible to define a value of R_{sh} by specifying two points in the cavity, as discussed in Chapter VI. The shunt resistance is then given by

$$R_{sh} = \frac{|\int \mathbf{E \cdot ds}|^2}{2 \times \text{average power loss}}.$$

From this value, equivalent values for L and C may be defined as follows,

$$Q = \frac{R_{sh}}{L\omega_o}, \qquad \omega_o = \frac{1}{\sqrt{LC}}.$$

Provided the limitations of these definitions are recognised a cavity may then be represented by the equivalent lumped circuit of Fig. 132.

Figure 132.—Equivalent circuit of a resonator.

In a similar manner, by determining the current flow across a specified line on the interior surface of the cavity, it is possible to represent a resonator in terms of a series loss resistance and a series inductance and capacity.

Coupling to a cavity is often represented by an ideal transformer with an impedance transformation ratio of K^2. Fig. 133

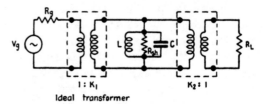

Figure 133.—Equivalent circuit of a resonator coupled to a generator V_g and to a load R_L.

illustrates a case where a cavity is employed as a band pass filter coupling a generator to a load as shown. The circuit may be simplified as shown in Fig. 134 which may then be

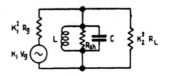

Figure 134.—Simplified form of Fig. 133.

analysed to determine its behaviour. As most cavities are usually employed in a resonant condition the circuit may be further simplified as shown in Fig. 135. In this circuit, the input and output couplings K_1 and K_2 are entirely arbitrary.

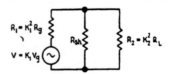

Figure 135.—Simplified circuit of Fig. 133 at the resonant frequency of the cavity.

They are often chosen to make the input impedance match that of the generator.

In this case,
$$R_1 = \frac{R_{sh} \cdot R_2}{R_{sh} + R_2}$$

$$R_2 = \frac{R_1}{1 - \dfrac{R_1}{R_{sh}}}$$

where $R_1 = K_1^2 R_G$ and $R_2 = K_2^2 R_L$.

The power coupled into the cavity is
$$P_{in} = \frac{V^2}{8R_1}$$

since the input impedance matches the generator.

The power to the load is then
$$P_{out} = \frac{V^2}{8R_1} \cdot \frac{\dfrac{1}{R_2}}{\dfrac{1}{R_{sh}} + \dfrac{1}{R_2}}$$

$$= \frac{V^2}{8} \cdot \frac{R_{sh}}{R_1} \cdot \frac{1}{R_{sh} + R_2} \cdot$$

Power is therefore transmitted through the cavity with an efficiency

$$\eta = \frac{P_{out}}{P_{in}} = \frac{1}{1 + \dfrac{R_2}{R_{sh}}}$$

$$= 1 - \frac{R_1}{R_{sh}} \cdot$$

If the unloaded Q is
$$Q_o = \frac{R_{sh}}{L\omega_o}$$

the Q when loaded by the input and output circuits is given by

$$\frac{1}{Q_L} = L\omega_o \left(\frac{1}{R_{sh}} + \frac{1}{R_1} + \frac{1}{R_2} \right)$$

$$= L\omega_o \frac{2}{R_1}$$

since $1/R_1 = 1/R_2 + 1/R_{sh}$.

Therefore

$$\frac{R_1}{R_{sh}} = \frac{2Q_L}{Q_o}$$

and

$$\eta = 1 - \frac{2Q_L}{Q_o}$$

As $Q_L \to Q_o/2$, the output coupling K_2 must be reduced in order that a match may be maintained at the input. The transmission loss of the cavity may be determined by measurements of Q and Q_L. The former is found from the bandwidth B of the cavity between the half power points with very loose input and output coupling

$$\text{i.e. } Q_o = \frac{f_o}{B}.$$

Similarly Q_L is determined by a measurement of the bandwidth B_L with the actual adjustment of the couplings, and for high Q cavities is given by

$$Q_L = \frac{f_o}{B_L}.$$

Other properties of the equivalent circuit may be derived for various coupling adjustments but the example should be sufficient to indicate the method.

The impedance looking into a cavity may be determined experimentally as shown in Fig. 136. With a knowledge of

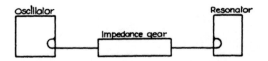

Figure 136.—Measurement of the input impedance of a resonator.

Q_o, Q_L and the coupled impedance, it is then possible to derive expressions for the degree of coupling.

7. Measurements at Microwavelengths

As this chapter is concerned with techniques, it is appropriate that mention be made of the techniques of measurements. It will not be a comprehensive survey ; rather, those methods in common use in this laboratory will be described.   Hence, the following remarks must be taken as indicative only of some of the methods which may be employed.

Frequency and Wavelength

The question at once arises, which of the two is to be employed. If we had an exact knowledge of the physical properties of various media it would be immaterial which were used.   This is not so, however, and for exact work it is preferable to decide on one or the other.   Frequency being the more fundamental, and being independent of the medium, is the desirable choice. However, in most laboratory measurements where accuracy is not paramount it is immaterial which unit is used, a value of 3×10^{10} centimetres per second for the velocity of wave propagation giving sufficient accuracy for the transformation.

Coaxial lines may be used for measuring wavelength by using a short length of line with a movable short circuiting plunger. By measuring the distance between two successive resonance points the wavelength may be determined.   A typical example is shown in Fig. 137.

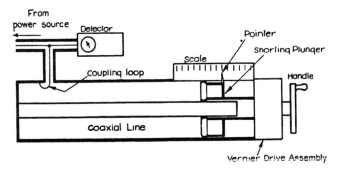

Figure 137.—Coaxial type wavemeter.   Resonance is indicated by a dip on the detector meter.

In the particular device shown, resonance is indicated by a dip in the crystal current reading.   The corrections for field distortion at the loop etc., are very small, and in many cases a direct reading of the distance will give sufficient accuracy.

Mention has already been made in section 6 of the method of tuning a cavity resonator. Fig. 131 shows a semi-coaxial type which may very conveniently be used as a wavemeter. Alternatively the end plate of a hollow cavity may be moved in or out to provide means of measuring wavelength. Such devices must be calibrated either by comparison with a coaxial type of instrument, or by comparison with the higher harmonics of a standard crystal oscillator. A crystal and meter are also coupled into the cavity in order to indicate the resonance point. Whilst the coaxial type gives a fundamental length measurement, it will be seen that the cavity type is only a comparison method and hence requires careful calibration if accuracy is to be achieved. Fixed wavelength resonators may be made and their resonant wavelengths determined by a length measurement. These are then useful as reference standards to provide check points in the various wavelength bands. Bracewell[7] describes a standard cavity of the type mentioned and also the construction and calibration of a cavity type wavemeter against a standard crystal.

In certain cases it is permissible to connect the wavemeter direct to the oscillator or signal source. For high accuracy however, this is not good practice, for the following reason. Most signal sources, particularly at microwave frequencies, contain only the oscillator with an output connection directly from it. As these oscillators do not supply much power, it is necessary to couple tightly to them which then means that a fairly large external impedance is reflected into the oscillatory circuit, thus in part determining the frequency of oscillation. If the load changes through connecting and disconnecting the wavemeter, then the oscillator frequency may change. Tuning through resonance on the wavemeter will cause large impedance changes and hence may exert a considerable effect on the frequency. In practice it is best.to isolate the oscillator with at least 10 decibels attenuation. Indeed it is preferable to use a directional coupler with a ratio of 100 to 1 to enable

[7] R. N. Bracewell, " L-band standard cavity," C.S.I.R. Radiophysics Laboratory, Report TI 134/2, September, 1944 ; and " L-band cavity wavemeter," C.S.I.R. Radiophysics Laboratory, Report TI 134/1, September, 1944.

the wavemeter to be left continuously in circuit. The directional coupler does not interfere with the main power used by the load and thus may be left in circuit for the duration of the measurements.

Power Measurements

Basically, most measurements of power at microwave frequencies involve the conversion of RF energy to some more easily measured form—usually heat. Any convenient method may then be used to determine the temperature rise of the element involved.

High Power Water Calorimeter Method (Fig. 138). A coaxial line (or waveguide) is shorted at one end and provided with a plug of " Mycalex " as shown to produce a section in which

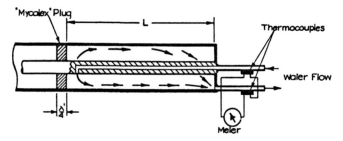

Figure 138.—Water calorimeter for the measurement of high RF powers.

water may circulate. The length L is chosen to be several wavelengths long, so that all the incident RF energy is absorbed. The flow of water is indicated by arrows, and thermocouples are attached to both the ingoing and outgoing water pipes to measure the temperature rise. The voltages generated by these elements are applied to a microammeter as shown. If the water flow, temperature difference, and thermal constant are known for the instrument, then the meter may be calibrated to read power in watts directly. Naturally there are several sources of error on the thermal side but these would be self evident when using the instrument and will not be discussed here. The dielectric constant of water is approximately 80 and it has a fairly high loss characteristic at ultra-high frequencies. To match to it from a coaxial line with air dielectric and characteristic impedance $Z_o = 50$ ohms, requires a quarter wave transformer of $Z = \sqrt{50 \times 6} = 17$ ohms. Therefore k (for same dimension) $= 8 \cdot 7$.

Mycalex is the only material available with high breakdown strength which approaches this value of dielectric constant. Locally available material has a dielectric constant of about 8 but some American firms have produced it with a value of $10 \cdot 7$. By making the water column long enough, very small reflections are obtained, and a match better than $0 \cdot 98$ can be obtained. Other devices exist for the measurement of high power, but space does not permit their inclusion here.

Low Power. Crystals, bolometers and thermistors are the main devices in common use. The crystal, which consists of a tungsten wire making contact with a small slab of silicon acts as a rectifier ; and a meter in its DC circuit will give a measure of the incident power. The calibration varies considerably with use however and hence the device is of limited application.

The bolometer and thermistor function by reason of resistance changes caused by the heating effect of the absorbed RF

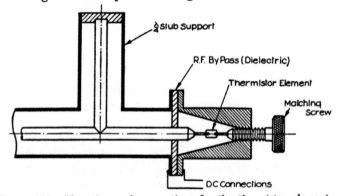

Figure 139.—Mounting and connections for the thermistor element.

power. Bolometers are made with a fine platinum wire whose sensitivity is of the order of several thousand ohms per watt. They are delicate and considerable care must be exercised in their use. Thermistors have a tiny bead made out of a mixture of metallic oxides (Mn_2O_3, SiO) and have a negative coefficient of resistance. Their sensitivity is equivalent to that of the bolometer, and at the same time they are much more robust.

The actual element to be used must be mounted in a suitable holder to enable it to absorb all the power and to enable DC connections to be made to it.

Such a device is shown in Fig. 139 with a matching screw to take account of variations in individual thermistors. It is fairly tolerant to frequency changes and requires few or no correcting factors.

A balanced bridge is used with this instrument to determine the power. As a rule, the instrument is calibrated on DC or 50 cycles AC and the effective range is from about 10 micro-watts to 4 to 5 milliwatts.

A typical circuit is shown in Fig. 140.

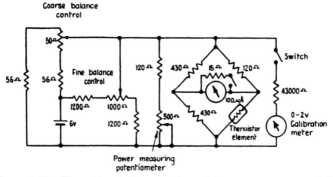

Figure 140.—Thermistor bridge circuit used to measure powers of 100 microwatts to 5 milliwatts.

For measuring high powers on a low power instrument a directional coupler (section 5*b*) or a known attenuator must be used. The attenuation may be either fixed (lossy cable or a lossy element in the feeder) or may be of a variable type. The variable types make use of the characteristics of a wave-guide operated at wavelengths longer than the critical value.

The propagation constant in a waveguide is given by equation 17 of Chapter VI.

$$\Gamma^2 = \left(\frac{2\pi}{\lambda_c}\right)^2 - \omega^2 \epsilon \mu \,.$$

If $(2\pi/\lambda_c)^2 < \omega^2 \epsilon \mu$, Γ is imaginary and so gives propagation with a real velocity and no attenuation. If, however, $(2\pi/\lambda_c)^2 > \omega^2 \epsilon \mu$, Γ is real and represents attenuation.

Taking the latter case, it can be shown that

$$\alpha = \frac{2\pi}{\lambda_c}\sqrt{1 - \left(\frac{\lambda_c}{\lambda}\right)^2} \text{ nepers per metre.}$$

This gives an attenuation device which may easily and correctly be calibrated by a length measurement and which moreover has the advantage of providing a linear variation when cali-

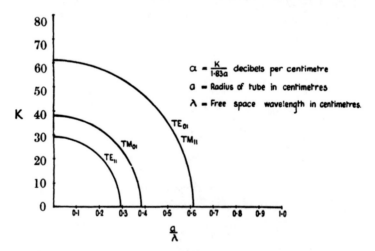

Figure 141.—Attenuation in a circular waveguide which is operated below its critical frequency.

brated in decibels. The most commonly used excitation is that in circular guides of TE_{11} or TM_{01} modes. In Fig. 141 a graphical means is given of determining the attenuation for any of several modes.

Fig. 142 shows a type of attenuator using the above mode in a circular guide, in which it is possible to obtain an attenuation from 20 to 100 decibels without difficulty. (It is not desirable to bring the loops too close together because higher order fields occur near the loop which upset the attenuation law). Unfortunately a minimum insertion loss of about

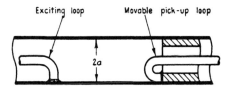

Figure 142.—Waveguide attenuator. Circular guide operated below its critical frequency.

20 decibels restricts the use of the attenuator. This attenuator, together with the thermistor meter, gives a fairly flexible system for measurements of power above one watt.

For very low power measurements (10^{-13} to about 10^{-5} watts) a receiver may be used as a comparison device ; a measurable amount of higher level power being attenuated by a known amount to the level of the power to be measured, and a comparison made with the unknown low power.

There are several precautions that must be observed in all power measurements. Firstly the device must be matched to the source of power. For this reason the advantage of standardising on a given impedance for all equipment will be immediately seen. Secondly, lossy elements must not be interposed between the source and the absorbing element unless their exact magnitude is known. Thirdly, the absorbing units are usually sensitive to changes in ambient temperature and so must be continually checked for correct adjustment.

Impedance Measurements

The theory of impedance measurements has been given in the previous chapter, and an indication of the type of equipment that may be used has already been given in section 2. Normally however, much more refined equipment is used in practice in order that a high degree of accuracy may be obtained.

Plate X (b) shows an impedance measuring gear that has been designed for use at wavelengths between 23 and 26 centimetres.

The measurement of impedance involves the accurate determination of the voltage or current node, and the voltage standing wave ratio S. The former is easily obtained by a scale on the instrument marked say in millimetres. The measurement of the latter depends to some extent on the magnitudes concerned, and the following remarks should indicate suitable methods for various cases.

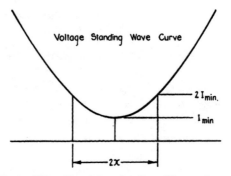

Voltage Standing Wave Curve

$2\,I_{min.}$

I_{min}

$2x$

Figure 143.—The determination of low values of S.

For high ratios of S (between 1·0 and 0·5) the crystal detector may be used with the probe assembly as previously described. Although the crystal has nominally a square law characteristic, it should always be checked if great accuracy is essential. For lower values of voltage standing wave ratio it is preferable to use a variable attenuator between the probe assembly and crystal and to maintain equal crystal current readings for node and antinode positions. By reading the change in attenuation the voltage standing wave ratio may then be determined.

For very low values of S (less than about 0·1) the following method may be used. Assume part of the voltage standing wave curve as shown in Fig. 143.

If x is the distance between the points where the detector current is twice its value at minimum, and assuming an approximate sine law curve, it may then be shown that the voltage standing wave ratio is given by,

$$S = \frac{2\pi x}{\lambda}$$

where λ is the wavelength in the measuring equipment.

A receiver whose detector characteristics are known is useful in making this measurement. It is connected to the probe assembly in place of the crystal. This method is of particular value when determining the complex dielectric constant of insulating materials.

CHAPTER VIII

AERIALS

THE development of ultra high frequency and microwave techniques during the last few years has given rise to a corresponding development in aerial design. The tendency has been to produce either very narrow beams or beams with a predetermined directional characteristic. To some extent this has occurred through modification of long wave techniques, e.g. the use of arrays of dipoles etc., but, because of the fact that it is possible at these frequencies to make aerials with linear dimensions large with respect to the wavelength, the problem has also been approached from the point of view of geometrical optics, and this has resulted in aerials using paraboloids, lenses, etc. A further type of aerial has developed out of the obvious use of an open ended waveguide as a radiator, namely the flared waveguide or horn aerial.

The theory of directional aerials at these frequencies is very similar to the theory of diffraction in optics. For instance, an array of point sources is analogous to a diffraction grating, the radiated field from a paraboloid fed by a point source corresponds to the optical case of Fraunhofer diffraction at an aperture, and that of a sectoral horn arises from Fresnel diffraction at its aperture.

The various types of aerials have been broadly grouped into the following three categories for the purposes of discussion.

(a) Elementary radiators, of which the half wave dipole is one of the most important types.

(b) Arrays of elementary radiators.

(c) Continuous surface radiators. These are aerials in which the radiation is produced by a continuous current distribution in a conducting sheet or by a continuous field distribution across an aperture, where the dimensions of the

conducting sheet or aperture contain at least several wavelengths (in order to distinguish between this type of radiator
and what has been termed an elementary radiator). An
example of this group is the paraboloid reflector.

It should be remembered that these three groups are by
no means distinct, but that the divisions have been drawn
for convenience only. In addition to the aerial systems
described under this grouping, other sections of the chapter
deal with various general and special aerial characteristics.
A brief summary of measuring techniques is also included.

1. General Aerial Properties

Directional Diagram of an Aerial

In order to define completely the properties of an aerial as a
directional radiator, it is necessary to know the radiated
power per unit solid angle in all directions and the polarisation
of the wave. The majority of applications require a plane
polarised wave which is, in general, the simplest type to
produce. In this chapter, aerials giving plane polarisation
only will be discussed. The plane containing the direction
of propagation and the electric vector is known as the E plane
and that containing the magnetic vector the H plane. If we
have an aerial situated at O (Fig. 144), we need to know the

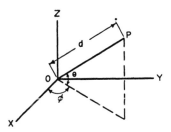

Figure 144.—Coordinate system for specifying the directional properties
of an aerial situated at 0.

radiated power per unit solid angle as a function of θ and ϕ.

Now power per unit solid angle is proportional to the field
intensity Φ at a fixed distance from O. Field intensity is
defined here as the power per unit area passing through a

plane at right angles to the direction of energy flow. Thus

$$\Phi = \frac{E^2}{\eta_o} \tag{1}$$

where Φ is in watts per square metre

$E \cdot$is the root mean square radiation electric field in volts per metre

$\eta_o = 120\pi$ ohms is the intrinsic impedance of free space.

Thus the directional properties of an aerial may be expressed in terms of either the field strength E or the field intensity Φ as a function of θ and ϕ.

To describe the directional distribution of E, the space factor $S(\theta,\phi)$ of an aerial is defined by

$$E = \frac{k}{d}S(\theta,\phi) \tag{2}$$

$$\text{and } S(\theta,\phi)_{max} \equiv 1 \tag{3}$$

where E is the radiation field in volts per metre

k is a constant depending on the total power radiated etc., and is derived on p. 222 and

d is the distance from the aerial in metres.

It follows from equation (1) that the directional distribution of Φ is then given by

$$\Phi = \frac{k^2}{d^2\eta_o} S^2(\theta,\phi)$$

$$= \frac{k^2}{d^2\eta_o} \Phi^*(\theta,\phi) \tag{4}$$

where $\Phi^*(\theta,\phi) = S^2(\theta,\phi)$ may be called the normalised intensity and $\Phi^*_{max} \equiv 1$. Either $S(\theta,\phi)$ or $\Phi^*(\theta,\phi)$ completely describe the directional properties of the aerial in three dimensions.

For many purposes it is sufficient to know the directional properties in one or two planes (in general the E and H planes) and we can then use a one dimensional space factor $S(\theta)$ or intensity $\Phi^*(\theta)$. The graph of $S(\theta)$ or $\Phi^*(\theta)$ against θ is known as the polar diagram of the aerial in the given plane. It must be specified, of course, whether field intensity or field strength is being plotted.

Because of the finite size of an aerial and also due to the presence of the induction field close to the aerial, the polar diagram is only independent of distance for large distances. However, in almost all applications we are concerned only with the distant field, and so in theoretical work the space factor is calculated in terms of the " field at infinity." A case where it is necessary to consider the field near a radiator is that of a feed for a paraboloid or lens. The polar diagram may be plotted in either polar or Cartesian coordinates, the latter being more convenient for very narrow beams. A typical diagram is shown in Fig. 145, plotted in both Cartesian

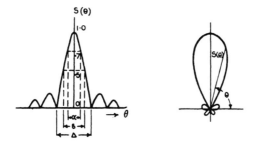

Figure 145.—Typical polar diagram having beamwidths α to half power points, δ to half field points, and Δ to zero field points, plotted in Cartesian and polar coordinates.

and polar coordinates. It is seen to consist of a main beam or lobe and a number of side lobes, which are generally kept as small as possible.

The beam width may be defined in any of the following ways,

(a) The width of the main lobe to half power points – α.

(b) The width of the main lobe to half field points = δ.

(c) The width of the main lobe between the first two minima in the polar diagram = Δ.

For convenience definitions (b) and (c) will be used.

Directivity and Gain

The directivity of an aerial is defined as the ratio of the power per unit solid angle radiated in the direction of maximum radiation to the average radiated power per unit solid angle

$$\text{i.e. Directivity} = \frac{4\pi\Phi_{max}}{\int\Phi d\Omega}$$

$$= \frac{4\pi}{\int\Phi^*d\Omega} \tag{5}$$

where Ω is the solid angle.

The gain of an aerial is defined relative to a standard radiator, and is the ratio of the power which must be supplied to the standard to produce a given field at a given distance in the direction of maximum radiation, to that which must be supplied to the aerial to produce the same field at the same distance. The standard aerial may be a hypothetical spherical (isotropic) radiator, a half wave dipole, or a short current element (infinitesimal dipole). The factors relating these three definitions of gain may be easily found, since the gain, G_s, of an infinitesimal dipole relative to a spherical radiator is 1·5 and that of a half wave dipole is 1·64. For an aerial in which there are no ohmic losses, the gain relative to a spherical radiator is equal to the directivity. An example for which this is not so is that of a rhombic aerial in which some of the power supplied to the system is dissipated in a resistive load. Another example is the non-resonant array of section 4.

Except in a few cases, the term *gain* will be employed in this chapter even when strictly directivity is meant, since in most of the aerials described ohmic losses are negligible. The symbols G_s, G_c, G_d will be used for gain relative to a spherical, infinitesimal, or half wave radiator respectively.

Radiation Resistance

The radiation resistance of an aerial referred to a given point in the system is equal to the resistance which, if inserted at that point, would dissipate the same energy as is actually radiated by the system. If the current at the reference point is I_o, then

$$R_r = \frac{\text{Radiated power}}{I_o^2}. \tag{6}$$

The concept of radiation resistance has its most important

application for radiators such as dipoles or slots, since in these cases a generator can be inserted at a definite point on the aerial and the resistive component of the load presented to the generator is equal to the radiation resistance referred to that point (neglecting ohmic losses). The most useful reference point is at the terminals of the aerial.

Bandwidth

Bandwidth is dependent on two considerations : (a) change in impedance of the aerial with frequency and consequent mismatch to the generator ; and (b) deterioration of polar diagram with frequency.

For most aerials, particularly those composed of resonant elements (a) is the more important, since impedance change with frequency is much more rapid than change of the polar diagram. However, in a particular case such as a long, endfed array of elements, frequency change will cause swinging of the beam, which may be undesirable.

In cases for which (a) is the major consideration, bandwidth is generally defined as the frequency range over which the voltage standing wave ratio[1] on the feeder is greater than some given amount, although in some cases, particularly half wave dipoles, the bandwidth can be defined by the equivalent Q of the radiator considered as a lumped resonant circuit. In this chapter the bandwidth Δf will be defined as the frequency range for which the voltage standing wave ratio > 0.7 i.e. the power standing wave ratio > 0.5. The relation between Δf and the Q of a system is given by

$$\frac{\Delta f}{f_0} = \frac{0.017}{Q} \tag{7}$$

where f_0 is the resonant frequency, and the voltage standing wave ratio is unity for $f = f_0$. A voltage standing wave ratio of 0.7 is equivalent to 3 per cent of power being reflected at the load.

The conditions which determine the bandwidth required of an aerial are : (a) allowable pattern deterioration ; (b) allowable standing wave ratio on the feeder ; (c) frequency

[1] For the definition or standing wave ratio see footnote 2, Chapter VI.

spectrum to be transmitted, e.g. in the case of pulsed radiation, television, or frequency modulation ; (d) tolerances on linear dimensions of the system. In general the smaller the band-width of an aerial i.e. the more highly resonant it is, the more critical the tolerances become.

Radiation Field

The intensity at a distance d from a spherical radiator is given by

$$\Phi = \frac{P_T}{4\pi d^2}$$

where P_T is the transmitted power.

Substituting from (1) we obtain

$$E = \frac{1}{d}\sqrt{30 P_T}. \tag{8}$$

Thus the field at a large distance d measured in the direction of maximum radiation, from an aerial of power gain G_s relative to a spherical radiator, is given by

$$E = \frac{1}{d}\sqrt{30 G_s P_T}. \tag{9}$$

The field in any direction relative to the aerial is

$$E = \frac{S(\theta, \phi)}{d}\sqrt{30 G_s P_T}. \tag{10}$$

The Theorem of Reciprocity

Consider two aerials, A and B having any orientation and displacement relative to one another. The theorem states that if a voltage V applied to A causes a current I at a given point in B, then the voltage V applied at this point in B will cause a current I in A at the point where V was originally applied.

The main use of this theorem is that it enables us to obtain the properties of a receiving aerial in terms of its properties as a transmitting aerial. Thus the polar diagram (i.e. the response to waves incident at varying angles), gain, and impedance of an aerial used for receiving are the same as when the aerial is used for transmitting. The main difference between receiving and transmitting aerials is that the latter must have a high power handling capacity, e.g. feeders must be of sufficient dimensions to avoid breakdown.

Aerials as Receivers of Waves

As stated above, the properties of an aerial as a receiver may be found from its properties as a radiator of energy. However it is sometimes useful to know how much power an aerial will absorb from an incident plane wave. This may be determined for any particular system, the simplest case probably being that of a current element, and once this is done all other cases may be evaluated if their gain relative to the particular aerial chosen is known.

Consider a simple aerial such as a current element or a half wave dipole situated in a field and delivering power to a matched load (i.e. load + ohmic resistance = radiation resistance). Then, of the power input to the system, one half is dissipated in the load plus ohmic resistance and one half in the radiation resistance, the latter portion being re-radiated.

The power absorbed by any aerial (i.e. the power dissipated in the load resistance) oriented for optimum reception may be shown to be

$$P_{Rmax} = \frac{G_s \lambda^2}{480\pi^2} E^2 \qquad (11)$$

where E = incident field (root mean square)

G_s = gain relative to an isotropic radiator.

For an aerial having any orientation relative to the incident wave

$$P_R = \frac{G_s \lambda^2}{480\pi^2} \Phi^*(\theta, \phi) E^2 \qquad (12)$$

$$= \frac{G_s \lambda^2}{480\pi^2} S^2(\theta, \phi) E^2. \qquad (13)$$

Absorption Cross Section (A_0)

Absorption cross section is defined as the area from which energy would have to be abstracted from the incident beam to equal the energy actually removed and absorbed in the load. It may be shown that

$$A_o = \frac{G_s \lambda^2}{4\pi} \, \Phi^*(\theta, \phi). \tag{14}$$

$$\text{Hence } A_{o_{max}} = \frac{G_s \lambda^2}{4\pi} \tag{15}$$

when the aerial has optimum orientation. If the actual cross sectional area of the aerial equals the absorption cross section then

$$G_s = \frac{4\pi A}{\lambda^2}.$$

It will be seen in the next section that this is the gain of an aerial which, when transmitting, has a field distribution uniform in phase and amplitude across an area A.

Re-radiation Cross Section (A_r)

Since currents are induced in an aerial by the incident wave, re-radiation can take place, and the re-radiation cross section is defined as the area of the incident wave front from which energy would have to be absorbed to equal the energy re-radiated. It is often erroneously stated, on the basis of a simple circuit analogy, that for any matched aerial the received energy is equally distributed between load resistance and radiation resistance, and so the re-radiation cross section equals the absorption cross section. If this were so, the black body concept, so useful in thermo-dynamics would be invalid.

A more exact circuit analogy is given in Fig. 146. The infinite transmission line of characteristic impedance Z_o represents free space which guides the wave to the aerial represented by the impedance Z. The impedance looking into the line at AA' represents radiation resistance and equals $\frac{1}{2}Z_o$, and so for a matched aerial we must have $Z = \frac{1}{2}Z_o$. It can be shown that in this case, of the incident power P, $P/2$ is absorbed and $P/2$ " re-radiated," half of this being reflected and half transmitted. This case would correspond to an array of dipoles without any reflecting sheet, and so for such an aerial $A_r = A_o$. It is also clear that the re-radiation gain equals the actual gain of the aerial.

If now we place a short circuit across the transmission line of Fig. 146 behind the load (say at BB') and adjust Z so that it again equals the impedance looking into the line at AA', then all the incident power is absorbed since in this case, the reflected waves cancel and there can be no trans-

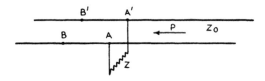

Figure 146.—Equivalent circuit for a receiving aerial with a plane wave incident on it.

mitted wave. A free space equivalent of this circuit is a resistive sheet of surface impedance η_o backed by a conducting sheet spaced $\lambda/4$ from it. Without the conducting sheet all of the incident power would not be absorbed. An equivalent aerial system is a rectangular array of dipoles backed by a reflecting sheet. Another less obvious example is any con‘ tinuous surface radiator which transmits all its power through an aperture in the form of a plane wave of uniform amplitude. In both of these cases all the incident power is absorbed (i.e. $A_r = 0$).

In the case of a paraboloid of gain factor less than unity (see section 5), a fraction of the incident power is re-radiated, but the re-radiation gain does not equal the aerial gain. Effectively the reflector may be considered as a matching transformer and if the gain factor is less than unity a mismatch exists and re-radiation occurs.

In the case of aerials such as horns which transmit a diverging wave, complete cancellation of the re-radiation clearly cannot occur ; that is the horn is not a perfect " matching transformer."

2. The Aerial Problem

The aerial problem may arise in two forms :

(*a*) Given a specific radiating system, find its directional characteristics, gain etc.

(*b*) Given a directional characteristic, find a radiating system which will produce it (subject of course, to additional requirements such as bandwidth, impedance, physical size etc.). This problem is discussed in section 8.

Problem (*a*) can be solved if a solution of the wave equation with the appropriate boundary conditions can be found. However, if as in the case of a thin half wave dipole, a suitable current distribution on a conducting surface, or in the case of a paraboloid, a suitable field distribution over an aperture can be assumed, the approximate solution becomes somewhat simpler.

3. Elementary Radiators

Elementary radiators form the basis for many other types of aerials (e.g. arrays, feeds for paraboloids, lenses, etc.). Two types, the half wave dipole and the half wave slot will be discussed in some detail.

Short Current Element ("Infinitesimal" dipole)

Consider a short current element of length δl (where $\delta l < \lambda$) situated at the origin and along the z axis of Fig. 144. The following results may be obtained :

$$S(\theta,\phi) = \cos \theta. \tag{16}$$

This is plotted in Fig. 147 in polar form.

$$R_r = 80\pi^2 \left(\frac{\delta l}{\lambda}\right)^2 = 790 \left(\frac{\delta l}{\lambda}\right)^2 \tag{17}$$

$$G_s = 1 \cdot 5 \tag{18}$$

and using equations (11) and (15) we obtain

$$A_{o\,max} = \frac{3}{8\pi} \lambda^2 = 0 \cdot 12 \lambda^2 \tag{19}$$

$$P_{R\,max} = \frac{\lambda^2 E^2}{320\pi^2} = 3 \cdot 2 \times 10^{-4} \lambda^2 E^2. \tag{20}$$

Because of the high reactive component of the impedance of short current elements and because it is in general more convenient to use a resonant dipole, they are rarely used at ultra high frequency except where there is a restriction on size as in the case of detector probes for impedance measuring gear etc.

The Half Wave Dipole

A thin conductor one half wavelength long behaves as a resonant element, and if the current distribution along it is known the radiation field may be calculated by considering it as an aggregate of current elements. This treatment leads to the following results :

$$S(\theta,\phi) = \frac{\cos\left(\frac{\pi}{2}\sin\theta\right)}{\cos\theta} \qquad (21)$$

where the dipole is centred at the origin of the coordinate system of Fig. 144 and lies along the z axis. The polar diagram

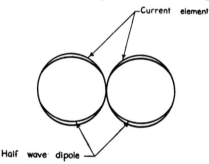

Figure 147.—Polar diagrams of a short current element and a half wave dipole in a plane containing them.

of a half wave dipole is plotted in Fig. 147 and may be compared with that of a short current element.

$$R_r = 73{\cdot}3 \text{ ohms} \qquad (22)$$

where R_r is referred to the centre of the dipole

$$G_s = 1{\cdot}64 \qquad (23)$$

$$P_{R_{max}} = 3{\cdot}46 \times 10^{-4} \ \lambda^2 \ E^2 \qquad (24)$$

$$A_{o_{max}} = 0{\cdot}13\lambda^2. \qquad (25)$$

If a matched load is inserted at the centre of the dipole, then the voltage induced across it by an incident field E is given by

$$V = \frac{\lambda E}{2\pi}. \tag{26}$$

Effect of Finite Thickness of a Half Wave Dipole. The main effect of finite thickness is to increase the bandwidth and to decrease the resonant length (see Fig. 152). Thus fat dipoles must be used when wide bandwidth is required. The effect on S, R_r, G_s, etc. is of the second order.

The Dipole as a Circuit Element. The assumption of a sinusoidal current distribution leads to useful results in the case of the thin half wave dipole, but it gives no indication of the properties of an actual dipole. In some other cases the treatment is useless. For instance in the case of an aerial one wavelength long and fed at the centre, the current at the centre becomes zero and the impedance infinite.

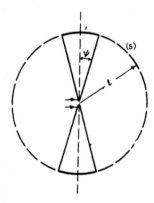

Figure 148.—Conical radiator with spherical ends and boundary sphere (S).

A method of overcoming this difficulty is used by Schelkunoff.[2] He considers a conical dipole of section given in Fig. 148, and assumes that the aerial and the surrounding

[2] S. A. Schelkunoff, *Electromagnetic Waves*, Van Nostrand, New York, 1943, pp. 441-79.

space are two waveguides consisting of (a), the aerial region
bounded by the conical surfaces and the hypothetical boundary
sphere (S) and (b), the space external to (S). The wave
approaching (S) is then partly transmitted and partly reflected,
the transmitted portion representing radiation. In order to
satisfy the boundary conditions at the conical surfaces and
on the sphere (S), it is necessary to consider a large number

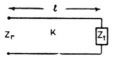

Figure 149.—Transmission line equivalent of a dipole of length 2*l* and
average characteristic impedance *K*.

of modes within and without (S). The aerial as far as the
principal wave is concerned may be considered as a transmission line terminated in an impedance Z_t, and having
characteristic impedance K (Fig. 149). Curves for Z_t are
given by Schelkunoff, so that if K is known, the input impedance Z_r can be determined.

For the conical aerial of Fig. 148

$$K = 120 \log_e \cot \frac{\psi}{2} \tag{27}$$

$$\doteqdot 120 \log_e \frac{2}{\psi} \quad \text{if } \psi \text{ is small.} \tag{28}$$

If the shape of the aerial is other than conical the same type
of treatment can be used (as long as the transverse dimensions

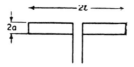

Figure 150.—Cross section of a cylindrical dipole.

are small) except that such aerials, unlike cones, are non-
uniform transmission lines. The quantity K becomes the

average characteristic impedance. For a cylindrical aerial (Fig. 150), the average characteristic impedance is given by

$$K = 120 \left(\log_e \frac{2l}{a} - 1 \right). \tag{29}$$

Curves are given by Schelkunoff for K, Z_r, Z_t, and for the resonant lengths of other shapes of dipole radiator.

Case of $l \doteq \lambda/4$. When l is just less than $\lambda/4$, half wave resonance occurs, the input impedance being resistive and equal to the radiation resistance. Since, in the region of half

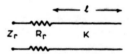

Figure 151.—Transmission line equivalent of a half wave dipole having radiation resistance R_r.

wave resonance, the change in the reactive component of Z_r with frequency is much greater than the change in its resistive component R_r we can represent the aerial by the equivalent transmission line circuit of Fig. 151. We have

$$Z_r = R_r + jX_r. \tag{30}$$

To a first order R_r is a constant equal approximately to 73 ohms and

$$X_r = -K \cot \frac{2\pi l}{\lambda}. \tag{31}$$

When $l = \lambda/4$ we can treat the dipole as a series resonant circuit, and using (31) the Q of a dipole is found to be

$$Q = \frac{\pi K}{4R_r} \doteq \frac{K}{100}. \tag{32}$$

For the cylindrical half wave dipole of Fig. 150

$$Q = 2 \cdot 76 \log_{10} \frac{\lambda}{2a} - 1 \cdot 20 \tag{33}$$

since $2l = \lambda/2$.

For a half wave conical dipole of half angle ψ

$$Q \doteqdot 2 \cdot 76 \log_{10} \frac{2}{\psi}. \tag{34}$$

The bandwidth is found from equation (7) to be

$$\frac{\Delta f}{f} = \frac{17}{K} \tag{35}$$

when the dipole matches the line at the resonant frequency.

Fig. 152 gives K, $\dfrac{\Delta f}{f}$, Q and percentage deviation p of the resonant length from $\lambda/2$ for a cylindrical dipole.

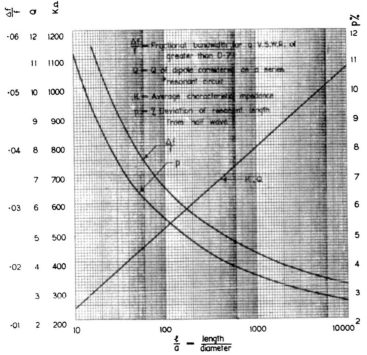

Figure 152.—Characteristic impedance, Q, and percentage deviation of resonant length from half wave, of a cylindrical dipole of length $2l$ and diameter $2a$ at half-wave resonance.

$$\left[2l = \frac{\lambda}{2} \left(1 - \frac{p}{100} \right) \right]$$

Methods of Feeding Dipoles

A dipole may be energised through a transmission line connected to its centre, and the feed is termed either resonant or non-resonant depending on whether or not standing waves exist on the line. It is in general desirable with high power radar transmitters to use a non-resonant feed since, for the same power, the maximum voltages appearing on the feeder are larger in the case of a resonant system.

Twin Line Feed. (a) *Resonant.* This consists of a twin line, an integral number of half wavelengths long connected between generator and dipole (Fig. 153). The impedance presented to the generator is thus the dipole impedance. The standing wave ratio on this feed depends on the ratio of characteristic

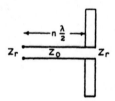

Figure 153.—Dipole with resonant feed consisting of a twin line an integral number of half-waves in length.

impedance Z_0 of the line to the dipole impedance. Such feeds find their major application as connecting links between the elements of arrays.

(b) *Non-resonant.* If no standing waves are to exist on the line it must be matched to the dipole, for instance by making Z_0 equal to the resonant dipole impedance. This, however, is in general difficult except when low powers are used and a line of very small spacing can be employed. If $Z_0 \neq Z_r$, then a matching network such as a quarter wave transformer (Fig. 154) or a system of stubs must be used. Methods of matching transmission lines to loads are described in Chapter VII.

One of the major disadvantages of twin lines at ultra high frequencies is that of radiation which may introduce losses, unwanted cross polarisation and other effects. The radiation from a non-resonant twin line of length at least twenty times the spacing is equal to twice that from a current element carrying the same current as the line and having a length equal to the line spacing.

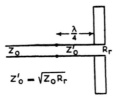

Figure 154.—Non-resonant twin line feed using a quarter-wave matching section.

Concentric Line Feed. Since it is quite feasible to make a concentric line having Z_o equal to 73 ohms, such lines may be matched directly to a dipole (Fig. 155). Since however, the dipole of Fig. 155 presents a balanced load and a concentric

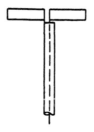

Figure 155.—Simple coaxial fed dipole.

line is unbalanced, currents will flow on the outside of the concentric line and will give rise to radiation, which may be undesirable. Thus some form of balance to unbalance transformer is needed. Some typical arrangements are shown in Fig. 156.

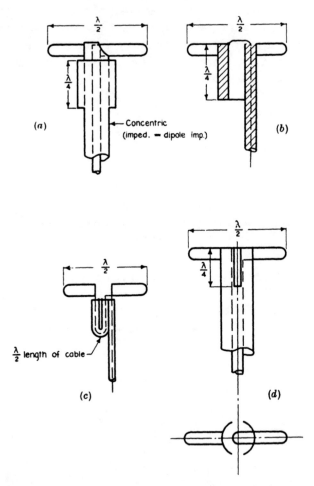

Figure 156.—Various feeds using balance to unbalance transformers to couple from an unbalanced coaxial transmission line to a balanced radiator. The device shown in (b) is known as the Pawsey stub.

Waveguide Feed. A dipole may be fed from a waveguide by means of a probe parallel to the electric field, a typical arrangement being shown in Fig. 157. A linear array of dipoles may be excited by spacing them along a waveguide.   The coupling to the guide and therefore the amount of energy abstracted from the guide can be adjusted by means of the probe.

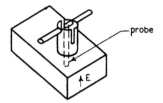

Figure 157.—Half-wave dipole excited from a waveguide by means of a probe situated parallel to the electric field.

Slot Radiators[3]

The concept of a slot in a conducting sheet as an efficient radiator is rather novel.   It may be shown that a narrow slot approximately one half wavelength long in an infinite conducting sheet and energised at its centre AA', (Fig. 158)

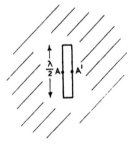

Figure 158.—Half-wave resonant rectangular slot in a conducting sheet.

behaves as an efficient resonant radiator of energy.   To a first approximation it behaves in a similar fashion to a twin transmission line shorted at each end and fed at the centre, thus making it a resonant element.   However, owing to the fact that the currents are not localised but spread out over the

[3] H. G. Booker, "Slot aerials and their relation to complementary wire aerials (Babinet's principle)," *J. Instn. Elect. Engrs.*, Vol. 93, Part IIIA, 1946, pp. 620-6.

sheet, the radiation from the slot is much greater than that from a two wire transmission line.

The properties of a slot may be determined from the equivalent properties of the complementary dipole by application of a modification of Babinet's principle which arises in optics.[3] Thus the field of a slot is obtained from that of the complementary dipole by interchanging electric and magnetic fields and reversing their direction on one side of the slot. That is, a slot is analogous to a hypothetical dipole consisting of a " perfect conductor of magnetic current " fed at its centre by a " magnetic current generator." A vertical slot has the same polar diagram and gain as a vertical dipole but produces a horizontally polarised field.

The radiation resistance of a narrow resonant slot is given by

$$R_{rd} \cdot R_{rs} = \frac{\eta^2}{4} \tag{36}$$

R_{rs} = slot resistance
R_{rd} = dipole resistance = 73 ohms.

Hence R_{rs} = 485 ohms.   (37)

The bandwidth of a rectangular resonant slot is equal to that of the complementary dipole and is about one half that of a cylindrical dipole of diameter equal to the slot width. Since the resonant length of a rectangular slot is slightly less than $\lambda/2$ and obeys a similar law to that of a half wave dipole (see Fig. 152) we have for a slot of length $2l$ and width b

$$Q \doteqdot 5 \cdot 5 \log_{10} \frac{\lambda}{b} - 2 \cdot 4 \tag{38}$$

$$\frac{\triangle f}{f} \doteqdot \frac{0 \cdot 03}{\log_{10} \dfrac{\lambda}{b} - 0 \cdot 43} . \tag{39}$$

The current induced in a matched load at the centre of a slot by a field E is given in terms of the voltage induced across a matched load connected to the complementary dipole by

$$I_s = \frac{2}{\eta} V_d. \tag{40}$$

Similarly $I_d = \frac{2}{\eta} V_s$ (41)

where the suffix " s " refers to the slot and " d " to the dipole.
From (26) and (40)

$$I_s = \frac{\lambda E}{\eta \pi}. \tag{42}$$

As in the case of a dipole a half wave slot may be represented
as a transmission line of the form shown in Fig. 159. Its
characteristic impedance K_s is given by

$$K_s K_d = \frac{\eta^2}{4} \tag{43}$$

where K_d is the characteristic impedance of the complementary
dipole.

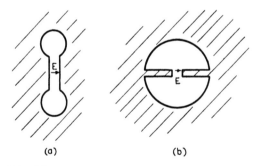

Figure 159.—Transmission line equivalent of a half-wave resonant slot
of characteristic impedance K_s and radiation resistance R_{rs}.

In the above, a rectangular slot only has been considered.
Some alternative shapes are shown in Fig. 160 (a) and (b).

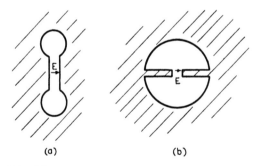

Figure 160.—Wide band slot radiators, (a) dumb-bell, (b) circular.

Application of Slots. Slots may replace dipoles in most applications, especially when polarisation perpendicular to that
of a dipole is required. A particularly promising application
which has not yet been exploited to any extent is to the design
of aircraft aerials of negligible wind resistance. It should be
possible to utilise the metal skin of the aircraft as the conducting sheet and produce an aerial which does not project outside

the plane and at the same time does not take up any space inside the plane. Although the theory assumes an infinite conducting sheet, the departure from the above results is small even for sheets of the order of one wavelength square.

Feeding of Slots. Slots may be fed by means of a transmission line connected to the centre of the slot in much the same manner as described for dipoles, except that, because of the higher impedance of the slots, the transforming devices will differ somewhat. One interesting feature, however, is that a slot presents a balanced load when fed at any point along its length. A graph of impedance against feed position for a typical slot is given in Fig. 161. Thus the slot can be used as its own

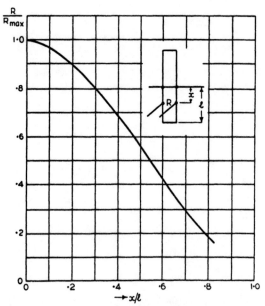

Figure 161.—Resistance of a resonant rectangular slot as a function of feed position.

matching transformer, and almost any twin or coaxial line can be matched directly by feeding the slot at the appropriate point. The effect of an off-centre feed on the polar diagram would be of the second order, since to a first approximation

the voltage distribution along the slot would still be sinusoidal.
Fig. 162 shows some typical arrangements. The simple case
of a slot in a metal sheet fed at its centre by a twin line is
illustrated in Fig. 162 (a). In order to suppress the radiation
on one side of the slot it may be placed in a resonant cavity
[Fig. 162 (b)]. The effect is to approximately double the slot
impedance.

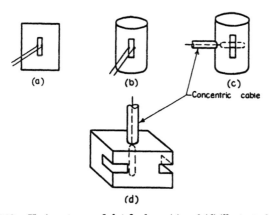

Figure 162.—Various types of slot feeds.   (c) and (d) illustrate how slots
        may be excited by placing them in resonant cavities.

Fig. 162 (c) shows how a resonant device such as a cavity
may be used as a balance to unbalance transformer, the cavity
being fed by a probe oriented so as to excite a wave with its
E vector at right angles to the slot. Fig. 162 (d) illustrates
two slots in a cavity excited by a single probe. This system
has been used in a beacon to produce a split beam, the cavity
being placed in a V reflector and the slots switched alternately
by shorting their centres.

Slots in Waveguides. Various slot arrangements in a waveguide
are shown in Fig. 163, together with a schematic diagram of
the current distribution in the guide walls for the TE_{10} mode.
Since the slots are narrow, slots AB and CD do not affect the
current distribution and so do not radiate. Slot EF intercepts
series current and so presents to the guide a series load, while

GH intercepts shunt currents and so presents a shunt load lumped at the centre of the slot. *GH* is a shunt slot with maximum coupling to the guide. This coupling may be reduced in either of the ways shown in Fig. 164.

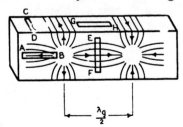

Figure 163.—Resonant slots in a waveguide. *EF* and *GH* are series and shunt slots respectively, while *AB* and *CD* are non-radiating slots.

The conductances presented to the guide by the slots of Fig. 164 (*a*) and (*b*) expressed as a ratio to the " characteristic conductance " of the guide are[4]

$$g = k \sin^2 \left(\frac{\pi x}{a} \right) \tag{44}$$

$$\text{and } g = k \sin^2\phi \tag{45}$$

respectively, where

$$k = 2 \cdot 09 \, \frac{\lambda_g}{\lambda} \cdot \frac{a}{b} \cos^2 \left(\frac{\pi}{2} \frac{\lambda}{\lambda_g} \right) \tag{46}$$

λ and λ_g are the free space and guide wavelengths respectively.

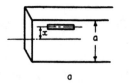 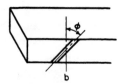

Figure 164.—(*a*) and (*b*) show how the coupling of a shunt slot to a wave-guide may be varied by either displacing it towards the centre of the broad face of the guide or by suitably orienting it in the narrow face.

If $g = 1$ and the guide is shorted one quarter wavelength from the end of the slot, all the energy supplied to the guide is

[4] W. H. Watson, "Resonant slots," *J. Instn. Elect. Engrs.*, Vol. 93, Part IIIA, 1946, pp. 747-77.

radiated. The system can be represented by a transmission line of unit characteristic impedance shunted by a conductance g lumped at the centre of the slot.

A slot such as AB (Fig. 163) can be excited if the field in its neighbourhood is distorted by a probe, for example, with a part of its length parallel to the electric vector (Fig. 165). The coupling of the slot to the guide can be varied by rotating the probe. Slots may be coupled in a similar fashion to concentric lines of large diameter.

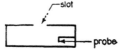

Figure 165.—Excitation of a symmetrically placed shunt slot through distortion of the electric field in a waveguide by means of a probe with a part of its length parallel to the field.

4. Arrays and Dielectric Rod Aerials
Linear Array With Uniform Amplitude Distribution

Consider a linear array of n point sources at a distance l apart and with progressive phase difference α between each element (Fig. 166). The space factor of the array in a plane containing

Figure 166.—Linear array of point sources with spacing l and phase difference α between successive elements.

the array is, (see Schelkunoff,[2] pp. 342-54)

$$S(\theta) = \left| \frac{\sin \dfrac{n\zeta}{2}}{n \sin \dfrac{\zeta}{2}} \right| \tag{47}$$

where $\zeta = \dfrac{2\pi l}{\lambda} \cos \theta - \alpha.$

The directions of the principal maxima are given by

$$\cos \bar{\theta}_p = \frac{\lambda}{l}\left(p + \frac{\alpha}{2\pi}\right) \qquad (48)$$

where $p = 0, \pm 1, \pm 2, \pm 3, \ldots\ldots\ldots\ldots$.

The criterion that the array has one principal maximum only is that

$$l < \lambda\left(1 - \frac{\alpha}{2\pi}\right), \ 0 < \alpha < \pi. \qquad (49)$$

If $l = \lambda(1 - \alpha/2\pi)$ then two principal maxima will occur, but as l is descreased one will diminish in amplitude. The direction of the main lobe is then given by

$$\cos \theta = \frac{\alpha\lambda}{2\pi l} \qquad (50)$$

corresponding to $p = 0$. By varying α, the beam can be made to " scan." This principle is used in the " Musa " array.

If it can be arranged that the space factor of the individual elements has a null in the direction of undesired maxima then larger spacings can be allowed. Directional diagrams of various arrays have been computed by numerous authors and a collection of such diagrams from various published papers is given by Brainerd and others.[5] For an array of other than point sources the space factor is found by multiplying (47) by the space factor of the individual elements.

With amplitude and phase distributions other than those described above it may not be easy to find an explicit expression for $S(\theta)$. However the radiation may be computed for a number of values of θ, either theoretically or by graphical vector addition of the radiation from the individual elements, enabling a graph of $S(\theta)$ against θ to be plotted.

End-fire Array

If we put $\alpha = 2\pi l/\lambda$, (e.g. by feeding the array from one end by a transmission line) then $\bar{\theta} = 0$, and the direction of maximum radiation is along the line of the elements. Such a linear array is called an end-fire array.

[5] J. G. Brainerd, G. Koehler, H. J. Reich, L. F. Woodruff, *Ultra High Frequency Techniques*, Van Nostrand, New York, 1943, pp. 419-24.

Equation (47) gives

$$S(\theta) = \frac{\sin \left(\dfrac{2\pi n l}{\lambda} \sin^2 \dfrac{\theta}{2} \right)}{n \sin \left(\dfrac{2\pi l}{\lambda} \sin^2 \dfrac{\theta}{2} \right)}. \tag{51}$$

For an end-fire array, with elements one quarter wavelength apart,

$$S(\theta) = \frac{\sin \left(\dfrac{n\pi}{2} \sin^2 \dfrac{\theta}{2} \right)}{n \sin \left(\dfrac{\pi}{2} \sin^2 \dfrac{\theta}{2} \right)} S_o(\theta) \tag{52}$$

where $S_o(\theta)$ is the space factor of the individual elements. The maximum element spacing of an end-fire array for a single principal lobe should not exceed $\lambda/4$. Equation (49) gives $l < \lambda/2$, but as l increases from $\lambda/4$ to $\lambda/2$, the radiation at $\theta = 180$ degrees increases until it equals the forward radiation.

The gain G relative to a single element of the array, is given by

$$G \doteq n \tag{53}$$

for large values of n.

A method of determining the gain of arrays with small numbers of elements is given by Terman.[6]

Beam Width of End-fire Arrays. The zeros in the polar diagram are found from (51) to occur for

$$\sin \frac{\theta_o}{2} = \sqrt{\frac{k\lambda}{2nl}}, \quad k = 1, 2, 3\ldots\ldots \tag{54}$$

Thus if n is large the width of the main lobe between zeros is

$$\Delta \doteq 2 \sqrt{\frac{2\lambda}{nl}} \text{radians} \tag{55}$$

$$\doteq 170 \sqrt{\frac{\lambda}{a}} \text{ degrees} \tag{56}$$

where a = overall length of the array.

The main disadvantage of the end-fire array is the difficulty of feeding a large number of elements spaced $\lambda/4$ or less apart.

[6] F. E. Terman, *Radio Engineers' Handbook*, McGraw-Hill, New York, 1943, p. 800.

Dielectric Rod Aerials

The use of a dielectric rod as a leaky waveguide radiator is of relatively recent origin and, although it is not an array in the sense of an aggregate of elementary radiators, it may be considered as the limiting case of an end-fire array and so has been included in this section.

If in (51) we let l tend to zero and keep a (i.e. nl) constant, we obtain a line distribution of uniform amplitude and with a progressive phase lag of 360 degrees per wavelength along the array. This gives

$$S(\theta) = \frac{\sin\left(\dfrac{2\pi a}{\lambda}\sin^2\dfrac{\theta}{2}\right)}{\dfrac{2\pi a}{\lambda}\sin^2\dfrac{\theta}{2}} \tag{57}$$

$$G_s = \frac{4a}{\lambda}\ (a \gg \lambda). \tag{58}$$

It is interesting to note that this is the same as that of a long end-fire array of point sources, $\lambda/4$ apart.

The dielectric aerial provides a method of realising the required phase distribution, although in general it will not give a uniform amplitude distribution. It consists of a tapered dielectric waveguide (Fig. 167) propagating a transverse electric wave and adjusted so that the velocity of the wave is equal to the free space velocity by shaping the rod, balancing the dielectric retardation against the waveguide acceleration.

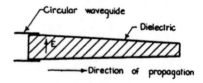

Figure 167.—Dielectric rod radiator.

Leakage from the rod, which may be fed from the wide end by matching it to a circular waveguide, causes it to act as an end-fire array. The power gain obtainable is slightly less than that given by (58) owing to tapering of the amplitude of the leakage field.

Broadside Linear Arrays

If in (47) we put $\alpha = 0$, then $\bar{\theta} = 90$ degrees i.e. the space factor has its maximum at right angles to the array. This array is known as a broadside array and its space factor is

$$S(\theta) = \frac{\sin\left(\dfrac{\pi n l}{\lambda} \cos \theta\right)}{n \sin \dfrac{\pi l}{\lambda} \cos \theta}. \tag{59}$$

The criterion that all secondary lobes are less than the main lobe is given by (49) to be

$$l < \lambda.$$

As l is increased from $\lambda/2$ to λ the radiation in the line of the array increases from zero to equal that of the main lobe. Thus it is desirable to keep l equal to or less than $\lambda/2$ except when the individual elements have a null along the line of the array (e.g. the linear array of dipoles of Fig. 168) and in this case the maximum allowable separation is one wavelength.

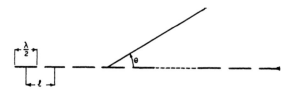

Figure 168.—Collinear array of half wave dipoles.

For a broadside array of n elements $\lambda/2$ apart

$$S(\theta) = \frac{\sin\left(\dfrac{n\pi}{2} \cos \theta\right)}{n \sin\left(\dfrac{\pi}{2} \cos \theta\right)} \tag{60}$$

$$\text{and } G \doteqdot n \tag{61}$$

for large n, where G is the gain over a single element of the array. For a continuous, in phase, line distribution of length a

$$S(\theta) = \frac{\sin \left(\dfrac{\pi a}{\lambda} \cos \theta \right)}{\dfrac{\pi a}{\lambda} \cos \theta}, \tag{62}$$

$$G_s = \frac{2a}{\lambda}. \tag{63}$$

Comparing this with (61), it is seen that a broadside array of elements $\lambda/2$ apart has the same gain as a continuous line distribution of the same length.

Beam Width. For large n, the width of the main lobe between zeros is

$$\Delta = \frac{2\lambda}{nl} \text{ radians} \tag{64}$$

$$\doteq \frac{120\lambda}{a} \text{ degrees} \tag{65}$$

where a = overall length. The total beam width to half field points on the polar diagram is given approximately by

$$\delta \doteq \frac{70\lambda}{a} \text{ degrees.} \tag{66}$$

Rectangular Arrays

The most general array is a three dimensional lattice of elements, and the characteristics of such arrays may be determined from the following rule.

Array of Arrays Rule. An aerial array may be considered as a combination of elementary arrays forming an array of arrays. The directional characteristics of such an array is given by the directional characteristics of the elementary array multiplied by the directional characteristics of the combination of elementary arrays, each elementary array being replaced by a point source located at its centre.

Consider a rectangular broadside array consisting of n columns and m rows of dipoles (Fig. 169). The space factor

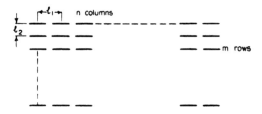

Figure 169.—Plane rectangular array of half wave dipoles.

in the horizontal plane (E plane) is

$$S(\theta) = \frac{\sin\left(\dfrac{\pi n l_1}{\lambda}\cos\theta\right)}{n\sin\left(\dfrac{\pi l_1}{\lambda}\cos\theta\right)} \cdot \frac{\cos\left(\dfrac{\pi}{2}\cos\theta\right)}{\cos\theta} \qquad (67)$$

where $\theta = 0$ along the array. In the vertical plane (H plane)

$$S(\phi) = \frac{\sin\left(\dfrac{\pi n l_2}{\lambda}\cos\phi\right)}{n\sin\left(\dfrac{\pi l_2}{\lambda}\cos\phi\right)} \qquad (68)$$

where $\phi = 0$ along the array. If $l_1 = l_2 = \lambda/2$ we obtain

$$G \doteqdot p \qquad (69)$$

where $p = nm =$ number of radiators and $G =$ gain over a single radiator.

Plane Reflector

If radiation from a broadside array is desired on one side only, a plane reflector can be placed behind the array. This has the effect of doubling the gain and giving a correspondingly narrower beam. The space factor of such an array is found by replacing the reflector by the image in it of the array, the image

being 180 degrees out of phase with the array. The spacing
of the elements from the reflector is not critical, but for very
close spacing the impedance of the elements will be very low.
In order to satisfy (49) the spacing should be $\lambda/4$ or less.

In Fig. 170 is plotted the radiation resistance of a $\lambda/2$
dipole situated at a distance d from and parallel to an infinite
reflecting sheet together with the directivity, D, of the system

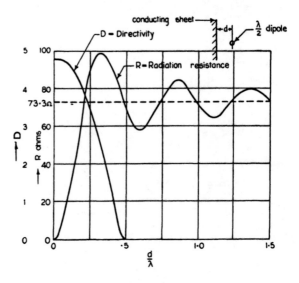

Figure 170.—Radiation resistance and directivity of a half-way dipole
at a distance d from, and parallel to, an infinite conducting sheet.

relative to that of an isolated dipole. The curve for gain
is substantially the same as that for directivity except at small
spacings, when it becomes less due to ohmic resistance. The
equation to the directivity curve is

$$D = \frac{R_\infty}{R}\left(2\sin\frac{2\pi d}{\lambda}\right)^2 \tag{70}$$

where $R_\infty = 73\cdot3\omega$ = radiation resistance of an isolated
dipole, and R = radiation resistance of the dipole and reflector.
If R equals the sum of the sum of the ohmic and radiation resistances presented at the dipole terminals, (70) gives the gain

of the system relative to a $\lambda/2$ dipole and in this case when $d \to 0$, $\dfrac{R_\infty}{R}$ remains finite. Thus the gain falls rapidly to zero.

The gain of a large rectangular array of n dipoles in front of a reflector is

$$G_s \doteqdot 2n \times 1\cdot64$$
$$\doteqdot 3\cdot28n$$

where $1\cdot64 =$ spherical gain of a dipole. The area of such an array when n is large is approximately

$$A = \frac{n\lambda^2}{4}$$

$$\text{giving } G_s \doteqdot \frac{4 \times 3\cdot3 \times A}{\lambda^2}. \tag{71}$$

It will be seen in section 5 that the gain of an aerial consisting of an aperture over which there is a uniform phase and amplitude distribution is

$$G_s = \frac{4\pi A}{\lambda^2}. \tag{72}$$

This agrees well with (71) indicating that such an array approximates to a uniform field distribution over an aperture.

Equations (71) and (72) show that in order to obtain a high gain a large area is required. It is possible, however, to obtain a higher gain for a given area than that given above for arrays of a small number of elements, by using small element spacing. This causes large mutual impedances and higher gain. For large arrays, mutual impedance effects tend to cancel and the expressions given above are valid.

Tapered Arrays

The long broadside linear array described above has side lobes of amplitude about 20 per cent of the main lobe. These may be reduced by tapering the amplitude of the current in the radiators from the centre to the edge of the array. This reduction is obtained at the expense of a wider main lobe and reduced gain.

Feeding of Broadside Arrays

The simplest method of feeding broadside arrays is by spacing the elements along a transmission line or waveguide. A typical case is illustrated in Fig. 171. Each alternate dipole is reversed to compensate for the 180 degrees change in phase along a $\lambda/2$ length of line. This is a resonant feed, since standing waves will exist on the feed line. An alternative to reversing the dipoles is to transpose the line between dipoles.

If the dipoles are spaced, say, every 200 degrees along the line instead of 180 degrees, the reflections back along the line from each element tend to cancel and the standing waves on

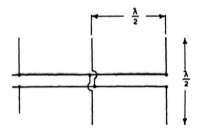

Figure 171.—Resonant-fed array of parallel half wave dipoles.

the line are reduced. The array is fed from one end, each element extracting a small fraction of the power, and the residue (generally about 10 per cent) is absorbed by a matched load at the other end. However since each element has to extract a progressively greater percentage of the power in order to maintain a given excitation, they must present a progressively smaller impedance to the line. This is generally difficult to arrange. Also the presence of the elements modifies the phase along the line and so the phase difference between the elements will vary from the desired amount. This effect however is usually small. A slight disadvantage is that if the elements are spaced every 200 degrees along the line (the line being transposed between radiators) there is a progressive phase difference of 20 degrees between radiators which will deflect the beam through an angle ψ from the normal to the array. This is found from (50) to be

$$\psi \doteqdot \frac{\lambda\alpha}{2\pi l} \text{ radians.} \qquad (73)$$

In the above example $\alpha = \dfrac{\pi}{9}$

$$l = \frac{5}{9}\lambda$$

giving $\psi \doteqdot 6$ degrees.

Any deviation from the operating frequency will cause α to change, and thus deflect the beam. This effect may be serious for long arrays having narrow beams.

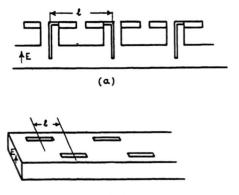

Figure 172.—Waveguide-fed arrays of (a) half wave dipoles and (b) half wave shunt slots.

Non-resonant feeds are more easily obtainable with wave-guides,[7] two typical examples being given in Fig. 172, which shows a waveguide array of probe coupled dipoles, alternate dipoles being reversed to compensate for the phase change along the guide and a similar array of slots, phase compensation

---

[7] A. L. Cullen and F. K. Goward, "The design of a waveguide-fed array of slots to give a specified radiation pattern," *J. Instn. Elect. Engrs.*, Vol. 93, Part IIIa, 1946, pp. 683-92.

being obtained by placing alternate slots on opposite sides of the centre line. Coupling is varied by adjusting the distance of the slot from the centre line [see equation (44)]. Slots may be excited by any of the methods described previously.

For non-resonant feeds the elements could be spaced, say, 200 degrees (in the guide) and the array terminated in a matched load. In this case, again, the beam will be deflected from the normal to the array. Because of the ease of varying the coupling of individual radiators, a non-resonant array is more easily obtainable in waveguide than conventional transmission line. A method of calculating the required element impedances is given in a report by Kaiser.[8]

If in Fig. 172, $l = \lambda_g/2$, (where λ_g = guide wavelength), and the guide is shorted $\lambda/4$ beyond the centre of the end radiator, a resonant system analogous to that of Fig. 171 is obtained, the only difference being that the element spacing is greater than one half wavelength in free space.

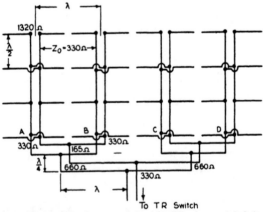

Figure 173.—Thirty-two element array of half-wave end-fed dipoles used with radar-set type LW/AW Mk.I showing the feeder system which uses 330 ohm twin transmission line throughout.

An example of the feeding of a typical rectangular array is given in Fig. 173. This array was used with the Australian

[8] T. R. Kaiser, "The design of a waveguide-fed slot array to produce a cosec[2] θ polar diagram," C.S.I.R. Radiophysics Laboratory, Report TI 223, January, 1946; A. L. Cullen and F. K. Goward.[7]

LW/AW Mk.I Radar set. The array consists of 8 banks each of four $\lambda/2$ dipoles backed by a wire mesh reflector. The dipoles are end fed as shown, the impedance at the feed point being 1320 ohms. Twin wire transmission line with $Z_o = 330$ ohms is used throughout.

Since the elements are spaced at $\lambda/2$ intervals along the line the input impedance to each of the banks A, B, C, D is 330 ohms. Banks A and B are connected by a twin line of length λ, so that the impedance at its centre is 165 ohms. This is transformed to 660 ohms by a $\lambda/4$ length of line and connected to the corresponding feed point of banks C and D by a line 2λ in length. This line is fed at its centre where the impedance is 330 ohms, so that it matches to the line from the transmitter.

The gain of this array is given by

$$G_d = 64$$

(i.e. twice the number of elements). The horizontal and vertical beam widths are found from (65) and (66) to be

$$\Delta_H = 30 \text{ degrees} \qquad \delta_H = 18 \text{ degrees}$$
$$\Delta_V = 60 \text{ degrees} \qquad \delta_V = 36 \text{ degrees}.$$

The feed to each individual bank of this aerial is resonant, since there are standing waves on the feeders.

Bandwidth of Arrays

The bandwidth of an array is difficult to estimate, due to mutual effects between elements, selectivity of feeders etc. As an example the bandwidth of a resonant array of 10 elements may be of the order of 1 per cent. The bandwidth of a long non-resonant array is always greater than that of a resonant array of the same length, since the selectivity of the feeding system is not as great.

Impedance of Array Elements

The input impedance of a dipole in either an end-fire or broadside array is modified by the close proximity of other resonant dipoles. It may be calculated from a knowledge of the self impedance of the element and the mutual impedances between it and all other elements. Tables of mutual impedances of $\lambda/2$ dipoles for various spacings and displacements are given by Terman,[6] (p. 779) and a more extensive table by Telecommunications Research Establishment.[9] If the number of elements is large, mutual effects tend to cancel and may be neglected.

Arrays of Dielectric Rods

The gain of a rectangular array of elements backed by a reflector is seen by (72) to be proportional to the area of the array. A somewhat higher gain may be obtained if each element of the array is itself an end-fire array. This implies that the absorption cross section of the array is greater than the actual area. The increase is greatest for small arrays. An example of this type is a broadside array of dielectric rods. Such a system has been developed by Bell Telephone Laboratories.

Yagi Arrays

The Yagi is a particular type of end-fire array consisting of a single fed element and a number of parasite elements. Currents are induced in the latter and the element spacings and parasitic impedances are arranged so that these currents produce fields which reinforce along the array.

If a parasitic element causes maximum radiation along the line from the driven element to the parasitic it is termed a director, and if in the opposite direction, a reflector. A dipole of length slightly less than the resonant length (i.e. with a capacitive impedance) behaves as a director, while if it is slightly greater (inductive) it behaves as a reflector. The optimum values of the lengths of such elements depend on the spacing, the length being adjusted so that the reactive component of the impedance causes the parasitic current to lag or lead the current in the fed element by the desired amount.

[9] Telecommunications Research Establishment, Mathematics Group, "Mutual impedance of half wave aerials," T.R.E. Report M3, January, 1941; R. A. Smith, *Aerials for Metre and Decimetre Wavelengths*, Cambridge University Press, 1949, pp. 10-1.

It is desirable to have the elements closely spaced in order to get high currents in the parasitics. The limit is set by selectivity and ohmic losses. Usually a spacing between 0.1λ and 0.25λ is used, and a compromise is made between physical size, gain, and impedance. Owing to the close proximity of the parasitic elements, the impedance of the driven element may be as low as 20 ohms. This may present feed difficulties which can, however, be overcome by using a folded dipole[10] as the driven element.

It is possible to obtain gain figures higher than expected from (58) for Yagi arrays having a small number of closely spaced radiators, owing to the lowering of the radiation resistance by mutual effects between the elements.

Since the current in each parasitic element depends on its mutual coupling to the other elements, calculation is difficult and the design of a Yagi is best performed experimentally.

5. Continuous Surface Radiators

This type of aerial has been defined earlier. The polar diagram may be computed if the current distribution over the surface is known, or, if the aerial has a plane aperture (e.g., a paraboloid or horn), from the field distribution across the aperture.

Radiation from a Plane Aperture, Having a Given Illumination[11]

Assuming that edge effects may be neglected, the polar dia gram may be calculated by applying the Kirchoff-Huygens' principle to the field across the aperture. The errors involved in the assumption of a discontinuity at the edge of the aperture are discussed by Stratton and Chu[12] and a more exact method

---

[10] D. S. Carter, " Simple Television Antennas," *R.C.A. Review*, Vol. 4, October, 1939, pp. 168-85.

[11] J. P. Ryan, "Design data relating directional diagram to aperture illumination of pencil and fan beam aerials," C.S.I.R. Radiophysics Laboratory, Report RP 241, February, 1945; S. Silver (Ed.), *Microwave Antenna Theory and Design* (Massachusetts Institute of Technology, Radiation Laboratory Series, Vol. 12), McGraw-Hill, New York, 1949, pp. 169-99.

of solving the diffraction problem is advanced. However, if the aperture is large with respect to the wavelength, the simple application of Huygens' principle gives results which agree closely with experiment.

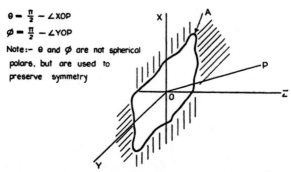

Figure 174.—Coordinate system for specifying the space factors (θ, ϕ) of an illuminated plane aperture in an opaque sheet situated in the XOY plane. The direction (θ, ϕ) is that of the vector OP which has direction cosines $\sin \theta$ and $\sin \phi$ relative to the X and Y axes.

The space factor for a plane aperture A (Fig. 174) may be shown to be

$$S(\theta,\phi) = \left| \int_A F(x,y)\, e^{j\frac{2\pi}{\lambda}(x\sin\theta + y\sin\phi)}\, dA \right| \qquad (74)$$

where $F(x,y) = f(x,y)e^{jg(x,y)}$ defines the aperture distribution in both amplitude and phase. The integral of (74) gives the amplitude and phase in the diffracted field and its amplitude gives $S(\theta,\phi)$. The space factor of (74) does not quite satisfy the definition of section 1 in that its maximum value, in general, is not unity. If the aerial produces a sufficiently narrow beam (i.e. if A is large) then we may write

$$\sin \theta \doteqdot \theta$$
$$\sin \phi \doteqdot \phi$$

and (74) becomes

$$S(\theta,\phi) = \left| \int_A F(x,y)\, e^{j\frac{2\pi}{\lambda}(x\theta + y\phi)}\, dA \right|. \qquad (75)$$

This assumption is justified for most microwave aerials. The form of (75) is in general more easily evaluated than (74).

The simplest aperture distribution is one which is uniform in amplitude and phase, i.e. $F(x,y) = 1$. The spherical gain of such a system will be denoted by G_{so} and it is given by

$$G_{so} = \frac{4\pi A}{\lambda^2} \qquad (76)$$

(where A is the aperture area), since for such an aerial, the absorption cross section equals the actual cross section (see page 224).

In the general case, when the direction of maximum radiation is along the normal to the aperture, it is clear that

$$G_s \propto \frac{S^2(0,0)}{\int\limits_A | F(x,y) |^2 \, dA},$$

since the numerator is proportional to the intensity along the normal and the denominator to the total power flowing through the aperture. If we put $\theta = \phi = 0$ in (74) and substitute we obtain

$$G_s \propto \frac{| \int F(x,y) \, dA |^2}{\int | F(x,y) |^2 \, dA}. \qquad (77)$$

The constant of proportionality may be found by putting $F(x,y)$ equal to unity and substituting from (76) giving

$$G_s = \frac{4\pi}{\lambda^2} \frac{| \int F(x,y) \, dA |^2}{\int | F(x,y) |^2 \, dA}. \qquad (78)$$

We may write

$$G_s = \gamma \, G_{so},$$

where

$$\gamma = \frac{1}{A} \frac{| \int F(x,y) \, dA |^2}{\int | F(x,y) |^2 \, dA}, \qquad (79)$$

and is known as the gain factor of the aerial. The factor γ is always less than unity, the case of uniform illumination being the optimum condition.

Rectangular Aperture

Consider a rectangular aperture of dimensions a and b in the x and y directions respectively, illuminated so that the aperture distribution may be expressed by

$$F(x,y) = F_1(x) \cdot F_2(y). \tag{80}$$

This gives

$$S(\theta,\phi) = \left| \int_{-a/2}^{a/2} F_1(x)e^{j\frac{2\pi}{\lambda}x\sin\theta} dx \right| \left| \int_{-b/2}^{b/2} F_2(y)e^{j\frac{2\pi}{\lambda}y\sin\phi} dy \right| \tag{81}$$

$$= S(\theta) \times S(\phi),$$

where

$$S(\theta) = \left| \int_{-a/2}^{a/2} F_1(x)e^{j\frac{2\pi}{\lambda}x\sin\theta} dx \right| \tag{82}$$

and

$$S(\phi) = \left| \int_{-b/2}^{b/2} F_2(y)e^{j\frac{2\pi}{\lambda}y\sin\phi} dy \right| \tag{83}$$

are the one dimensional space factors in the planes $\phi = 0$ and $\theta = 0$ respectively. Thus we need only consider the one dimensional polar diagrams.

Similarly we may write

$$\gamma = \gamma_1 \cdot \gamma_2 \tag{84}$$

where γ_1 and γ_2 are the vertical and horizontal one dimensional gain factors respectively.

$$\gamma_1 = \frac{\left| \int_{-a/2}^{a/2} F_1(x)dx \right|^2}{a \int_{-a/2}^{a/2} \left| F_1(x) \right|^2 dx}. \tag{85}$$

Graphs of $S(\theta)$ and gain factors for a number of distributions $F(x)$ are given by Ryan.[11]

The polar diagram is determined completely in amplitude and phase by

$$S(\theta)e^{j\alpha(\theta)} = \int_{-a/2}^{a/2} F(x)e^{j\frac{2\pi}{\lambda}x\sin\theta}\,dx. \tag{86}$$

This may be written

$$S(\theta)e^{j\alpha(\theta)} = \int_{-\infty}^{\infty} F(x)e^{j\frac{2\pi}{\lambda}\theta x}\,dx \tag{87}$$

since $F(x) = 0$ for $a/2 < x < -a/2$ and for sufficiently large apertures, as θ increases $S(\theta)$ rapidly becomes zero, thus we may replace $\sin\theta$ by θ.

If in (86) we write

$$\frac{2\pi}{\lambda}\sin\theta = \psi$$

and

$$S(\theta)e^{j\alpha(\theta)} = \sqrt{2\pi}\,T(\psi),$$

we have

$$T(\psi) = \frac{1}{\sqrt{2\pi}}\int_{-\infty}^{\infty} F(x)e^{jx\psi}\,dx.$$

An equivalent statement is that $T(\psi)$ is the Fourier transform (Stratton[12] pp. 285-9) of $F(x)$ and so correspondingly,

$$F(x) = \frac{1}{\sqrt{2\pi}}\int_{-\infty}^{\infty} T(\psi)e^{-jx\psi}\,d\psi$$

$$-\frac{1}{\lambda}\int_{-\infty}^{\infty} S(\theta)e^{j\alpha(\theta)}e^{-j\frac{2\pi}{\lambda}x\sin\theta}\cos\theta\,d\theta.$$

Therefore

$$F(x) \doteq \frac{1}{\lambda}\int_{-\infty}^{x} S(\theta)e^{j\alpha(\theta)}e^{-j\frac{2\pi}{\lambda}x\theta}\,d\theta \tag{88}$$

from which the aperture distribution corresponding to a given polar diagram may be determined (see section 6).

Uniformly Illuminated Rectangular Aperture

From (86) we obtain for the space factor in the plane $\phi = 0$ (Fig. 174)

$$S(\theta) = \frac{\sin\left(\dfrac{\pi a}{\lambda} \sin\theta\right)}{\dfrac{\pi a}{\lambda} \sin\theta} \qquad (89)$$

$$\doteq \frac{\sin\dfrac{\pi a\theta}{\lambda}}{\dfrac{\pi a\theta}{\lambda}}$$

which is the same as that for the continuous, in-phase, linear array of p. 246.

From equation (76)

$$G_s = \frac{4\pi ab}{\lambda^2},$$

where a and b are the sides of the aperture.

Equation (89) gives the beam widths in the plane $\phi = 0$ to be

$$\Delta \doteq 120\frac{\lambda}{a} \text{ degrees between the first two minima,}$$

$$\delta \doteq 70\frac{\lambda}{a} \text{ degrees to the half field points,}$$

amplitude of first side lobe = 21 per cent.

The important results are that gain is proportional to area and that beam width is inversely proportional to linear dimensions.

Circular Aperture With Uniform Illumination

It may be shown that

$$S(\theta) = \frac{J_1\left(\dfrac{\pi a}{\lambda} \sin\theta\right)}{\dfrac{\pi a}{\lambda} \sin\theta}. \qquad (90)$$

Also $G_s = \dfrac{\pi^2 a^2}{\lambda^2}$ from equation (76).

The beam widths are

$$\Delta \doteqdot 140\frac{\lambda}{a} \text{ degrees}$$

$$\delta \doteqdot 85\frac{\lambda}{a} \text{ degrees}$$

and the amplitude of the first side lobe = 13 per cent, where a is the diameter of the aperture.

In Fig. 175, $S(\theta)$ is plotted against $\frac{\pi a}{\lambda}\sin \theta$ for a rectangular and a circular aperture with uniform illumination and for a rectangular aperture with illumination $f(x) = \cos \frac{\pi x}{a}$ which is more nearly the form realised in practice. This distribution has a gain factor of 0·811. It is seen that the

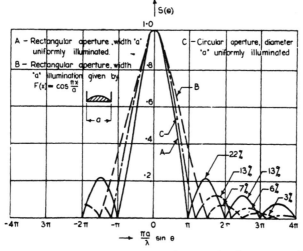

Figure 175.—Polar diagrams corresponding to simple aperture distributions for rectangular and circular apertures.

circular aperture gives a wider beam and smaller side lobes than the rectangular aperture, as also does the cos ($\pi x/a$) distribution. However, in the latter case, there is also a loss in gain for a given aperture area. The curves for the rectangular aperture apply also to large rectangular broadside arrays.

The case of a circular aperture having a circularly symmetrical amplitude distribution given by

$$f(r) = 1 - \left(\frac{4r^2}{a^2}\right)^t,$$

where r is distance from the centre of the aperture, is evaluated by Ryan[11] for $0 < t < 2$.

Parabolic Reflectors

The use of a parabolic reflector to obtain an equiphase distribution across a plane aperture is very common. Such reflectors may be either paraboloids of revolution with circular apertures or parabolic cylinders with rectangular apertures. The former may be cut so that the projected aperture is rectangular in shape. The source of excitation or feed in the case of the paraboloid must be an effective point source and in the case of a parabolic cylinder it must act as an effective line source.

If the field distribution in the aperture of the parabolic reflector is known by calculation or measurement, the radiation characteristics may be calculated by the principles and results discussed in the previous sections.

Paraboloids. For the theoretical behaviour of paraboloids with various feeds the reader is referred to the papers listed below.[13]

The condition of uniform phase across the aperture of a paraboloid is automatically obtained by using a small (effectively point) source at its focus ; however, a uniform amplitude distribution is never obtained owing to the peculiar polar diagram which would be required from the source. If the focus were in the plane of the aperture, four times more energy would have to be directed at the edge than at the centre.

---

[13] C. C. Cutler, "Parabolic antenna design for microwaves," *Proc. I.R.E.* Vol. 35, 1947, pp. 1284-94; E. G. Brewitt-Taylor, "A detailed experimental study of the factors influencing the polar diagram of a dipole in a parabolic mirror," *J. Instn. Elect. Engrs.*, Vol. 93, Part IIIA, 1946, pp. 679-82: H. T. Friis and W. D. Lewis, "Radar antennas," *Bell Syst. Tech. J.*, Vol. 26, 1947, pp. 219-317; R. Dabord, "*Reflecteurs et lignes de transmission pour ondes ultra-courtes*," *Onde Elect.*, Vol. 11, 1932, pp. 53-82.

Feeds for Paraboloids. The conditions which the paraboloid feed must satisfy are : (*a*) it must act sensibly as a point source, and (*b*) it must not obstruct a large portion of the aperture.

Location of Feed. The effective phase centre of the feed must be located at the focus. The position for most feeds must be found experimentally since the exact phase centre will not be known. Once the aperture has been fixed the choice of focal length depends on the directivity of the feed.

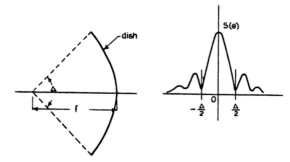

Figure 176.—Polar diagram of a paraboloid feed. Beam width is adjusted so that the main lobe just fills the dish.

A general rule is that the main lobe of the primary pattern should just fill the reflector (Fig. 176). The side lobes should fall outside the reflector since, in general, they will be opposite in phase to the main lobe. If the feed has a " back lobe " in phase with the forward lobe, the focal length should be made an odd number of quarter wavelengths in order that it should reinforce the reflected radiation, and for a back lobe 180 degrees out of phase with the forward lobe, *f* should be an integral number of half wavelengths. This is important for feeds such as dipoles which have a large amount of back radiation, but not for feeds such as horns etc. If it is desired to keep *f* small, then the feed diagram must be broad.

The amplitude distribution across the aperture may be determined approximately from the feed diagram, which should be measured at about a distance *f* from the source.

The effect of the tapered distribution is to lower the gain, at the same time widening the beam and reducing side lobes. By using a suitable aperture distribution side lobes may be entirely eliminated at the expense of a broader beam.   Fig. 177

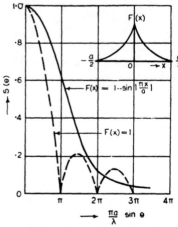

Figure 177.—Illustrates how a suitably tapered aperture illumination produces a diagram with no side lobes.

illustrates such a case, where the distribution is given by $F(x) = 1 - \sin \left| \dfrac{\pi x}{a} \right|$. The diagram corresponding to $F(x) = 1$ is given also for comparison.

The simplest feed is a single dipole and it is shown theoretically by Darbord[13] that the optimum dish is one with its focus

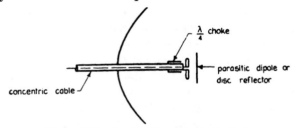

Figure 178.—Paraboloid with dipole feed.

in the aperture plane.   However, owing to the small degree of illumination at the edges, it may be economical to have the focus a short distance in front.   The gain factor of such a system is less than 0·5 since only one half of the dipole radiation falls on the reflector. It may be improved by using a parasitic dipole or a metallic plane or cylindrical reflector behind it (Fig. 178).

Typical feeds are shown in Figs. 179 and 180. Other possible feeds are, slots, small arrays of dipoles or slots, di-

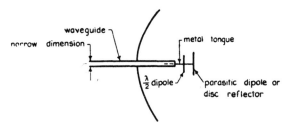

Figure 179.—This dipole feed differs from that of Fig. 178 in that the dipole is excited from a waveguide.

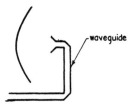

Figure 180.—Paraboloid with a horn or open waveguide feed.

electric rods etc. A paraboloid with good illumination has a gain factor of the order of 0·7.

Parabolic Cylinders

A parabolic cylinder fed by an " in phase " line source will produce a constant phase distribution across its aperture (Fig. 181). The line source may consist of a linear array of dipoles, slots, or dielectric rods, or alternatively of the cheese or pill box aerials described below. The design of the transverse section of the cylindrical parabola as far as illumination, focal length, etc. are concerned depends on the same considerations as for the paraboloid of revolution, but the problem in this case is one-dimensional.

The ends of the parabolic cylinder may be closed by parallel plates without materially affecting the performance if the electric vector is parallel to the axis of cylinder.   If however the electric vector is parallel to the end plates, waveguide propagation may take place within the cylinder and allowance may have to be made for this, especially in short cylinders.

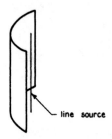

Figure 181.—Parabolic cylinder fed by a line source.

Cheese Aerial

This is the name given to a parabolic cylinder closed at its ends by parallel plates and fed with the electric vector parallel to the plates.   Fig. 182 shows a cheese fed by a sectoral horn (see p. 271) excited in the TE_{10} mode.

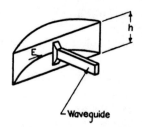

Figure 182.—Cheese aerial with a horn feed.

Unless the cheese is sufficiently narrow many modes of propagation may occur within it ; however, by using a horn feed of aperture equal to the cheese height, the lowest TE mode only is propagated and the amplitude distribution across the short dimension of the aperture follows a cosine law corres-

ponding to this mode. If the width of the horn is made
approximately 2/3 of the height of the cheese and placed
symmetrically, the primary pattern across the cheese mouth is
that of Fig. 183 (a). This may be subjected to a Fourier

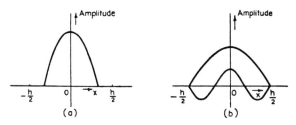

Figure 183.—(a) Primary illumination across the small dimension of the
cheese aperture. (b) Approximate analysis of this primary illumination
into two Fourier components.

analysis and effectively resolved into a " first harmonic "
and a "third harmonic " curve, plus higher harmonics of
smaller amplitudes [Fig. 183 (b)]. Thus, substantially two
modes of propagation will occur within the cheese.

If now the focal length is adjusted so that after reflection
at the parabola the two modes arrive in the plane of the opening with opposite phase, the amplitude distribution across the
aperture will be that of Fig. 184. This distribution, apart

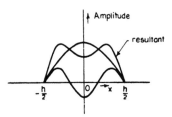

Figure 184.—Secondary illumination at the cheese aperture after reflection
in the cheese.

from the dip in the centre is approximately uniform, and in
actual practice the dip is not very pronounced, probably owing
to the presence of higher modes.

Satisfactory illumination across the long dimension of
the aperture is obtained since it is fed by a wave diffracted

at the narrow dimension of the horn, which causes a very wide beam having its phase centre almost in the aperture. This distribution will of course be tapered. The effective phase centre for illumination of the narrow dimension of the cheese is near the apex of the horn. If the height of the cheese is small, it may be fed by an open ended waveguide.

Pill Box Aerial[14]

If a parabolic cylinder closed at both ends is fed by a wave having its electric field parallel to the cylinder, only one wave, the TEM, can be propagated, and the phase velocity between the plates equals the free space velocity. A typical pill box with waveguide feed is shown in Fig. 185. The distribution across the narrow dimension in this case is constant.

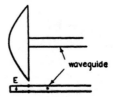

Figure 185.—Pill-box aerial with waveguide feed.

Both the pill box and the narrow cheese provide a good approximation to a line source with phase centre just inside the aperture, producing a very narrow beam in their own plane and a wide beam in the plane at right angles.

Bandwidth of Reflector Type Aerials

Since the feed can be made with a relatively large bandwidth, the bandwidth of a reflector type aerial is considerably greater than that of an array. It is however slightly less than the bandwidth of the feed since some of the reflected energy is picked up by the feed and causes standing waves on the feeder. This standing wave can be eliminated at one frequency by a matching adjustment on the feeder, but will be present for other frequencies owing to the change in phase of

[14] H. T. Friis and W. D. Lewis, "Radar antennas," *Bell Syst. Tech. J.*, Vol. 26, 1947, pp. 219-317.

the reflected radiation reaching the feed, with frequency. The bandwidth of a reflector type aerial is less for small paraboloids owing to the smaller focal length.

Lens Aerials

Dielectric Lenses.[15]   As in the case of a paraboloid, the lens is a development from geometrical optics.   It may be a solid of revolution or cylindrical, and in the latter case may be enclosed between parallel plates and fed by a horn in a similar fashion to the cheese and pill box.   The lens contours are not spherical as is usual in optics, but are designed to give freedom from spherical aberration for a point source at the focus.

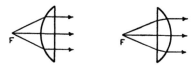

Figure 186.—Simple dielectric lenses used to produce a constant phase distribution across an aperture from a point source.

The lens must be designed so that a point source at the focus gives parallel emergent rays, i.e. constant phase over the aperture (Fig. 186).   For a constant phase illumination, the electrical path lengths to the aperture may differ by an integral number of wavelengths.   Thus the lens may be " stepped,"

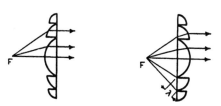

Figure 187.—Stepped dielectric lenses.

each step corresponding to one wavelength change in path length.   Two typical stepped lenses are shown in Fig. 187. If the lens aperture is large it may be necessary to step the lens for each two or more wavelengths' change in path length.

[15] J. Brown, "Microwave lenses," *Electronic Engng.*, Vol. 22, Part I, 1950, pp. 127–31.

In general, the feed conditions are similar to those for a paraboloid but, because of the difficulties of illuminating stepped lenses at large angles, the feed in this case needs to be somewhat more directional than for a paraboloid.

Advantages of the lens over the paraboloid are : (a) the feed does not obstruct the aperture ; (b) there is little energy reflected back to the feed. Thus the bandwidth of a lens aerial is effectively determined by that of the feed and by the chromatic aberration of the lens. Chromatic aberration is in general negligible for a simple lens but may be serious in the case of a large stepped lens. Reflections from lens surfaces may be eliminated by a $\lambda/4$ thickness of dielectric whose refractive index is the geometric mean of that of air and the lens material. The gain factor of a lens aerial is of the same order as that of a paraboloid of the same size with similar illumination.

Metal Plate Lenses.[16] A medium consisting of parallel metal plates with spacing of the order of waveguide dimensions behaves as a dielectric of refractive index less than unity to a wave polarised with its electric vector parallel to the plates.

If the plates have spacing b, then the phase velocity between the plates for the lowest TE mode is given by

$$v = \frac{c}{\sqrt{1 - \dfrac{\lambda^2}{4b^2}}} \qquad (91)$$

where λ is the free space wavelength and c is the free space velocity and so the refractive index is

$$n = \sqrt{1 - \frac{\lambda^2}{4b^2}}. \qquad (92)$$

It is possible to give n any value between zero and unity ; however for small values of n the system becomes extremely

[16] W. E. Kock, "Metal-lens antennas," *Proc. I.R.E.*, Vol. 34, 1946, pp. 828-836· J. Brown, "Microwave lenses," *Electronic Engng.*, Vol. 22, Part II, 1950, pp. 227-31.

frequency selective. Since $n < 1$, a concave metal plate lens is equivalent to a convex dielectric lens. It has been shown that the plates may be replaced by suitably spaced wires without effectively changing the characteristics.[17] Lenses may be employed to straighten wave fronts at the aperture of other types of aerials (e.g. horns). Because of the dependence of n on λ, metal plate lenses suffer greater chromatic aberration than dielectric lenses.

Electromagnetic Horns[18]

The polar diagram from an open ended waveguide is in general very broad owing to the small dimensions of the aperture. The diagram may be improved by increasing these dimensions, but it becomes increasingly difficult to suppress unwanted modes in the guide, the presence of which may cause the pattern to deteriorate.

This objection may be overcome by using a waveguide of normal dimensions and flaring it out in either the E or H plane to the desired aperture. Because of the uniform change of section unwanted modes will not be excited, except at the beginning of the flare, where the dimensions are such that they will be attenuated. Such a device is known as a sectoral horn, and the lowest modes which may be propagated are the TE_{10} and the TE_{01} [Fig. 188 (a) and (b)]. Because of the flare, the wave propagated in the horn will not be plane, but will diverge ; thus the phase distribution across the aperture will not be constant. This is analogous to the optical case of Fresnel diffraction at an aperture.

For both the TE_{10} and the TE_{01} horn there will be an optimum radial length r_0 for a given aperture. This may be seen qualitatively as follows. If r_0 is kept constant and the aperture increased, the gain will increase owing to the increase in aperture. However as the beam cannot be narrower than

---

[17] See Kock.[16]
[18] See Schelkunoff,[2] pp. 360-5; L. J. Chu and W. L. Barrow, "Electromagnetic horn design," *Elect. Engng.*, Vol. 58, 1939, pp. 333-8; W. D. Oliphant, "The electromagnetic horn," *Electronic Engng.*, Vol. 21, Part I, 1949, pp. 255-8, and Part II, pp. 294-9.

the horn angle, at some point the gain will commence to decrease. In the case of the TE_{10} wave this occurs when the phase difference between the fields at the edge and centre of the aperture is 90 degrees.

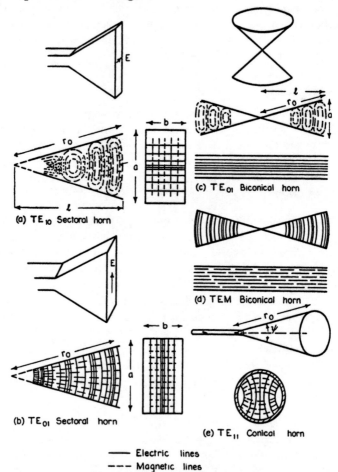

(a) TE$_{10}$ Sectoral horn

(b) TE$_{01}$ Sectoral horn

(c) TE$_{01}$ Biconical horn

(d) TEM Biconical horn

(e) TE$_{11}$ Conical horn

——— Electric lines
--- Magnetic lines

Figure 188.—Field configurations in sectoral, conical, and biconical electromagnetic horns.

Sectoral Horn Excited by the TE_{01} Mode [Fig. 188 (b)].

The assumption that the aperture distribution is the same as that which would exist at that point in an infinite horn of the same flare angle leads to the following expression (Schelkunoff[2] pp. 360-5) for the field distribution across the broad dimension of the horn,

$$F(x) = e^{-j \frac{\pi x^2}{\lambda r_0}}.$$

Substituting this in equation (85) gives the corresponding gain factor,

$$\gamma_a = \frac{2 \lambda r_0}{a^2} \cdot \left[C^2\left(\frac{a}{\sqrt{2 \lambda r_0}}\right) + S^2\left(\frac{a}{\sqrt{2 \lambda r_0}}\right) \right] \quad (93)$$

where a = horn aperture in the broad dimension, $C(x)$ and $S(x)$ are the Fresnel integrals

$$\text{i.e. } C(x) = \int_0^x \cos \frac{\pi t^2}{2} \, dt$$

$$S(x) = \int_0^x \sin \frac{\pi t^2}{2} \, dt.$$

For a horn of given depth, the gain is proportional to $a\gamma_a$, and the optimum condition may be found by maximising $a\gamma_a$ with respect to a, keeping r_0 constant. This gives

$$a = \sqrt{2 \lambda r_0} \quad (94)$$

i.e. the optimum radial length for a given aperture a is

$$r_0 = 0 \cdot 25 \frac{a^2}{\lambda}.$$

Substituting l, the distance from the apex to the centre of the aperture, in (94) gives

$$r_0 - l = 0 \cdot 25 \lambda$$

i.e. the phase difference between the fields at the edge and centre of the aperture is 90 degrees for an optimum horn.

Substituting (94) in (93), the gain factor of the optimum horn is found to be

$$\gamma_a = 0 \cdot 80.$$

The field distribution across the narrow dimension is uniform in phase, and follows a cosine law in amplitude giving

$$\gamma_b = 0 \cdot 81$$

where γ_b is the one dimensional gain factor in the narrow dimension of the aperture. The overall factor γ of such a horn is given by

$$\gamma = \gamma_a \gamma_b = 0 \cdot 65.$$

Thus the gain of an optimum TE_{01} horn is

$$G_s = 8 \cdot 2 \frac{ab}{\lambda^2}. \tag{95}$$

Sectoral Horn Excited by the TE_{10} Mode [Fig. 188 (a)]. A similar treatment to that of the TE_{01} horn gives

$$F(x) = \cos \frac{\pi x}{a} \, e^{-j \frac{\pi \, x^2}{\lambda \, r_0}}$$

and

$$\gamma_a = \frac{\lambda r_0}{a^2} \{ [C(u) - C(v)]^2 + [S(u) - S(v)]^2 \} \tag{96}$$

where

$$u = \frac{1}{\sqrt{2}} \left(\frac{\sqrt{\lambda r_0}}{a} + \frac{a}{\sqrt{\lambda r_0}} \right)$$

$$v = \frac{1}{\sqrt{2}} \left(\frac{\sqrt{\lambda r_0}}{a} - \frac{a}{\sqrt{\lambda r_0}} \right).$$

Maximising $a\gamma_a$ with respect to a gives the optimum relation between radial length and aperture :

$$a = 1 \cdot 78 \sqrt{\lambda r_0}, \tag{97}$$

$$\text{i.e. } r_0 = 0 \cdot 32 \frac{a^2}{\lambda}$$

$$\text{and } r_0 - l = 0 \cdot 4\lambda.$$

Substituting (97) in (96) we obtain $\gamma_a = 0 \cdot 63$ for the optimum horn. The field distribution across the narrow dimension of the aperture is uniform in both phase and amplitude, giving $\gamma_b = 1$ and $\gamma = \gamma_a \gamma_b = 0 \cdot 63$. Thus the gain of such a horn is given by

$$G_s = 7 \cdot 92 \frac{ab}{\lambda^2}. \tag{98}$$

It will be seen that for a given area of aperture, both types of sectoral horn have approximately equal gain when the dimensions are optimum ; however the radial length of the TE_{01} horn will be less than that of the TE_{10}.

In Fig. 189 γ_a is plotted against $a/\sqrt{r_0\lambda}$ for TE_{01} and TE_{10} horns, together with graphs of $G_s(\lambda/r_0)^{\frac{1}{2}}$ for horns of one wavelength depth ($b = \lambda$). For other horns, G_s should be multiplied by b/λ.

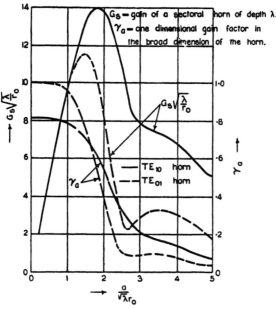

Figure 189.—Gain and gain factor of sectoral horns as a function of horn dimensions. The curves for gain factor apply also to biconical horns.

Pyramidal Horn. If the horn is flared in both the E and H planes, the gain factor in the dimension parallel to the E vector is given by (93) and in the dimension parallel to the H vector by (96). If the flare in one plane is slight it may be treated as a sectoral horn of the same aperture. For an optimum flare in both the E and H plane, the overall gain factor is equal to 0·50, and

$$G_s = 6\cdot33\frac{A}{\lambda^2} \qquad (99)$$

where A is the aperture area.

Conical Horn.[19] For a TE_{11} conical horn [Fig. 188 (e)] fed by a circular waveguide, the optimum dimensions are given by

$$r_o = \frac{0 \cdot 3 \lambda \cos \tfrac{1}{2}\psi}{1 - \cos \tfrac{1}{2}\psi}. \tag{100}$$

Biconical Horn.[20] Fig. 188 (c) and (d) illustrate the form of the electromagnetic field in a biconical horn for the TE_{01} and the TEM mode respectively. It will be noted that the former corresponds to the TE_{10} mode in a sectoral horn, and the latter to the TE_{01} mode ; thus the optimum dimensions are the same in both cases.

If we consider a meridian plane only, the gain factor corresponding to the aperture distribution is the same as for the corresponding sectoral horn and may be found from equations (93) and (96) or from Fig. 189 for given horn dimensions. The gain is then given by[21]

$$G_s = \gamma_u \cdot \frac{2a}{\lambda} \tag{101}$$

since $2a/\lambda$ is the gain corresponding to uniform phase and amplitude aperture distribution [see equation (63)].

For optimum horns,

TEM	TE_{01}	
$a = \sqrt{2\lambda r_0}.$	$a = 1 \cdot 78 \sqrt{\lambda r_0}.$	(102)
$G_s = 1 \cdot 6 \dfrac{a}{\lambda}.$	$G_s = 1 \cdot 26 \dfrac{a}{\lambda}.$	(103)

Bandwidth of Horns. Effectively the horn may be regarded as a tapered matching section between the guide and free space, thus if the taper is not too rapid the impedance of the horn will not vary rapidly with change in frequency.

[19] G. C. Southworth and A. P. King, " Metal horns as directive receivers of ultra short waves," *Proc. I.R.E.*, Vol. 27, February, 1939, p. 95.

[20] W. L. Barrow, L. J. Chu and J. J. Jansen, " Biconical electromagnetic horns," *Proc. I.R.E.*, Vol. 27, December, 1939, p. 769.

[21] Barrow, Chu and Jansen have plotted gain relative to a current element against flare angle for various values of r_0/λ ; however, the figures for gain are 4·8 times higher than expected from (101) i.e. some 3 or 4 times higher than that of an in-phase, uniform amplitude, line source of the same aperture.

Combination of Lenses with Horns. It has been seen that horns have gain factors considerably less than unity because of the diverging wave at the horn aperture. By the use of lenses of the type described previously the aperture distribution may be made uniform in phase, and satisfactory horns can be constructed whose dimensions differ considerably from the optimum (e.g. short horns with large flares).

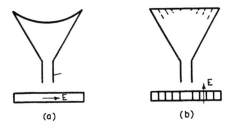

Figure 190.—Combined horn and lens systems to produce constant phase across the horn aperture.

Fig. 190 shows two arrangements. That of (a) is for a TE_{01} horn, in which the horn structure itself is used as a lens owing to the higher phase velocity in the horn than in free space, and that of (b) shows a metal plate lens designed to straighten the wave front in a TE_{10} horn.

6. The Design of Aerials

The problem of designing an aerial which has a specified space factor has arisen in radar equipments of the GCI (Ground Control Interception) and ASV (Aircraft to Surface Vessel) types. In both of these cases the basic requirement is for a radar to give coverage such that the echo intensity is independent of range up to a given range. It may be shown that to satisfy this requirement we must have $S(\theta)$ proportional to cosec θ over a given range of θ in the vertical plane. This type of aerial is known as a cosec^2 θ aerial since the energy radiated at an elevation θ is proportional to cosec^2 θ. The directional diagram required is shown, plotted in both polar and Cartesian form in Fig. 191 (a) and (b) respectively. The cosec^2 θ pattern is produced between the angles θ_1 and θ_2.

In most cases the problem is one dimensional, a particular distorted polar diagram being required in one plane only. Two methods of solution are available, based on geometrical optics and diffraction theory respectively.

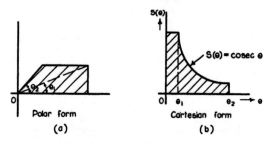

Figure 191.—Polar diagram of a " cosec^2 θ " aerial plotted in (a) polar and (b) Cartesian coordinates. $S(\theta)$ = cosec θ for $\theta_1 < \theta < \theta_2$ and $S(\theta) = 0$ for $\theta < 0$.

Geometrical Optics

(a) Standard Optical System and Distributed Feed. A typical system is shown in Fig. 192. A tapered feed is employed producing a fanned beam which approaches a cosec^2 θ pattern. By suitable choice of feed a wide range of secondary patterns

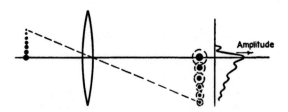

Figure 192.—Illustrating how a distorted polar diagram may be obtained by using a distributed feed, tapered in amplitude, at the focus of a lens.

may be obtained. The secondary pattern will be influenced by all of the errors of the optical system, the main effect being coma. This may be reduced by placing the sources in the feed along a curve instead of a straight line.

The feed may be an array of dipoles, slots, or other radiators, and the focusing system, which for convenience has been shown as a lens may equally well be a paraboloid.

(b) Localised Source and Modified Optical System. Distorted polar diagrams have been obtained by the use of a localised feed and a modified parabolic reflector. The design is based on a one-one correspondence between the intensity distribution in the feed pattern and in the secondary pattern, the reflector being modified in such a fashion as to give the desired form of the latter.

Before modification, the reflector may be a paraboloid of revolution in which case the feed should be an effective point source. In this case the section of the paraboloid in the plane in which the modified diagram is required is distorted. Alternatively, a cylindrical reflector may be used whose cross-section is a modified parabola, fed by a line source, e.g. an array of dipoles or slots, a pill box etc.

It should be remembered that the actual polar diagram will only approach the desired diagram if the aperture is many wavelengths in linear dimension and the system approaches an optical one.

An appropriate method of calculating the space factor of modified reflector systems is to compute the aperture distribution from a knowledge of the feed diagram and to use (82) to compute the secondary pattern.

Diffraction Theory
Aperture Distribution to Give a Specified Polar Diagram. The aperture distribution $F(x)$ corresponding to a given field pattern $S(\theta)e^{j\alpha(\theta)}$ may be determined from (88), since $S(\theta)e^{j\alpha(\theta)}$ is the Fourier transform of $F(x)$.

In general it is desired to produce a pattern which is specified by $S(\theta)$ and any suitable phase distribution $\alpha(\theta)$ may be chosen. This allows an infinite number of aperture distributions for a single polar diagram, and the amplitude of each of these distributions will not fall to zero for a finite value of x except in special cases. It should be possible

however to choose a distribution, the amplitude of which
falls to a small value for a finite value of x. It may then be
amputated at this point without appreciably affecting the
diagram. The effect of this in general is to introduce unwanted side lobes and to smooth out any discontinuities in
the polar diagram because of the diffraction effect. The
effect is smaller, the larger the aperture.

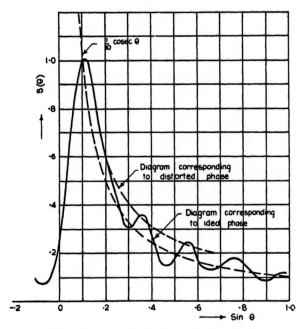

Figure 193.—Polar diagrams for a thirteen element slot array giving
$S(\theta) \doteqdot \frac{1}{10} \operatorname{cosec} \theta$ for $\sin \theta > \cdot 1$.

Aperture distributions which give $S(\theta) = \operatorname{cosec} \theta$ have been
extensively studied and a number of such distributions are
given in amplitude and phase in a Telecommunications Research Establishment report.[22] All of these distributions
require infinite amplitude for $x = 0$ and the effect of using

[22] P. M. Woodward, "A method of calculating the field over a plane
aperture required to produce a given polar diagram," *J. Instn. Elect. Engrs.*,
Vol. 93, Part IIIA, 1946, pp. 1554-8.

finite amplitude is to cause a departure from the cosec θ law for small angles. The case of a triangular polar diagram has been studied by Cullen.[7]

It has been seen (section 4) that an array of slots or dipoles suitably spaced behaves as a uniform " in phase " line distribution. In a similar fashion, an array of elements with suitable phasing and amplitude may be used to approximate a continuous aperture distribution (subject, of course, to the limitations on spacing discussed in section 4).

The required phase distribution is obtained by suitably spacing the elements along a waveguide or transmission line, and the amplitude distribution by adjusting the coupling to

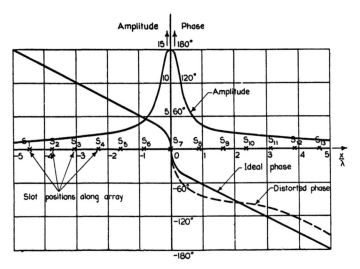

Figure 194. Ideal and actual aperture distributions corresponding to the polar diagrams of Fig. 193.

the guide or line. This is most easily effected in the case of guide by using shunt slots spaced from the centre line or by probe fed dipoles. It is usual to choose a mean element spacing of about 200 degrees and to terminate the array with a matched load in order to reduce the standing waves on the feeder (i.e. a non-resonant array). A method of determination of element conductance and spacing along the array is given

in a report by Kaiser.[8]   Coupling of elements to the guide
causes a distortion of the phase distribution along the array
which may be serious if any elements are coupled tightly.

A waveguide-slot array of 13 elements to give a $\cosec^2 \theta$
diagram has been designed and its characteristics are given
in Figs. 193, 194 and 195.   The full curves of Fig. 194 show

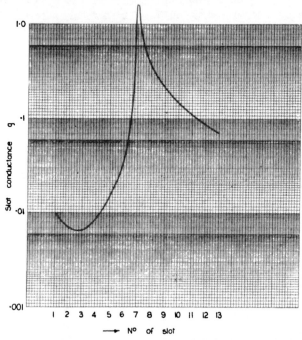

Figure 195.—Slot conductance as a function of slot position for the
thirteen element slot array, giving the amplitude distribution of Fig. 194.

the required aperture distribution, and the dotted curve the
distortion of the phase distribution due to the coupling of the
elements to the guide.   Fig. 195 gives the normalised conductance of each slot as a function of its position along the
guide.   This conductance distribution is calculated from the
aperture distribution.   The desired polar diagram ($\frac{1}{10} \cosec \theta$),
the diagram due to the ideal, but finite, aperture distribution,
and that due to the distorted aperture distribution are shown in

Fig. 193. The latter two curves have been computed from the phase and amplitude of the radiation from each individual element.

The required spacing of the slots from the centre line of the guide may be calculated from (44). Such an array may be combined with a cylindrical paraboloid to give a narrow beam in the other plane, or even with a shaped reflector if it is desired to obtain a distorted diagram in two planes at right angles.

7. Rapid Scanning Methods

In order to accurately estimate the direction of a target it is possible to use an aerial which provides a narrow beam and which scans in either two or three dimensions depending on whether azimuth or elevation alone, or both, is to be determined. With scanning systems, high gain aerials are employed which tend to increase the range of the set. This in turn is reduced by the introduction of scanning losses, since for very rapid scans the number of pulses reflected from the target during a single sweep may be quite small.

Two types of scanning are used, electrical and mechanical. The former describes all systems in which the aperture is held fixed while the beam is caused to swing by changing the phase distribution, and the latter those systems in which the aerial as a whole is scanned. Since mechanical scanning is mostly a problem of mechanical design, electrical scanning only will be considered here. For a general account of scanning systems, the reader is referred to the literature.[23]

Moving Feed Fixed Aperture Scanning

If a paraboloid or lens feed is displaced from the axis through a distance t, the beam scans through an angle approximately equal to t/f where f is the focal length. However, if the scan is more than a few beam widths the phase distribution across the aperture becomes non-linear and the diagram deteriorates.

---

[23] E. G. Schneider, "Radar," *Proc. I.R.E.*, Vol. 34, 1946, pp. 556-9; H. T. Friis and W. D. Lewis, "Radar antennas," *Bell Syst. Tech. J.*, Vol. 26, 1947, pp. 219-317.

The side lobe on one side of the main lobe increases and the gain decreases. The deterioration is due to coma in the focusing system and may be minimised by either using large focal lengths or by modifying the focusing system. The former requires a directive feed and large feed displacement, and may be undesirable. In general, however, a simple focusing system designed to be free from coma suffers badly from spherical aberration, and it is necessary to use a zoned reflector consisting of a number of elements free from spherical aberration situated on a coma free surface and spaced so that radiation from the individual elements reinforces. Alternatively a combination of two simple optical systems such as the Schmidt (reflector plus correcting lens) and the Schwarzchild (double reflector) systems may be used. A discussion of the design of mirrors for scanning systems is given by Gooden.[24]

The path of the beam in scanning systems of this type may be conical or linear. The locus of the field in the former case is a circle in the focal plane producing a beam which scans in a cone, and in the latter, a straight line in the focal plane. Instead of an oscillating feed, a number of switched fixed feeds spaced along a line in the focal plane may be used.

One of the disadvantages of a system using a feed moving along a line in the focal plane is that of providing a mechanical structure capable of allowing a large acceleration at the end of the feed traverse. This may be overcome by means of a rolled parallel plate system. Fig. 196 (a) shows a waveguide-fed parallel plate horn feeding in turn a parabolic mirror. In the plane of the horn the effective phase centre of the feed is at the mouth of the waveguide and so the section of the mirror in this plane has focal length f, while in the plane at right angles the phase centre is at the mouth of the parallel plates and the mirror section in this plane has focal length f'. A pencil beam is produced which can be made to scan by moving the waveguide along the line AA'. An equivalent of the system of Fig. 196 (a) is that of Fig. 196 (b) where A_1A_1' is the mirror

[24] J. S. Gooden, "The use of geometrical optical theory in the design of mirror shapes for scanning aerials," C.S.I.R. Radiophysics Laboratory, Report RP 242, February, 1945.

image of AA' in the reflecting strip BC. This system may now be rolled about an axis parallel to BB' so that AA' becomes a complete circle [Fig. 196 (c)], which is then the locus of the feed. A linear " sawtooth " scan is thus produced. This method is possible because parallel plate systems may be folded without effectively altering their characteristics as long as the plates remain parallel.

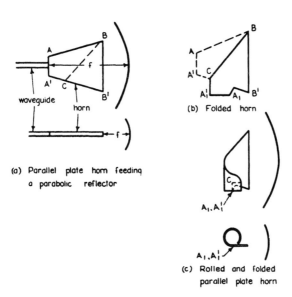

(a) Parallel plate horn feeding
a parabolic reflector

(b) Folded horn

(c) Rolled and folded
parallel plate horn

Figure 196.—The development of a rolled and folded horn scanner from a parallel plate system, producing a linear saw-tooth scan from a rotating waveguide feed.

Variable Path (Foster) Scanner

Fig. 197 represents a pill-box parallel-plate system fed by a waveguide inserted at F. This produces a constant phase distribution across AA' and since the electrical path length for all rays PQ is equal (to d say), a constant phase distribution is produced across BB'. If now by some means the path length d' of a ray distance x from AB is made to vary linearly with x giving $d' = d + bx$, the constant phase front at BB'

is deflected through an angle θ given by $\tan \theta = b/\lambda$, and the beam is deflected through an equal angle. Thus by providing a means of varying b the beam may be swung through large angles without distortion.

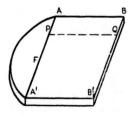

Figure 197.—Pill-box with the parallel plates extended beyond the aperture of the parabola.

This is accomplished very neatly in a scanner developed by J. S. Foster at Massachusetts Institute of Technology.[23] The construction of this scanner is illustrated in Fig. 198. It consists of a conical structure [Fig. 198 (*a*)] the cross section of which is given in more detail in Fig. 198 (*b*). As the rotor

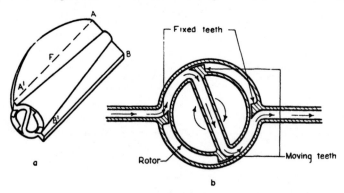

Figure 198.—Schematic diagram of the variable path (Foster) scanner. Rotation of the inner cone produces a linear saw-tooth scan.

is turned the path length increases by an amount dependent on the distance from the edge. In order that the rotor may rotate through a full 360 degrees the reflecting barriers consist of teeth such that those on the rotor will pass through those on the stator. These teeth are sufficiently closely spaced (i.e. $\ll \lambda/2$) to act as almost perfect reflectors.

Rapid scans of the order of ± 20 degrees without distortion or appreciable loss of gain are obtainable with this system. The system of Fig. 198 when used in conjunction with a parabolic cylinder will produce a scanning pencil beam.

Musa Systems

It has been seen (section 4) that a linear array of elements with a progressive phase difference α between adjacent elements produces a beam making an angle θ with the normal given by

$$\sin \theta = \frac{\alpha \lambda}{2\pi l} \qquad \text{where } l \text{ is the element spacing.}$$

The beam can be made to scan by varying α. This is similar in principle to the variable path length scanner described above, and large angle scans with negligible distortion are obtainable.

One obvious way of accomplishing the required phase variation would be to space the elements along a transmission line or waveguide and to use a frequency modulated signal. One difficulty in this, of course, is to provide a satisfactory automatic frequency control system in the receiver. If the frequency is kept constant the phase may be varied by means of line lengtheners or phase changers in the feeds to the individual elements. Such a device has been developed by Bell Telephone Laboratories using a broadside dielectric rod array for a ship gunnery radar.

If the radiating elements are spaced along a waveguide the progressive phase difference α may be varied by altering the wavelength in the guide, which can be done without changing the frequency. A method which has been employed is to space parallel resonant rings along the guide. The effective wavelength in the guide depends on the orientation of the rings, and by connecting the rings through a gear system which keeps all of the rings parallel while rotating, the beam may be caused to scan.

The same effect is accomplished in a different fashion in the Eagle scanner. The guide wavelength for the TE_{10} mode is related to the free space wavelength by

$$\lambda_g = \frac{\lambda}{\sqrt{1 - \left(\dfrac{\lambda}{2a}\right)^2}}$$

where a is the width of the guide. In the Eagle scanner, λ_g, in a waveguide array of dipoles is varied by using a waveguide of variable width a. It is necessary of course to have a well designed choke system to prevent undue leakage of radiation from the guide. It is possible to get scans of up to $\pm$ 30 degrees with a minimum of distortion.

8. Measuring Techniques

For most applications the properties of an aerial which must be known are its polar diagram, gain, impedance, and bandwidth. Also, if experimental work is being carried out on a system, it is often useful to know the field distribution across its aperture. This section briefly describes some methods of measuring these quantities.

An important fact not immediately obvious is that the form of the aperture distribution and polar diagram of the aerial are independent of the matching between aerial and transmitter (or receiver). Thus it is not important in the measurement of either of these quantities to have a carefully matched system.

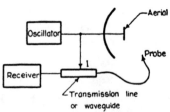

Figure 199.—Schematic set-up for experimental determination of the aperture distribution of an aerial in both amplitude and phase.

Aperture Distribution

The amplitude distribution may be determined by moving a probe which is coupled to a receiver (e.g. a crystal and meter) across the aperture. To determine phase, some arrangement such as that shown in Fig. 199 is required. The phase distribution may be determined from the position of the input I to cancel the pickup from the probe. An alternative is to fix the position of I and to use a variable length of line or guide in the feed to the probe.

Polar Diagram

For polar diagram measurements the system to be tested may be used either as a transmitting or receiving aerial, the latter being generally more convenient.

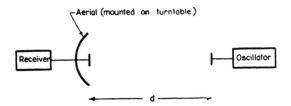

Figure 200.—Schematic set-up for determination of the polar diagram of an aerial.

The aerial is set up on a turntable and connected to a calibrated receiver. In order to produce an effectively plane wave at the aerial, an aerial fed by an oscillator is set up at a large distance, d (Fig. 200). By the theorem of reciprocity the receiver and oscillator of Fig. 200 may be interchanged without altering the polar diagram. The turntable is rotated and either the voltage or power input to the receiver is plotted against θ. The former is proportional to $S(\theta)$ and the latter to $\Phi(\theta)$.

A good experimental criterion for an effectively plane wave at the aerial is that the phase from the centre to the edge of the aerial aperture should not change by more than $\pi/4$. This means that for an aperture of width a, we must have

$$d > \frac{a^2}{\lambda}. \tag{104}$$

The measurements should be carried out over a plane surface free from obstructions which might give spurious reflections.

Automatic Polar Diagram Recorder. In order to eliminate the time factor involved in taking large numbers of polar diagrams during aerial experiments an equipment has been developed in the laboratory for presenting polar diagrams on a cathode ray screen. This consists basically of a potentiometer, geared to the turntable with a voltage applied across it and connected to the X deflection of a cathode ray tube, the Y plates being fed by the output from a receiver. X deflection is proportional to the angle through which the turntable is rotated and the Y deflection to receiver output, which may be calibrated in terms of receiver input. By using the set-up of Fig. 200, the polar diagram is traced on the cathode ray tube screen as the turntable is rotated. A persistent screen tube is employed in the equipment.

By substituting a recording meter, the drum of which rotates with the turntable, for the cathode ray tube it would be possible to make an equipment to give a permanent plot of the polar diagram. Equipments of this type have been used by many laboratories.[25]

Gain

It is safe to say that at the present time there is no completely satisfactory method of measuring aerial gain. The methods available are the following.

(a) By Direct Comparison. The set-up of Fig. 200 is employed. Making sure that the aerial is carefully matched to the receiver, the peak power input to the receiver is measured. The aerial is then replaced by a standard aerial such as a half wave dipole which is carefully matched to the receiver and the peak power input is again measured. The ratio of these two quantities gives the ratio of the gains of the two aerials, and if that of the standard is known the other may be determined. Calibrated horns are often used at centimetre wavelengths as secondary standards.

This method is quite satisfactory if it can be carried out in a location remote from the earth and other reflecting bodies.

[25] C. C. Cutler, "Microwave antenna measurements," *Proc. I.R.E.*, Vol. 35, 1947, pp. 1462-71.

The presence of the earth often cannot be ignored, however, especially with low gain aerials and the different effect of earth reflections on the two aerials may give rise to large errors in gain, especially if the properties of the standard (gain, polar diagram, etc.) differ greatly from those of the aerial under test. This often makes the half wave dipoles an unsuitable standard for use with directive aerials. The ideal would be a set of secondary standards covering a large range of gain in relatively small steps.

(b) An Indirect Method. The aerial is matched to an oscillator and placed in front of a plane sheet. Reflected radiation which effectively originates from its image in the sheet is picked up by the aerial and causes standing waves on the feeder.

If the voltage standing wave ratio is S when the aerial is distant d from the reflector the gain is given by

$$G_s = 8 \; \frac{1 - S}{1 + S} \frac{d}{\lambda}. \tag{105}$$

In deriving this formula it has been assumed that no energy is re-radiated by the aerial. However this is not so, and in general some of the energy picked up by the aerial is re-radiated, reflected from the sheet, and picked up again by the aerial, thus modifying the standing waves on the feeder. Pippard, Burrell and Cromie[26] have shown experimentally, that this re-radiation may introduce serious errors into the method. This difficulty may be overcome by taking a series of measurements at varying distances from the sheet. It will be found that the graph of G_s against d is oscillatory, the mean value being the correct value of G_s.

Unfortunately, although this method looks quite promising, it will only give satisfactory results for aerials whose gain lies in a relatively restricted range, owing to limitations on the distance of the aerial from the reflector ($d \geqslant a^2/2\lambda$ where

[26] A. B. Pippard; C. J. Burrell and E. E. Cromie, "The influence of re-radiation on measurements of the power gain of an aerial," *J. Instn. Elect. Engrs.*, Vol. 93, Part IIIA, 1946, pp. 720-2; L. D. Lawson, "Some methods for determining the power gain of microwave aerials," *J. Instn. Elect. Engrs.*, Vol. 95, 1948, pp. 205-9.

a is the width of the aerial), on the sensitivity of the standing wave measuring gear, and because a finite reflecting sheet must be used.

(c) By Integration of the Polar Diagram.   If $\Phi^*(\theta,\phi)$ is known for all angles θ, ϕ, the gain may be determined from equation (5). It is difficult, however, to measure a polar diagram in three dimensions, and $\Phi^*(\theta,\phi)$ is in general known only in the planes $\theta = 0$ and $\phi = 0$.

In special cases this is sufficient information for an estimation of the gain. For a three dimensional polar diagram which is a solid of revolution, it may be shown that

$$G_s = \frac{2}{\int_{-\pi/2}^{\pi/2} S^2(\theta) \cos \theta d\theta} \tag{106}$$

where $S(\theta)$ is the space factor in a plane containing the axis of revolution, θ being measured from the normal to the axis.

If $S(\theta)_{max}$ occurs for $\theta = 0$, and if the beam is narrow

$$G_s \doteqdot \frac{2}{\int_{-\pi/2}^{\pi/2} S^2(\theta) \, d\theta}. \tag{107}$$

This method can be applied to aerials such as linear arrays or biconical horns. The integrations of (106) and (107) may be performed graphically from the experimental polar diagram. Graphical integration methods of determining the gain of some paraboloid reflectors from the polar diagram have been developed.[27] Whereas methods (a) and (b) measure the actual gain of the overall system including feeders, graphical integration gives the directivity, and so can only be used for gain determination when ohmic losses and losses in the feeder

[27] L. D. Lawson[26]; J. A. Saxton, "Determination of aerial gain from its polar pattern," *Wireless Engr.*, Vol. 25, 1948, pp. 110-6.

system are negligible (e.g. it does not apply to a non-resonant array in which any significant portion of the power is absorbed in a load).

Impedance

The simplest method of measuring aerial impedance is to insert a section of slotted line or guide in the feeder and to measure the standing waves. The aerial is usually adjusted for no standing waves at a spot frequency, but at other frequencies standing waves will be present. The bandwidth may be determined by plotting standing wave ratio against frequency. The technique of impedance measurement is described more fully in Chapter VI.

CHAPTER IX

AERIAL DUPLEXING

An aerial duplexer or TR/RT switch is an electronically operated switching device, which enables the same aerial to be used for both transmitting and receiving in pulsed radar equipments. This type of duplexer should be distinguished from the duplexer of telegraphy which permits signals to be sent simultaneously in opposite directions over the same line.

The elimination of one aerial represents a considerable saving of space and equipment for the same performance. Normally the maximum possible aerial performance is required for a given available space. With a duplexer, the common aerial may have twice the area and hence twice the power gain of each separate unit. An increase in the overall power gain of the aerial system of 4 times or 6 decibels is thus theoretically possible, although in practice the figure is reduced somewhat by losses in the duplexer.

The advantages of aerial duplexing were realised during the early development of radar on metre wavelengths. Tentative experiments were made at the Telecommunications Research Establishment in England prior to 1939, but the work was discontinued before reaching a satisfactory solution. Subsequent investigations at the Radiophysics Laboratory into the design of the radio frequency circuits for such a device showed the importance of appropriate impedance transformation, and in May 1940 a successful unit was operated with a low power transmitter.[1]

The operation of an elementary switch may be represented symbolically as shown in Fig. 201. During the transmitted pulse, the switch connects the aerial to the transmitter and the receiver is disconnected. The latter is necessary to prevent loss of transmitter energy and, more important, to prevent

---

[1] J. L. Pawsey and H. C. Minnett, "Early Australian developments in the use of a single aerial for transmission and reception," C.S.I.R.O. Radiophysics Laboratory, Report RPR 51, January, 1947.

damage to the receiver. In the intervals between transmitted pulses, the aerial is connected by the switch to the receiver and the transmitter is disconnected so that it does not absorb any of the echo signals received by the aerial.

In practice, these two switching operations are performed by separate switching units. The aerial duplexer thus consists of a TR or transmit-receive switch, which disconnects the

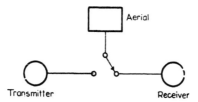

Figure 201.—Elementary aerial switch permitting the use of a common aerial for transmitting and receiving.

receiver at high power levels and an RT or receive-transmit switch which disconnects the transmitter at low power levels. Hence the term TR/RT switch is an alternative to duplexer. It should be mentioned that in some cases, it is possible to dispense with the RT unit to disconnect the transmitter ; the duplexer then consists only of the TR switch (see next section).

1. The Fundamental Method

Fig. 202 shows symbolically the circuit which forms the basis of nearly all aerial duplexers. The feeder system is assumed to be loss free and to consist of any of the usual transmission lines, open wire, coaxial or waveguide. The TR and RT units are represented by single-pole, single-throw switches S_1 and S_2 respectively which are also assumed loss free.

During the transmitted pulse both switches are closed so that points Q_1 and Q_2 are short-circuited. These short circuits reflect as open circuits across the shunt junctions P_1 and P_2 owing to the impedance inverting property of a quarter wavelength of transmission line. Hence the transmitter power flows to the aerial without reflection at P_2 and P_1 and there is no leakage past Q_1 to the receiver. The transmission line sections P_1Q_1 and P_2Q_2 are resonant but absorb no power as the line has been assumed loss-free. All the generator power thus reaches the aerial.

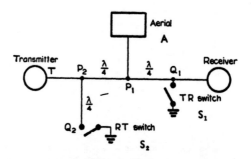

Figure 202.—The basic circuit used for aerial duplexing. Single conductor transmission line is used to represent the open wire, coaxial or waveguide feeder system. The idealised switches S_1 and S_2 are normally open, so that energy received by the aerial flows only to the receiver; they close during the transmitted pulse to protect the receiver and to direct the energy to the aerial.

At the end of the transmitted pulse both switches are opened and remain open until the beginning of the next pulse. The open circuit at Q_2 transforms to a short circuit across P_2 and an open circuit across P_1. The echo energy received by the aerial then flows to the receiver without reflection by the infinite impedances shunted across P_1 and Q_1. Also none of the energy reaches the quiescent transmitter. In this condition the section $P_1 P_2 Q_2$ is resonant and as before does not dissipate any power.

Elimination of the RT Switch

In some radar equipments, it is possible to simplify the duplexing system by elimination of the RT switch. This is due to the fact that the radio frequency impedance of many transmitting tubes in the quiescent state is a highly reactive mis match to the feeder, although the output impedance matches the feeder during the pulse. Hence, if the RT switch is omitted, echo energy flowing from the junction P_1 into the branch P_1T will be almost entirely reflected at the transmitter and will return to P_1 with a phase which depends on the phase of the reflection and the length P_1T. If the latter is chosen correctly, the input impedance of P_1T at P_1 may be made infinite provided no reflection loss occurs at T. The transmitter itself, therefore, acts as the RT switch.

In this arrangement the complete line section P_1T is resonant and in practice has to be reasonably short to avoid losses and to keep the frequency selectivity within specified limits. It is sometimes possible to fix the length P_1T at the desired value, if the properties of all transmitting tubes to be used in the particular set are sufficiently alike. Otherwise, it is made variable by introducing a telescopic section or more commonly by arranging for the junction P_1 to slide along the line AT.

Series and Shunt Junctions

The junctions P_1 and P_2 have been assumed in the above discussion to be of the shunt type represented symbolically in Fig. 203 (a). A series type junction is sometimes used and this can be represented as shown in Fig. 203 (b). The

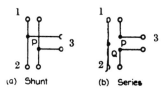

(a) Shunt   (b) Series

Figure 203.—Symbolic representation of two possible forms of transmission line tee junction : (a) shunt type, (b) series type.

operation of a duplexer with series type junctions at P_1 and P_2 is essentially similar to that already described except that the switches S_1 and S_2 have to be a half wavelength from their respective junctions instead of a quarter wavelength. This is because the series junction of Fig. 203 (b) requires a short

circuit across the terminals P if energy is to flow without
reflection between the branches 1 and 2. In the shunt case
[Fig. 203 (a)], an open circuit was needed across P. Similarly
reflection-free transmission between branches 1 and 3 requires
a short circuit across Q in the series case and an open circuit
across P in the shunt case. Examples of both types of junction
are described in section 4.

2. Electronic Switches

Practical Considerations

The action of the ideal switches S_1 and S_2 is approximated
in practice by such non-linear elements as vacuum diodes,
spark gaps and gas discharge tubes. These devices are non-
conducting at sufficiently small power levels and therefore
have a high impedance. At high power levels they are conducting and their impedance falls to a low value. Thus the
" opening and closing " of the switches is automatically
controlled by the power level on the feeder system. Since
in practice, electronic switches cannot achieve zero impedance
in the " closed " condition, and since in the " open " condition
circuit losses prevent the impedance from being infinite, an
aerial duplexing system involves a certain loss of power, while
the power leakage into the receiver is never quite zero.

In most cases it is desirable to insert an impedance transforming network between the switching element and the points
across which it is connected in the feeder system, (i.e. Q_1 or Q_2
in Fig. 202). This is because it is usually necessary to transform the two impedance levels of the switching element to
values which are more suitable for efficient operation relative
to the characteristic impedance of the feeder. Moreover
the impedance network is useful in tuning out the stray reactances which are associated with practical forms of switching
element. This unit of the electronic switch is discussed in
more detail in section 3.

Switch Requirements

Each of the two switches is required to close in a time small
compared with the transmitter pulse duration (0·25 to 20

microseconds) and to open in a period of the same order as the pulse duration. This cycle of operations is repeated at the pulse repetition frequency of the radar equipment (50 to 5000 cycles per second).

In the case of the crystal converters used in centimetre wave sets, the attenuation produced by the TR switch between the receiver and transmitter feeders at high power levels must be between 60 and 70 decibels. This ensures that the leakage of transmitter power to the receiver is reduced to a level which will not " burn out " the crystal (see Chapter XII, section 3). At longer wavelengths, where the first stage of the receiver is a more robust tube converter or radio frequency amplifier, the requirement is less stringent.

The losses of the complete duplexing unit must be less than 6 decibels to compete with the system of separate aerials and the unit must operate effectively over the frequency band covered by the radar equipment. It must therefore be either tunable or broadband.

Obviously, the design and construction of the switch will be considerably influenced by the intended operating wavelength.

Diode Switches

Historically, the thermionic diode was one of the first tubes to be employed as a switch in the metre wavelength radar equipment then in service. Early in 1940, for example, the British television-type diodes EA50 and D1 were used by the Radiophysics Laboratory in low power, 1½-metre equipments.[1] Several such tubes were used in parallel in higher power British sets. Eventually as transmitter power increased, special diodes with high emission cathodes had to be developed to handle the heavier current pulses. The radio frequency design of these tubes was also improved by reducing the length of electrode connections to a minimum (Fig. 204).

Diodes are inherently fast in operation, but are limited to fairly low-power equipments.   A small DC potential is necessary to bias the tube beyond cut-off so that it has a high impedance at the low power levels of received echoes.

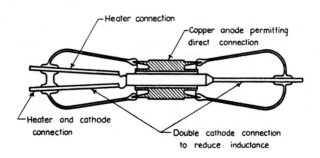

Figure 204.—Thermionic diode designed for use at $1\frac{1}{2}$ metres as an electronic switching tube.   Lead inductance is reduced as much as possible.

Spark Gap Switches

Spark discharges at or near atmospheric pressure have been used from time to time as an alternative to diode switches. The tungsten discharge gaps may be either exposed or enclosed in a glass envelope.   An early example of the enclosed construction is (i in Plate XI (a).   A filling of argon at a fairly high pressure (400 to 500 millimetres) together with a glass sleeve over the gap ensures that a spark discharge rather than a glow discharge takes place.   The latter type of operation was found to be characterised by slow deionisation of the gap so that the switch was slow to open.   Although the spark gap switching element is simple to use and manufacture, it is characterised by a high maintaining voltage (high conducting impedance) and fairly rapid deterioration of the electrodes. Its application has been confined to cases where the requirements on leakage power are not too severe (as with receivers using tube converters or radio frequency amplifiers) and where frequent maintenance is possible.

(a) Three types of gas discharge switching tube. The spark discharge tube (i) and the glow discharge tube (ii) are both designed for 1½ metres. Tube (iii) is a glow discharge tube (type CV83) for centimetre wavelengths; it incorporates a DC discharge electrode to provide a source of ions.

(b) Resonant cavity with CV83 gas discharge tube in position. A part of the cavity attached to a mounting flange has been removed to show the interior. The discharge gap, tuning plugs and input coupling window in the flange are visible. An output coupling loop may be inserted through the hole at the rear of the cavity (see p. 308).

PLATE XI.

Glow Discharge Tubes

The type of switch which has received the most attention
utilises a glow discharge between electrodes in a low pressure
gas tube. These switches are universally used for centimetre-
wave radar and will probably supersede other types at the
longer wavelengths. It is now possible to design tubes which
meet all the present requirements, are simple to manufacture
and have adequate life. Owing to the low gas pressure, the
maintaining voltage across the gap is quite low and deterioration of the electrodes is almost negligible. It is necessary to
provide a source of ions in the region of the gap to ensure that
the tube strikes with the first application of transmitter power.
Otherwise when the radar equipment is first switched on,
several transmitter pulses may occur before a discharge commences. By careful choice of the gas filling, it has been found
possible to increase the deionisation rate in the gap to a value
which is high enough for most purposes.

The gas discharge tube (ii) in Plate XI *a* was designed
some years ago at the Radiophysics Laboratory for use with
1½-metre equipments, the power of which could not be handled
by the available diodes.[2] The inert gas filling was at a pressure
of 10 to 20 centimetres of mercury and traces of radioactive
material within the tube provided an initial source of ions in
the region of the tungsten gap. Deionisation of this particular
tube, after the discharge, was assisted by means of a moderate
DC potential across the gap.

At centimetre wavelengths, the form of the gas discharge
tube first developed and still in widespread use.is shown at (iii)
in Plate XI (*a*). Two copper discs *D* are shaped to form the
radio frequency discharge gap *G* and are sealed into a glass
envelope containing the gas filling. The parts of the discs
external to the envelope are intended to clamp into metal
fittings to form a resonant cavity which acts as the impedance
transformer. The similarity of this construction to that of
many types of reflex klystron is evident. Indeed, the early

---

[2] M. I. G. Iliffe and H. R. Oliphant, " Interim report on the manufacture
and performance of glow gaps for common aerial working," C.S.I.R. Radiophysics Laboratory, Report RP 139/1, March, 1942.

models were evolved from such klystrons by omitting the electron gun and filling with gas.

An auxiliary electrode A, usually called the " keep-alive " electrode, provides a source of ions. The application of a DC voltage between this electrode and the adjacent copper disc causes a continuous discharge. The ions thus generated diffuse into the region of the radio frequency discharge gap G and ensure instantaneous striking when the equipment is first switched on. The discharge also ensures consistent operation during the ionising period and in this respect a negative potential on the auxiliary electrode has been found to give the best results.

The maintaining potential of the auxiliary discharge is usually 350 to 450 volts and the striking potential a few hundred volts higher. A series resistance limits the discharge current to a value sufficiently low (usually less than 1 milliampere) to prevent chemical reactions which cause the gas filling to disappear or "clean up." Trouble has sometimes been experienced with relaxation oscillations in this DC circuit owing to the capacity in parallel with the auxiliary gap (Fig. 205). Such oscillations impair the receiver protection pro-

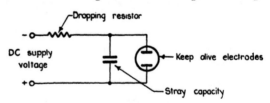

Figure 205.—The circuit of the auxiliary electrode, including the stray capacity across the discharge gap ; with certain values of the circuit constants, relaxation oscillations are possible.

perties of the TR switch and the DC circuit must be carefully designed to avoid them.

The best gas filling has been the subject of much experiment. Probably the most common filling is water vapour at a pressure of 5 to 20 millimetres, chosen chiefly because it has a rapid deionisation rate. Sometimes hydrogen is also added. The factors on which depend the choice of the gas and its pressure are discussed more fully in section 5.

3. Impedance Transformers

Theory

In the discussion of the fundamental operation of an aerial duplexer, the switches were idealised and assumed to place either an infinite or zero impedance across the feeder. Suppose that the corresponding impedances of an electronic switch are pure resistances R and r respectively. Then the higher the value of R compared with the characteristic impedance Z_o of the feeder, the more nearly will ideal operation be approached and the smaller will be the loss of received signal. Similarly, the resistance r represents a better " short circuit " and gives greater receiver protection in the case of the TR switch, the smaller its value compared with Z_o. It is thus clear that the ratio R/r may be taken as a figure of merit for a switch and should be as large as possible.

Many practical switches with a desirably high figure of merit have too large a value of r compared with Z_o for adequate re-

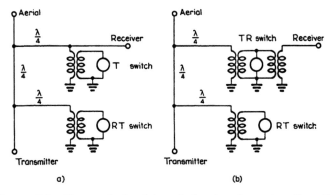

Figure 206.—Two methods of introducing impedance transformation between the switches and the feeder system. The arrangement used with the TR switch in (b) is very common at microwave frequencies.

ceiver protection. For this reason it is normal to introduce an impedance transformation between the switch and the feeder to reduce the value of r (and R) relative to Z_o. It may be shown that the ratio R/r is unchanged whatever the transformation ratio provided the transformer is loss free. In practice, dissipation in the latter decreases the ratio slightly.

Fig. 206 (a) and (b) show two methods of introducing the required transformation. In Fig. 206 (a), the transformer is inserted between the switch and the feeder, so that the impedance levels of the switch are transformed to lower values at the feeder. The equivalent arrangement shown in Fig. 206 (b) is now very common. Here the feeder impedance on each side of the switch is transformed to a higher value, so that the switch operates across a higher impedance level. The action of both circuits is essentially the same but the leakage power from the second, when a gas discharge tube is used, is less critically dependent on the adjustment of the circuit.

In the above discussion, the switch has been represented at high power levels by a small linear resistance r. The non-linear action of the electronic discharge may be allowed for by defining r as an effective value, which dissipates the same power as the discharge with the observed voltages on the system. Actually, the two impedance levels at the external terminals of a practical switch have complex values owing to the effect of connecting leads and other sources of stray impedance. However, it is possible to use the transformer to resonate the system, so that the impedance values connected across the feeder are both pure resistances. The complete impedance network may then be regarded as introducing transformation between the feeder and the discharge gap of the electronic switch. Later in this section, it will be apparent that in microwave duplexers, the radio frequency circuits of the impedance transformer and the discharge tube are very closely merged.

Lumped Circuit Transformers

At wavelengths of a few metres and above, lumped circuits are compact and have low loss. One type of circuit is shown in Fig. 207. C_1 tunes the circuit to resonance when the switching tube is non-conducting, and C_2 is a tuning adjustment for the conducting condition.   The position of the tap on the inductance controls the transformation ratio.   Many other lumped-circuit impedance-transforming networks may of course be devised.

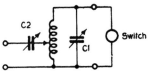

Figure 207.—An impedance transformer using lumped circuit elements and suitable for use at wavelengths longer than a few metres.

Transmission Line Transformers

At a wavelength of about a metre, transformers using transmission line circuit elements become more efficient.[3]   Open-wire units have been used extensively in the form shown in

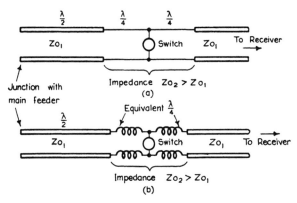

Figure 208.—(a) An open wire impedance transformer consisting of a half wavelength of high impedance line across the centre of which the switch is shunted.   (b) A similar unit in which the conductors of the transforming section are coiled for compactness.

Fig. 208 (a).   Here the switching tube is connected across the centre of an approximate half-wavelength of open-wire line of higher impedance than the rest of the feeder system.   An

[3] F. E. Terman, " Resonant lines in radio circuits," *Electrical Engineering*, Vol. 53, July, 1934, p. 1046 to final page.

impedance transformation therefore occurs between the input
terminals and the switch, and between the switch and the
output connections, so that the transformation belongs to the
type shown in Fig. 206 (b). To make a more compact unit,
the conductors of the high-impedance section may be coiled
as in Fig. 208 (b). This increases the characteristic impedance
of the section and consequently the transformation ratio but
the losses of the circuit tend to be greater.

With coaxial feeder systems, a coaxial-type transformer is
more convenient and becomes essential at wavelengths where
the radiation loss from open wires is prohibitive. Fig. 209 is a

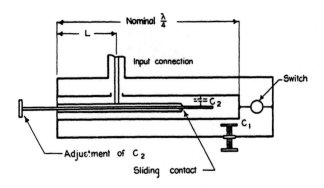

Figure 209.—A coaxial, quarter-wave transformer. The dimension L
determines the transformation ratio and C_1 and C_2 are tuning adjustments.

typical coaxial line design, which has the equivalent circuit of
Fig. 207.   The switching tube connects across the high impedance end of the coaxial unit, which is an approximate quarter-
wavelength long.   The capacity C_1 resonates the circuit as
before and the series capacity C_2 is most conveniently obtained
by the adjustable-length coaxial line inside the inner conductor
as shown.   The distance L of the input connection from the
short-circuited end of the unit controls the transformer ratio.

In cases where the transformation ratio required is low, a
double quarter-wave transformer has been used with success.

Resonant Cavity Transformers

The impedance transformer of Fig. 209 may be regarded as a quarter wave coaxial resonator operating in the *TEM* mode. At a wavelength of about 10 centimetres other forms of resonant cavity having lower losses than the coaxial one are feasible. Fig. 210 shows a form of cylindrical cavity transformer which is

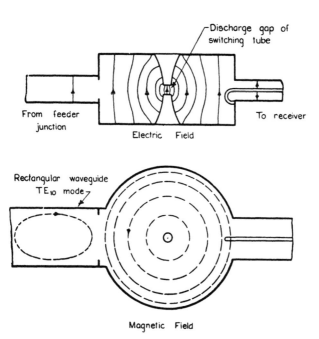

Figure 210.—A resonant-cavity impedance transformer for centimetre wavelengths. A discharge tube of the type(iii)in Plate XI (a) is an essential part of this transformer. Both iris and loop coupling to the cavity are illustrated.

commonly used in conjunction with a gas discharge tube of the type (iii) described in section 2 and shown in Plate XI (a). The discs and discharge gap of the tube clamp into the cavity of which they form a part. When the gap is non-conducting, the cavity is tuned to resonance in the dominant mode by means of plugs around its circumference. The electromagnetic field then has the distribution indicated. Coupling to the cavity may be either by window from a waveguide or by loop from a coaxial line. Both methods are shown in the diagram. Transformation occurs between the input coupling and the discharge gap and between the gap and the output coupling. When a discharge occurs at high power levels, the cavity is detuned and behaves as a very low impedance across the input window or loop. Plate XI (b) shows a gas discharge tube in position in a cavity, part of which is removed to show the interior. The discharge gap, tuning plugs, input coupling window and keep-alive electrode are visible.

A recent American trend has been to combine the tube and cavity into an integral unit. The cavity with discharge gap is formed in a copper block. One of the cones of the gap is mounted on a flexible diaphragm, so that the gap spacing may be varied by an external screw adjustment, thus tuning the cavity. The gas filling is contained in the cavity and the coupling holes are sealed by glass windows. The unit is clamped between waveguide flanges and discarded at the end of its useful life.

Broadband TR Cells

Under certain conditions, a magnetron may generate a very small fraction of its total power at frequencies which are harmonics of the fundamental frequency. Transmitter energy at the fundamental frequency is strongly attenuated by the action of the low-impedance gas discharge in the TR cavity before reaching the receiver. However, the attenuation of harmonic energy may be quite small for the shape of cavity described above, so that the leakage level of harmonic power is sufficient to impair the crystal mixer performance.

A form of waveguide TR unit which overcomes this trouble is shown in Fig. 211. Reflectionless glass windows are used to confine the gas filling to a section of waveguide thus forming a cell which may be removed from the waveguide transmission line for replacement. In this cell, resonant irises with closely spaced discharge gaps are located as indicated. At low power levels, the resonant irises transmit the signal energy without reflection and as they are low Q devices, the cell is an inherently

broadband unit and requires no tuning adjustment.   The losses
are also very low.   At high power levels, each iris discharges
and reduces the leakage power, including harmonic power, to a
very low level.   Two irises have been found sufficient to protect
the receiver, but three irises with staggered resonant frequencies are sometimes used to give satisfactory operation over
a still wider band.

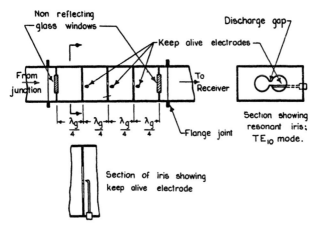

Figure 211.—Broadband TR cell using three stagger-tuned resonant
irises each having a narrow discharge gap at the centre.   Non-reflecting
glass windows seal off the gas cell and " keep-alive " electrodes are
provided for each gap.

4. Typical Duplexing Units

The electronic switching elements and transformers described
above are combined as TR and RT switches in duplexing
units, which are basically identical with the circuit of
Figs. 202 and 203.   The operation of the examples given below
should be clear from the discussion of fundamental system
operation in section 1.

Fig. 212 (a) is an open-wire design for a wavelength of
200 megacycles per second using spark gaps or glow discharge
tubes.   The balanced output from the TR unit is transferred
to the coaxial receiver feeder by means of a balance to unbalance transformer (see Chapter VIII, p. 234).

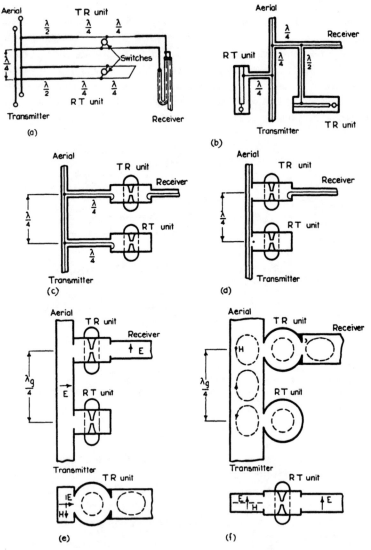

Figure 212.—Typical aerial duplexing units. (a) Open-wire duplexer for 200 Mc/s. (b) Coaxial duplexer for 200 Mc/s. (c) Microwave unit with coaxial shunt junctions. (d) Microwave duplexer in which coaxial series junctions are used. (e) A microwave waveguide unit with series type junctions (TE_{10} mode). (f) Waveguide unit with the cavities coupled to the narrow side of the rectangular waveguide (TE_{10} mode).

The coaxial duplexer of Fig. 212 (b) is also for 200 megacycle operation and may use similar switching tubes. Shunt junctions are shown and the quarter-wave coaxial impedance-transformers are inserted between the tubes and the feeder system according to the circuit of Fig. 206 (a). Cavity transformers are used in the microwave unit of Fig. 212 (c) in which the impedance transformation is of the type described by the circuit of Fig. 206 (b). A series type junction to a coaxial line may conveniently be made by coupling the branch impedance in series with portion or all of the current in the outer conductor of the line. Fig. 212 (d) shows an arrangement of this type in which the half wavelengths of line between the cavities and their respective series junctions are eliminated. This enables the cavities to be mounted directly on the main coaxial feeder as shown, so that a compact unit results. Each cavity is coupled by means of a window to a portion only of the current in the feeder in order to produce the required impedance transformation. If the RT switch is omitted, the length of transmission line between the TR cavity and the transmitter may be varied by arranging for this cavity to slide along the feeder. A slot in the latter allows coupling to the cavity window.

Two waveguide systems using the TE_{10} mode in rectangular tube are illustrated in Fig. 212 (e) and (f). The series type junctions in Fig. 212 (e) are obtained by window-coupling the cavities to the broad face of the guide. It is also possible to couple the cavities to the narrow side of the guide as shown in Fig. 212 (f).

Magic Tee Duplexer

A recent American development in aerial duplexing technique uses a scheme in which the leakage powers from two TR switches are cancelled by the use of the magic tee junction. This method differs somewhat from the fundamental one discussed in section 1. Magic tee properties are possessed by many types of junction, but one of the most common is that described in Chapter VII. This is simply a four-way waveguide junction having both E- and H-plane branches sym-

metrically arranged as indicated in Fig. 213 (*a*). As a result
of this symmetry, the junction has properties, certain of
which may be briefly restated as follows. Equiphase waves
are excited in branches 1 and 2 by a wave in the *H* branch,
and opposite phase waves by a wave in the *E* branch. There
is no coupling between the *E* and *H* branches. Conversely,

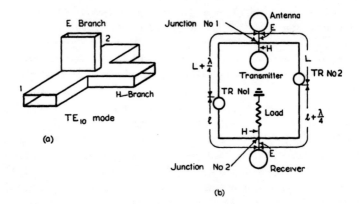

Figure 213.—A duplexer in which the leakage power from two TR units
is cancelled by using two ' magic tee " junctions (*a*) in the circuit shown
in (*b*).

if the waves in 1 and 2 are reflected they will couple to the
H branch if they reach the junction in phase, and to the
E branch if 180 degrees out of phase.

The cancellation duplexer is shown in Fig. 213 (*b*) in which
the branches of the magic tee are shown in the one plane for
convenience. Power from the magnetron excites in-phase
waves in the adjacent branches and these are reflected by the
discharging TR cavities. Owing to a difference of half a
wavelength in the total path length travelled, the waves
arrive back at the junction in opposite phase and hence couple
to the *E* branch, which leads to the aerial. Power leaking
through the TR cavities, however, arrives at the second
junction in phase and couples via the *H* branch to a dummy
load. If the leakage pulses are identical no power reaches
the receiver.

Echoes received by the aerial excite out of phase waves in 1 and 2 which are still out of phase when they reach the second junction and hence couple into the E branch and then to the receiver. The RT function in this circuit is performed by the first magic tee since there is no transfer of received energy from the E to the H branch.

5. Performance Characteristics

Equivalent Circuits

The operation of a duplexing unit may be analysed at both high and low power levels by considering the equivalent

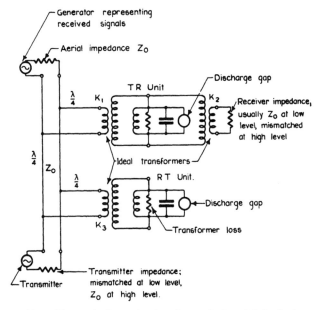

Figure 214.—The equivalent circuit of a typical aerial duplexing unit. Ideal transformers of ratio K are used to represent the impedance transformation.

lumped circuits. An electronic switch is often replaced by a circuit in which the impedance transformation is referred to the discharge gap and is represented by an ideal transformer of impedance ratio K. The tuning and losses of the circuit are allowed for by a parallel resonant circuit and a shunt resistance respectively across the gap.   Fig. 214 is the circuit of a duplexing unit in which this representation is used.   If the switching tube is a diode, the impedance of the discharge may be taken as a small equivalent resistance as discussed in section 3. The discharge of a gas tube on the other hand may be replaced by a generator of constant voltage equal to the maintaining voltage of the discharge, since the latter is practically independent of transmitter power.

The operation of the TR and RT switches may be investigated separately.   Considerably simplified circuits may then be drawn for low level and high level operation.   Such circuits have been analysed by Pawsey[1] in the case of a diode switch and by Samuel, Clark and Mumford[4] in the case of a gas discharge switch.

An important general result of the analysis is that a compromise must be made between leakage-power level and loss of received signal   The latter is due mainly to ohmic and dielectric losses in the switching tube and the transformer. When the receiver protection requirements have been satisfied (see next section), the low-level loss of a duplexing unit is between 1 and 2 decibels.   Somewhat lower losses for the same protection are obtained with the broad band TR cell described in section 3, p. 308.   A figure of $\frac{1}{2}$ to 1 decibel may be realised.

Leakage Power

The power leaking through a TR switch during the transmitted pulse is one of the most important factors in duplexer design. The leakage pulse from a gas discharge tube has some characteristic features which result from the properties of the discharge.   Fig. 215 shows the shape of the envelope of a typical pulse.   The front edge of the pulse has a " spike," which is due to the finite time required for the gas near the discharge gap to ionise.   This transient period is very short and, although not known with certainty, is thought to be of the order of 10^{-8} seconds or less.

[4] A. L. Samuel, J. W. Clark, and W. W. Mumford, "The gas discharge transmit-receive switch," *Bell Syst. Tech. J.*, Vol. 25, 1946, pp. 48-101.

When the gas has had time to ionise, a steady maintaining potential is established across the gap and this persists until the cessation of the pulse. This maintaining potential is independent of the transmitter power and depends only on the nature and pressure of the gas filling. It contributes one of the components of the leakage power in the " flat " of Fig. 215. A second component becomes apparent with high-power transmitters. This component is found to be proportional to the transmitter power and is present even when the discharge gap is bridged by a perfect short circuit. It is due to a form of direct coupling in which power leaks past the

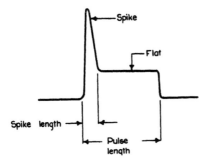

Figure 215.—The shape of the RF envelope of the leakage pulse from a gas discharge tube. The " spike " at the start of the pulse is due to the finite ionisation time of the gas.

short circuit shunted across the path to the receiver. The reduction of the direct coupling to a minimum is a matter of circuit design and is independent of the gas filling.

The type of receiver imposing the most stringent requirements on leakage power is the crystal-converter superheterodyne, which is used almost exclusively in centimetre radar. In order not to impair crystal performance, it is usually considered desirable to limit the peak power in the flat to less than 100 milliwatts, so that with a 400 kilowatt transmitter the degree of protection required is 66 decibels. With careful attention to gas filling and cavity design this can be achieved.[5]

[5] A. H. Cooke, G. Fertel and N. L. Harris, "Electronic switches for single-aerial working," J. Instn. Elect. Engrs., Vol. 93, Part III A, 1946, pp. 1575-84.

The maximum temperature of the crystal contact during the spike depends on the duration of the latter as well as on its peak power. This is because the duration is comparable with the thermal time constant of a crystal contact so that there is insufficient time for equilibrium to be reached. Hence the spike energy rather than its peak power is the important factor. It is thought that with present levels of flat leakage, the spike energy is the factor mainly responsible for crystal "burn-out" and it should be restricted to a value less than 0·3 to 2·0 ergs per pulse depending on the type of crystal used.

Recovery Time

At the end of the transmitted pulse the ions in the gap region are removed by molecular processes at a speed which depends on the nature and pressure of the gas. Usually those gases which possess a low maintaining potential are slow to deionise. It is important that this deionisation process should be fast enough to prevent significant attenuation of the shortest range echo which the radar has to detect. The period required after the transmitted pulse for the transmission efficiency of the TR switch to reach a certain percentage of maximum (usually defined as 50 per cent) is called the recovery time.

Life of a Gas Tube

The life of a gas discharge tube depends on the rate of clean-up of the gas inside the tube. The resulting reduction in pressure usually causes an increase in recovery time and leakage power. How much the DC "keep alive" discharge current contributes to the clean-up appears to be uncertain, but normally this current is kept to a minimum. When water vapour is used as a filling, it gradually breaks up into hydrogen and oxygen and the latter oxidises the copper parts of the tube. Some work in America has therefore been directed to preoxidation of these parts and attempts have also been made to seal into the tube a source of water vapour such as treated silica gel.

Choice of Filling in a Gas Tube[6]

The nature and pressure of the filling of a gas tube is a compromise between factors which may be summarised briefly as follows :

(a) Low maintaining voltage for small leakage power.

(b) Quick ionisation for small " spike " energy.

(c) Quick deionisation for short recovery time.

(d) Reasonably long life.

Up to the present, the filling of a tube has been selected on an empirical basis. Much work is needed to elucidate the fundamental mechanisms of radio frequency gas discharges and to provide adequate design data.

6. Measuring Techniques

Methods of measuring the leakage power, losses and recovery time of an aerial duplexer are of interest owing to the special nature of the techniques which have been developed.

Leakage Power

In the case of a diode switch, the leakage pulse is a reduced version of the transmitter pulse and may have an amplitude of 10 volts or so. If the wavelength is in the metre range, the amplitude may be measured with a peak voltmeter.

At centimetre wavelengths, the fundamental method involves the measurement of power averaged over a repetition period by means of a bolometer or thermistor bridge of the type described in Chapter VII. However, the measurement is complicated by the various leakage pulse components from the gas discharge switch as discussed in section 5. A means whereby these components can be separated has been shown.[7] The curve of Fig. 215 may be taken to represent the variation

[6] L. D. Smullin and C. G. Montgomery (Eds.), *Microwave Duplexers* (Massachusetts Institute of Technology, Radiation Laboratory Series, Vol. 14), McGraw-Hill, New York, 1948, pp. 139-223.

[7] A. L. Samuel, J. W. Clark and W. W. Mumford, "The gas discharge transmit-receive switch," *Bell Syst. Tech. J.*, Vol. 25, 1946, pp. 48-101.

of peak leakage power during the pulse, where the term peak refers to an average over several radio frequency cycles. We then write

$$\overline{P}_R = EF + P_F F\delta + A\overline{P}_T$$

where $\overline{P}_R$ is the total leakage power averaged over a repetition period, $\overline{P}_T$ is the average transmitter power, P_F is the peak power in the flat period due to the finite discharge voltage, E is the energy in each spike, F is the pulse repetition frequency, δ is the pulse duration and A is the direct coupling attenuation factor.

The first component is the average spike power, the second the average flat power due to the finite discharge voltage and the third the average flat power due to direct coupling. The quantities E, P_F and A are independent of F, δ and $\overline{P}_T$ and may be evaluated by measuring the change of $\overline{P}_R$ as F, δ and $\overline{P}_T$ are separately varied. Thus

$$A = \frac{d\overline{P}_R}{d\overline{P}_T},$$

$$P_F = \frac{1}{F}\frac{d\overline{P}_R}{d\delta},$$

$$\text{and } E = \frac{d\overline{P}_R}{dF} - P_F\delta$$

$$= \frac{d\overline{P}_R}{dF} - \frac{\delta}{F}\frac{d\overline{P}_R}{d\delta}.$$

By cancellation of the flat components of the leakage pulse by means of a pulse of equal amplitude and opposite phase the spike alone may be obtained. An attenuated portion of the transmitter pulse is suitable for this purpose. An average power measurement of the spike and a knowledge of F then give an alternative method of calculating the spike energy E.

By launching spike pulses down a long transmission line terminated by a short circuit, the spike duration may be determined. The length of the line is increased until the standing wave ratio at the input end becomes unity. This occurs when the start of the spike, after reflection from the far end, arrives back at the input just as the end of the spike is leaving. The spike duration may therefore be calculated if the velocity of propagation is known. It is of the order of 10^{-8} seconds, which is too short for the envelope to be viewed on an oscilloscope owing to the difficulty of video amplification.

Losses

The overall low-level loss of a duplexing unit may be found by an insertion measurement. The aerial is replaced by a signal generator and the power output of the latter is noted for a convenient receiver setting. The generator is then connected directly to the receiver and the new output for the same indication is found. If corrections are made for feeder losses external to the duplexer, the difference in the two readings is a measure of the low-level loss of the latter. A similar method may obviously be used to evaluate the loss of transmitted power which occurs during the transmitter pulse. An oscillator with sufficient power output to cause the switch gaps to discharge is connected in place of the transmitter and a power measuring device substituted for the aerial.

It has been shown from equivalent circuit analysis that the low-level loss of an individual switch may be expressed completely in terms of the loaded and unloaded Q's of the circuit. These may be found from bandwidth measurements using standard techniques. This method of determining the low-level loss is especially useful at centimetre wavelengths with cavity-type switches.

Recovery Time

One method of measuring this quantity is to connect the duplexer into a radar equipment with a Class A indicator. The output of a low-level pulsed oscillator is injected into the aerial feeder. If the transmitter is switched off, a series of unsynchronised pulses of equal amplitude appear on the time base. When the transmitter is operating, the oscillator pulses immediately following the transmitter pulse are attenuated. The envelope of the tops of these pulses is the recovery curve and its duration may be determined by using the range measuring facilities of the radar.

In another method, the oscillator is triggered by the transmitter after a certain interval which may be varied by means of a calibrated delay network. Thus the pulse may be moved along the time base away from the transmitter pulse until its amplitude recovers to the defined percentage of the maximum. The delay interval is then equal to the recovery time.

CHAPTER X

RECEIVERS

THE function of a radar receiving system is to provide a visual display on an indicator such as a cathode ray tube, from which a non-technical operator can determine the position and nature of bodies from which echoes are being received. In this chapter, which is intended as an introduction to the more detailed treatment of receiver principles given in later chapters, we shall consider the cathode ray tube, timing unit and other auxiliaries as part of the receiving system. This is in conformity with conventional television practice. We shall also assume that the reader is familiar with the basic features of cathode ray tubes, time bases and superheterodyne receivers.

1. Characteristics of Receiving Systems

Requirements of the System

The nature of the information to be displayed and the accuracy with which it is to be determined depend on the operational function of the radar set. In many cases, the range and bearing of a moving target such as an aircraft or surface vessel are required with considerable accuracy. In other cases, as in navigation and blind bombing, the relative intensities of echoes from buildings and topographical features of the terrain are important.

In all cases the received power is small, and must undergo considerable amplification before it can be applied to the indicator tube. Means must also be provided for accurate ranging or the precise determination of the time delay between the emission of the transmitted pulse and the arrival of the echo from the target. Since the velocity of electromagnetic waves is approximately 328 yards per microsecond, this delay is one microsecond per 164 yards, or 10·7 microseconds per statute mile. In practice, measurement of time intervals ranging from one or two milliseconds to a few microseconds is involved.

Since the transmitter emits pulses at intervals which may vary from 20 milliseconds to 500 microseconds, a cathode ray tube provided with a precision time base accurately synchronised to the transmitter, is an obvious method of solving the ranging problem. After amplification and demodulation in a receiver the incoming echoes may be applied to the cathode

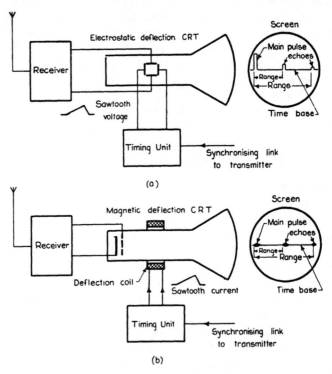

Figure 216.—The basic elements of a radar receiver.

ray tube causing either deflection or intensity modulation of the electron beam as already mentioned in Chapter II. The basic elements of the two systems, are depicted in Fig. 216 (a) and (b) respectively. The first type, known as a Class A display, utilises an electro-statically focused and deflected cathode ray tube in which the bearing of the target is obtained

from the azimuthal position of the aerial as shown on a suitable
mechanical indicator, while the range is of course proportional
to the distance between the echo and the main pulse on the
face of the cathode ray tube.

It will readily be appreciated that this display is very
inefficient from the point of view of traffic handling capacity
since the aerial must be stopped and set on the bearing of the
target before data can be obtained. For this reason the plan
position indicator described in Chapter II was developed
and is now in almost universal use. Several modifications and
refinements have been used, but the combination of intensity
modulation, long afterglow screen and sweeping of the time
base in synchronism with the aerial movement is usually
employed. This development has not only solved the traffic
handling problem but has provided the operator with a pictorial
display capable of easy interpretation. The PPI display
is particularly impressive when fitted to an airborne radar
of high resolving power (see Plate I). Many other types
of display have been developed for radar purposes, but they
are all basically similar to those already described and will
not therefore be referred to in detail.

Comparison With Television Receivers

At this stage it is interesting to compare the requirements
of high quality television and radar receivers. In radar the
time base is automatically initiated each time the transmitter
emits a pulse, whereas in television the time base of the receiver
normally runs freely but is synchronised by means of impulses
inserted in the television carrier wave. However, linearity
and accurate synchronisation of the time base are necessary
in both cases, and the problems associated with intensity
modulation are also comparable. The active horizontal
scanning period employed in typical television receivers is of
the order of 65 microseconds, corresponding to a radar range
of only 6 miles. The time available for the return of the
scanning spot after each sweep is only 10 microseconds, however, whereas in radar we usually have at least 300 microseconds. This adequately ensures the restoration of circuit

equilibrium before the subsequent cycle. In both cases, a blanking voltage is applied to the cathode ray tube grid during the inactive interval. The spot is thus rendered visible only during the active interval, that is when useful information is being received.

The basic difference in radar practice is that we are primarily concerned with obtaining quantitative data regarding the position of echoing objects, the most important being range. In television the problem is good qualitative reproduction of a moving scene ; hence a very short persistence phosphor is required on the cathode ray tube screen and minor irregularities in linearity of the time base may not have serious consequences.

Treating now the reproduction of the signal, we note that rapid changes in signal level are frequently occurring and that the times available for imparting the information to the radar observer or television viewer are roughly of the same order. The intelligence is therefore distributed over a considerable frequency band, necessitating a wide frequency response characteristic in the receivers in both cases. However, the signal to noise ratio requirements differ considerably. In radar we can obtain useful information if the received signal power is of the same order as the noise power in the input stages of the receiver whereas in television, signal to noise ratios below twenty to one result in unsatisfactory pictures. Another important difference lies in the permissible non-linear distortion. In television, while some compression in the overall range of signal intensities is permitted, non-linear distortion is in general detrimental. In radar such distortion or even overloading is of small consequence, provided the circuits can recover rapidly. In fact when using intensity modulated displays it is advantageous to limit all signals to a predetermined level after demodulation. This greatly improves the display as otherwise groups of strong echoes would cause patches of high intrinsic brilliancy to appear on the screen. Burning and defocusing are then prone to occur and the visibility of adjacent small targets is impaired.

Other common features will be apparent after reading Chapters XII and XIV, but enough has been said at this stage to indicate analogous and differing requirements.

We may therefore regard a radar receiving system as being composed of three distinct units, namely the cathode ray tube which displays the information, a timing unit which initiates the sequence of circuit operations essential to accurate range measurement, and a receiver whose function is to amplify and demodulate the received signals before they are applied to the cathode ray tube.   In practice all these units may be housed in one receptacle or they may be separated, depending upon such factors as portability, land or airborne location, etc. Plates XII, XIII (a) and XIII (b) are photographs of typical receiving units.

2. The Cathode Ray Tube

The cathode ray tube has received a great deal of attention in the past few years which has led to important improvements, but not to any fundamentally new techniques.   Since we have assumed the reader is familiar with the basic theory and constructional features of pre-war tubes, only general design innovations will be mentioned here.

The major requirements are robustness, adequate screen illumination and afterglow, small spot diameter and uniformly good focusing over the full tube face.   These characteristics should be obtainable with moderate accelerating potentials, otherwise additional circuit complexity is introduced to compensate for the reduced deflection sensitivities, quite apart from additional power supply problems.   Finally, there must be adequate screen area for plotting purposes, although increasing the screen diameter does not necessarily improve the resolution.

Classification of Types

Tubes in general use may be classified into two main groups depending upon whether the focusing and deflection are electrostatic or magnetic.   The former are used mainly for Class A displays, test equipment and monitoring, and are therefore usually fitted with medium persistence screens (about 2 milliseconds). They do not differ in any great detail from pre-war types except that in some cases additional acceleration is imparted to the beam after deflection. This produces higher screen illumination without appreciable reduction in deflection sensitivity.

The magnetically deflected types are commonly used in intensity modulated displays and are usually characterised by a special screen construction that produces an afterglow of the order of 30 seconds which is essential for efficient plan position presentation. They have greater beam current densities and better illumination and focusing properties than the electrostatic types when operated at similar anode potentials. Since the focusing and deflection fields are obtained from external coils a simpler tube construction results. Production variations in electrical characteristics are in turn lessened due to the reduction of tube elements. Many types are available, but we shall confine our attention here to the American types 12DP7, 7BP7, and 5FP7, illustrated in Plate XIV (a). These tubes have given excellent service in radar and are similar to one another except for size. Their overall length is relatively short and the tube faces are made flat by using pressed instead of blown glass in the case of the 5FP7 and 7BP7 types. Practically the full diameter can then be used for accurate ranging and plotting. A double layer screen, sometimes termed a cascaded phosphor, is utilised to achieve the long afterglow characteristics. The sulphide layer adjacent to the glass produces an orange phosphorescence lasting many seconds when excited by ultra violet radiation. Directly behind the first layer lies a sulphide which emits a short persistence blue light with an ultra violet component when excited by the impinging electron beam. This light is partially absorbed by the glass-backed layer which then emits the long persistence phosphorescence. An orange filter placed over the tube face attenuates any residual blue light from the second layer. Short persistence operation, but at reduced efficiency, may be obtained by using a blue in lieu of an orange filter.

The upper rack houses the superheterodyne receiver, below which is
situated a five inch cathode ray tube and the timing circuits. The
other units are power supplies, switch panels, etc. (see p. 325).

PLATE XII.—A land-based air-warning receiver.

facing p. 326]

(a) The plan-position indicator of a naval surface radar search set, type A276 (see p. 325).

[By courtesy of Amalgamated Wireless Australasia Ltd.

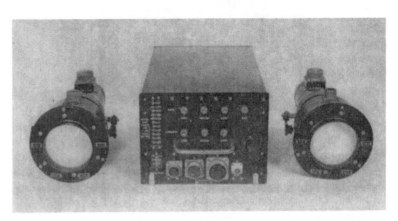

(b) An airborne receiver. The central unit contains the intermediate and video frequency amplifiers as well as the cathode ray tube timing circuits. Separate cathode ray tubes are provided for the pilot and the observer (see p. 325).

PLATE XIII.

The sole disadvantage of magnetically operated tubes lies in the extra complexity of the deflecting circuits. The correct current waveforms for magnetic deflection are, in general, more difficult to produce than the voltage waveforms used in electrostatically deflected tubes. In addition the weight of the focusing coil and deflection yoke amounts to several pounds. For these reasons some electrostatic tubes are fitted with long persistence screens and used for intensity modulated displays in lightweight airborne equipment.

Some excellent electron gun structures have been developed. These include an auxiliary anode which has essentially the same function as the screen grid of a pentode, and reduces the dependence of beam current intensity and cut-off bias on the potential of the accelerating anode. This auxiliary anode is also useful for blanking purposes. Accelerating potentials ranging from 4 to 6 kilovolts are used and in many cases the spot may be focused down to a diameter of 0·3 millimetres. In the majority of applications a 5- or 7-inch screen suffices and if operated under conditions of low ambient illumination good sensitivity is obtainable.

Special Types

When a larger screen is necessary for detailed plotting and simultaneous viewing by several observers under normal lighting conditions a special tube, known as a skiatron[1] [Plate XIV (b)] is used with the optical projection scheme shown in Fig. 217. Its screen diameter seldom exceeds three inches and the active agent in the screen material is usually potassium chloride which has the property of changing its colour when bombarded by electrons. Echoes are revealed, therefore, not as small light sources on the screen, but by patches of different colour, a deep magenta in this instance. They are readily visible under strong external illumination and are suitable for projection. The decay characteristic depends upon several factors, chiefly the screen temperature. A typical skiatron may be operated between 30 and 40 degrees centigrade

[1] P. G. R. King and J. F. Gittins, "The Skiatron or Dark-Trace Tube," *J. Instn. Elect. Engrs.*, Vol. 93, Part IIIA, 1946, pp. 822-31.

and illuminated by 10,000 foot-candles of mercury light. Under these conditions the decay period is about 10 seconds but it may be varied from a few seconds to a few days with suitable temperature control.

Requirements also exist for viewing short persistence tubes in daylight without a visor. A good example of a suitable tube

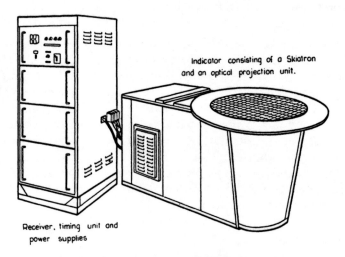

Indicator consisting of a Skiatron and an optical projection unit.

Receiver, timing unit and power supplies

Figure 217.—A receiver incorporating a skiatron for large-scale plotting and simultaneous viewing by several observers.

is the VCR526,[2] [Plate XIV (c)] in which adequate illumination has been obtained by the use of a high beam current density of about 200 microamperes, together with a post deflection acceleration potential of 2500 volts and a willemite screen. The tube is compact, having an overall length and screen diameter of 10 and $2\frac{3}{4}$ inches respectively.

Research has recently been directed to the production of a screen material that not only fulfils these daylight requirements but is also suitable for night viewing without impairing the sensitivity of the dark-adapted eye. In the latter case

[2] J. Sharpe, " The development of the VCR526 cathode ray tube for daylight viewing," General Electric Company (Wembley) Research Laboratory, Report 8645, March, 1945.

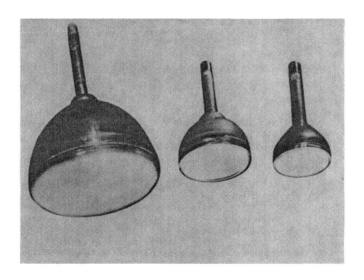

(a) Typical magnetically focused and deflected cathode ray tubes used in PPI indicators (Types 12DP7, 7BP7 and 5FP7).

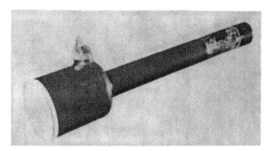

(b) A skiatron or colour-trace cathode ray tube, suitable for use in an epidiascope type of projector.

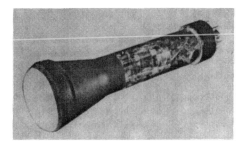

(c) The VCR526 cathode ray tube, which is suitable for daylight viewing without a visor.

PLATE XIV.

facing p. 328]

a ZnS.CdS Ag phosphor was used with a red filter for night viewing.

A good account of recent developments in cathode ray tubes has been given by Jesty, Moss and Puleston.[3]

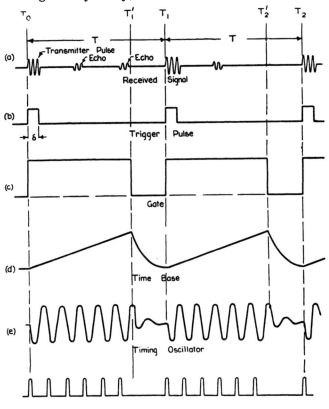

Figure 218.—The basic timing operations in a radar receiver (refer text).

3. The Timing Unit

This unit includes circuits for ranging, beam deflection and the generation of waveforms to accomplish rapid electronic switching. Coincident with each transmitted pulse, a sequence

[3] L. C. Jesty, H. Moss and R. Puleston, "War-time developments in cathode-ray tubes for radar," *J. Instn. Elect. Engrs.*, Vol. 93, Part III A, 1946, pp. 149-68.

of accurately timed operations is initiated, as shown in Fig. 218. Confining our attention to one pulse repetition period $T_0 T_1$ the interval $T_0 T_1'$ corresponds to the time delay of an echo at the maximum range of the system and represents the active period during which useful information is received and the timing circuits are operative. The pulse repetition period should exceed this time by a few hundred microseconds to ensure the restoration of circuits to their quiescent state before the subsequent cycle $T_1 T_2$. Coincident with the start of the transmitter pulse at T_0 is a trigger pulse (*b*), which initiates the following set of waveforms and controls their accurate relative timing. Firstly a rectangular pulse (*c*) termed a gate, which controls the duration of the active period $T_0 T_1'$ and at the same time switches on the cathode ray tube beam for this interval ; secondly a linear time base (*d*) which deflects the beam ; and thirdly a sinusoidal oscillation (*e*) whose period is accurately defined. In the case illustrated a calibration " pip " (*f*) is produced each time the sine wave crosses the zero axis in the decreasing direction. The interval between the emitted pulse and the receipt of the echo may then be measured by reference to this known time scale. Circuit details, refinements and variants are given in Chapters XIV and XV.

The unit may also contain circuits from which azimuth and sometimes elevation data are obtained. In a practical system several range scales may be provided by switching the gate length, time base velocity and timing oscillator frequency. Range may be measured to within 25 yards without undue circuit complexity or special operational skill.

4. The Superheterodyne Receiver

The limitations of vacuum tubes as amplifiers at ultra-high frequencies, due to the increased input conductance arising both from cathode lead inductance and transit time of the electrons, are well known.[4] At these frequencies the inter-

---

[4] M. J. O. Strutt and A. van der Ziel, " The causes for the increase of input admittance of modern high frequency amplifying tubes on short waves," *Proc. I.R.E.*, Vol. 26, August, 1938, pp. 1011-32 ; and W. R. Ferris, " The input resistance of vacuum tubes as ultra-high frequency amplifiers," *Proc. I.R.E.*, Vol. 24, January, 1936, pp. 82-105.

stage loading becomes so severe that the amplification may fall below unity above a critical frequency which is dependent upon the valve construction. Hence in radar receivers the superheterodyne principle has found almost universal application, most of the amplification taking place at a frequency lower than that of the input signal. The following sections discuss only the general consideration affecting receiver performance ; details of amplifiers, mixers, etc., are given more fully in subsequent chapters.

Bandwidth Requirements

The bandwidth and selectivity requirements differ greatly from those of the conventional broadcast receiver since the radar transmitter is, in effect, pulse amplitude modulated. Fourier analysis of a rectangular pulse, as indicated in Fig. 219,

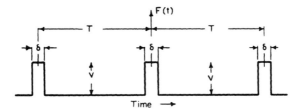

Figure 219.—A recurrent rectangular pulse.

yields the spectrum shown in Fig. 220,[5] where the abscissae represent the order of the harmonic and the ordinates their relative amplitudes. To preserve generality the harmonics have been expressed in terms of T/δ, where T is the pulse repetition period and δ is the pulse duration. For the ideal pulse shown an infinite number of harmonics exists. A radio frequency carrier modulated by this type of pulse will therefore radiate a large number of side bands, spaced at intervals equal to the pulse repetition frequency and symmetrically disposed on either side of the carrier. The corresponding side band energy distribution up to the second zero is shown in Fig. 221. Note that a large portion is contained in a band

[5] E. A. Guillemin, *Communication networks*. Wiley, New York, 1931, Vol. 2, Chapter XI.

2/δ megacycles per second wide, centred on the carrier frequency f_o, and that the position of the zeros is $f_o \pm 1/\delta$ mega-

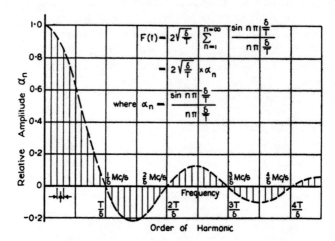

Figure 220.—Fourier analysis of the pulse shown in Fig. 219.

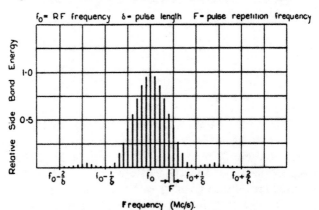

Figure 221.—Relative side-band energy radiated from a transmitter whose carrier frequency (f_0) is modulated by the rectangular pulse shown in Fig. 219.

cycles per second and $f_o \pm 2/\delta$ megacycles per second, independent of the repetition period T.

If perfect reproduction of the pulse were essential the receiving circuits would have to pass all the pulse component frequencies with negligible amplitude and phase distortion. In practice the permissible distortion depends upon the operational requirements of the radar set. The distortion produced in a network can be calculated if its phase and amplitude characteristics are known.[5] Such treatment however is beyond the scope of this book. It will suffice to present some semi-empirical design rules which are applicable to several stages of single tuned, isochronous circuits as commonly employed in radar because of their suitability for production methods.

Where maximum sensitivity is the paramount consideration, the bandwidth between the half-power points of the receiver frequency response characteristic is usually chosen to be

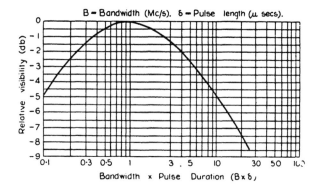

Figure 222.—The effect of bandwidth on the visibility of small signals in the presence of noise. The optimum bandwidth is the reciprocal of the pulse duration.

$1/\delta$ megacycles per second, where δ is the pulse duration in microseconds. Thus for a pulse of one microsecond duration the bandwidth would be one megacycle per second. Small signals viewed on a cathode ray tube against a background of noise are most easily seen under these conditions. The effect of departures from the optimum is shown in Fig. 222. Note that the maximum is broad, a considerable advantage in some cases where other practical considerations necessitate departures from the optimum. Radar systems use pulse durations ranging from 20 microseconds to 0·1 microseconds. The longer pulses are used for search radars and the shorter in high discrimination sets. Receivers, therefore, have bandwidths ranging from 50 kilocycles per second to 10 megacycles per second. The majority fall within the range of 0·25 to 2 megacycles per second.

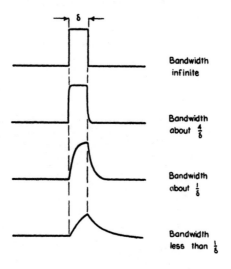

Figure 223.—The effect of receiver bandwidth on the reproduction of an echo whose power is considerably greater than receiver noise.

When high ranging accuracy is required it is essential that a clearly defined reference point on the echo, such as its leading edge, be provided. The transition time, usually taken as the time required for the receiver output pulse to rise from 10 to 90 per cent of its maximum amplitude, must then be considerably smaller. Under these circumstances the bandwidth may be of the order of $4/\delta$ megacycles per second. Fig. 223 depicts the appearance of a strong echo after demodulation for a variety of bandwidth conditions. The subject is treated in greater detail in Chapter XIII.

Sensitivity

Since sensitivity is of major importance in most applications considerable effort has been expended in improving receiver techniques. From equations (4) and (5), Chapter II, we note that the ultimate free space range d_o is proportional to $(P/p_o)^{\frac{1}{4}}$ where P is the peak transmitter power and p_o the minimum detectable signal power of the receiver. Thus a two to one reduction in p_o increases d_o by about 19 per cent. Such a reduction in p_o has the same overall effect as doubling the transmitter power.

The ultimate sensitivity of the receiver is limited by " noise " arising from thermal agitation in conductors and by the irregular nature of electronic emission in vacuum tubes. The fluctuation in the anode current of pentodes is greater than that of triodes due to " screen partition noise " arising from the random division of the cathode current between the screen and the anode. These noise currents cause fluctuating voltages to be developed across grid-cathode impedances, and after amplification, result in noise modulation of the electron beam of the indicator. In general a small echo can only be distinguished if its power is of the same order as that due to noise. Since noise is uniformly distributed over an infinite frequency spectrum, the noise power output of the receiver decreases linearly with bandwidth. However, as already indicated, at values below $1/\delta$ megacycles per second the severe pulse distortion which occurs outweighs the advantage of decreased noise power output. It will be assumed therefore in subsequent discussions that the optimum bandwidth B, or $1/\delta$ megacycles per second is employed.

In the specification of receiver sensitivity it is usual to compare the signal to noise power ratio of the receiver with that of a fictitious ideal receiver assumed to be noise-free and having an infinite input impedance. If the ideal receiver is connected to a signal generator of electromotive force v_a and resistance R_a (Fig. 224), the output noise is caused solely by thermal agitation voltages set up in R_a. It is easily shown that when

v_a is adjusted for an output signal to noise power ratio of unity, the available power P of the generator is given by

$$P = kT_a B$$

where k is Boltzmann's constant ($1 \cdot 37 \times 10^{-23}$ joules per degree centigrade), T_a is the temperature of R_a (degrees Kelvin), and B is the receiver bandwidth (cycles per second). When a similar procedure is applied to a practical receiver it

Figure 224.—A diagram illustrating the measurement of the power sensitivity of the ideal receiver.

is found that the available power of the generator must exceed $kT_a B$ by a factor N. This is defined as the noise factor of the receiver. The additional noise is produced in the valves and the resistive components of the interstage impedances.

The use of the available power of the signal generator in this definition rather than the generator voltage applied to the receiver input terminals as in broadcast receiver practice is preferable for several reasons. Power is more general and fundamental than voltage and the thermal limit, or datum level, is $kT_a B$ which is the available noise power of any resistance at temperature T_a.[6] Again it would be necessary to measure the input impedance of the receiver before its terminal voltage could be calculated, for under conditions of optimum sensitivity, the generator and receiver impedance are not necessarily matched.[7]

When the receiver is connected to an aerial, v_a and R_a become the induced signal voltage and radiation resistance respectively. Burgess[8] has shown that, although radiation

[6] J. B. Johnson, " Thermal agitation of electricity in conductors," *Physical Review*, Vol. 32, July, 1928, pp. 97-109.

[7] E. W. Herold, " An analysis of the signal to noise ratio of ultra high frequency receivers," *R.C.A. Review*, Vol. 6, January, 1942, pp. 302-31.

[8] R. E. Burgess, " Noise in receiving aerial systems," *Proc. Physical Society*, Vol. 53, May, 1941, pp. 293-304.

resistance is not itself a source of thermal agitation noise, the Johnson [6] formula ($P = kT_a B$) can be correctly applied when the aerial is in radiative equilibrium with its surroundings. In practice we can usually assign the ambient temperature to R_a. The reader is referred to Burgess's paper for further discussion on the correspondence between thermal agitation noise and black body radiation phenomena.

In laboratory measurements of noise factor the signal to noise power ratio at the second detector is measured using a CW signal generator, but in practice the sensitivity at the cathode ray tube is the more important criterion. Since the signal at this stage is transformed into a small bright spot, the ultimate sensitivity depends on physiological factors. Under ideal conditions when the noise background is perfectly uniform the eye can detect a signal whose level is sufficient to produce a 2 per cent increase in brightness over the background level. Payne-Scott has shown [9] that under these conditions we can detect a signal whose power is 15 to 18 decibels below the noise power, depending upon the type of detector and the " transconductance " of the cathode ray tube. However in practical radar systems the noise background is far from uniform and in addition other factors tend to prevent this limit being attained. Some of the system parameters influencing the visibility of small signals in intensity modulated displays are pulse repetition frequency, pulse duration, bandwidth, time base speed and the radiation pattern and speed of rotation of the aerial. The properties of the cathode ray tube itself including those of the screen and the electron gun structure, are also important. Although many existing radar systems can detect signals whose powers are of the same order as noise in the input circuits, it is preferable to calculate the sensitivity at the cathode ray tube making use of the basic theory and charts in Payne-Scott's paper.

[9] Ruby Payne-Scott, "The visibility of small signals on radar P.P.I. displays," *Proc. I.R.E.*, Vol. 36, 1948, pp. 180-96.

Other Receiver Considerations

The overall amplification of the receiver must be adequate to raise the fluctuating noise potentials in the input circuit to a level of sufficient magnitude to modulate the electron beam of the cathode ray tube; otherwise signals of the same magnitude as noise will not be discernible. Voltage gains ranging from 10^6 to 10^8 are required depending upon the noise factor, bandwidth and cathode ray tube sensitivity, while provision for the aging of tubes and for production tolerances is made by providing ten to twenty decibels reserve amplification, a manual gain control permitting adjustment to the optimum level.

The intermediate frequency employed usually lies between 15 and 60 megacycles per second, a value of 30 megacycles per second being very common. Some of the reasons dictating the employment of such high frequencies are the width of the frequency band containing significant side band components and the ease of separation of intermediate and video frequency components after detection. On the other hand excessively high intermediate frequencies should be avoided owing to high grid input admittances, alignment problems and mechanical difficulties in the provision of the necessary transmission line tuning.

Radio frequency amplification is seldom employed at wavelengths shorter than 25 centimetres. Instead the signal is changed immediately to the intermediate frequency by the use of a crystal converter as described in Chapter XII. At longer wavelengths triodes and pentodes give useful radio frequency amplification and improve the signal to noise ratio of the receiver (Chapter XIII). Triodes are also useful in this range as local oscillators and mixers. Klystron local oscillators are normally employed in the centimetre range of wavelengths, although disc seal triodes such as the GL446 are also used. A detailed discussion of local oscillators and mixers will be found in Chapters XI and XII respectively.

The reflex klystron tube is useful when automatic tuning is necessary. Automatic frequency control (AFC) is usually employed at frequencies above 2000 megacycles per second, since the local oscillator and magnetron frequencies both vary to some extent. Without AFC the intermediate frequency generated will not be effectively amplified since it may depart from its nominal value. The overall receiver sensitivity may show signs of deterioration if this departure is as great as one quarter of the overall bandwidth. For example, assuming a signal frequency of 3000 megacycles per second and a bandwidth of 1 megacycle per second, stability better than 1 part in 12,000 is required. In such cases AFC is imperative. The usual method is to employ a separate AFC mixer fed by a small fraction of the transmitter power (see Chapter XII, p. 403). After conversion and amplification this power is fed to a frequency discriminator and integrating circuit followed by a DC amplifier. The output or error voltage of this amplifier is then applied to the reflector of a klystron local oscillator. The dependence of frequency on the potential of this electrode (see Chapter XI) is utilized to maintain the correct intermediate frequency. If excessive frequency drifts occur, this simple method may be ineffective. In such cases additional circuits have been designed to search automatically for the correct local oscillator frequency, after which the normal AFC circuit again assumes control and switches off the searching circuit.

The special problems of AFC in pulse reception have been dealt with by Moxon and others.[10]

Other units of a radar receiver such as the second detector, video amplifier, cathode follower, etc. are in the main conventional. Special circuits and techniques are given in succeeding chapters.

Typical Receivers

Typical receivers in schematic form are shown in Figs. 225 and 226. The former is suitable for wavelengths of the order of $1\frac{1}{2}$ metres while the latter is commonly used in the microwave region. The separation of the units as shown is purely

[10] L. A. Moxon, J. Croney, W. G. Johnston and C. A. Laws, "Some automatic control circuits for radar receivers," *J. Instn. Elect. Engrs.*, Vol. 93, Part IIIA, 1946, pp. 1143-58.

for constructional and operational convenience and is not of basic importance, although some advantage is gained in reducing instability. There are of course variants and other

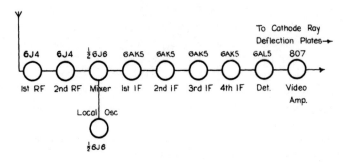

Figure 225.—Typical form of a superheterodyne receiver suitable for the 1·5 metre range of wavelength.

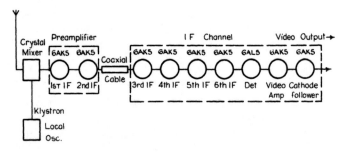

Figure 226.—A typical microwave (25 centimetres to 3 centimetres wavelength) superheterodyne receiver.

special features incorporated in many radar receivers but these are beyond the scope of this book.

Fig. 227 illustrates the values of sensitivity typical of modern receivers under field conditions ; better performance is possible under controlled laboratory conditions. Curve (*a*) is the noise factor expressed in decibels above kT_aB while curve (b) is the noise power of the complete receiver referred to the input terminals for an overall bandwidth of one megacycle per second.

While radar has been responsible for the accelerated development of sensitive wide band microwave receivers, nothing

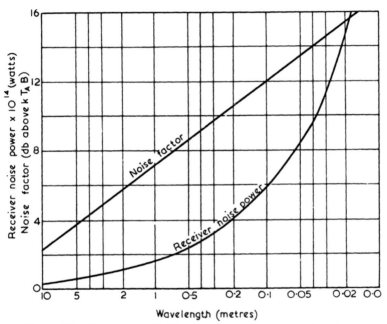

Figure 227.—Average sensitivities of radar receivers operated under field conditions.

as important as the discovery of the resonant cavity magnetron has resulted. The most notable advances have been in crystal mixers, low voltage klystrons, disc seal triodes and miniature pentodes and triodes suitable for high gain, wide band amplifiers.

BIBLIOGRAPHY

L. A. Moxon, *Recent Advances in Receivers*, Cambridge University Press, 1950.

S. Van Voorhis, *Microwave Receivers*, McGraw-Hill, New York, 1948.

CHAPTER XI

LOCAL OSCILLATORS

FOR frequencies up to a few hundred megacycles per second the now well-known " acorn " type triode, or a valve of similar design, with a lumped or a twin wire oscillatory circuit, is used as the local oscillator in the radar receiver. At higher frequencies however the more conventional triode oscillators will no longer operate, and valves and circuits of special design must be employed. In this chapter we shall be concerned with these special oscillators which have been developed to take over where the earlier designs fail, and to cover the frequency band known as the microwave region.

In the field of microwave transmitting valves the magnetron holds undisputed sway, but for the local oscillator, and also for test oscillators and signal generators, the requirement is somewhat different. It is that of an easily tuned, low power, continuous wave oscillator. When, as is almost invariably the case at microwavelengths, the local oscillator is operating with a crystal mixer, a power output of not less than 20 milliwatts is necessary.[1] For application as a test oscillator or signal generator, the oscillator should be capable of producing about 200 milliwatts. The requirements just outlined are satisfied by disc-seal triode oscillators or, especially for the very short wavelengths, by klystrons.

1. Frequency Limitations of Triode Oscillators

The factors limiting the extent to which the frequency of a conventional triode oscillator may be increased have been treated extensively in the published literature.[2] For this reason the subject will not be considered in detail here, and we shall merely recapitulate the main effects and point out the measures which have been taken to overcome them in the development of a triode oscillator which will operate in the microwave region.

---

[1] The power requirements of local oscillators are discussed in Chapter XII.

[2] See for example, R. I. Sarbacher and W. A. Edson, *Hyper and Ultra-high Frequency Engineering* (Wiley, New York, 1943) and the comprehensive bibliography included therein.

The three main frequency-limiting effects are : (*a*) inadequacy of the external oscillatory circuit, (*b*) excessive interelectrode capacity and lead inductance of the valve itself, and (*c*) finite electron transit time.

The External Oscillatory Circuit

As the operating frequency is increased, circuits made up of lumped inductance and capacity must give place to circuits with distributed reactance. For the longer decimetre waves a section of twin transmission line with adjustable short circuit is practicable, but at higher frequencies the physical dimensions of such a system become too small to be handled effectively, and the radiation losses become excessive unless the line is very carefully shielded. The problem is effectively solved by the use of a section of coaxial transmission line or of a resonant cavity, of which the former may be regarded as a special case. Such circuits are self-shielding, since radio frequency currents flow only on the internal surfaces. Moreover, if the walls are silver- or copper-plated the ohmic losses in these circuits are very small. The microwave triode should therefore be designed for easy adaptation to such circuits.

Capacity and Inductance of the Valve Structure

These reactances of necessity form part of the oscillatory circuit, and ultimately limit the extent to which the resonant frequency may be increased. The commonly used coaxial electrode structure with wire leads is unsuitable from this point of view. However, the difficulty has been effectively overcome in the microwave triode with the adoption of a parallel plane electrode structure, as we shall see in the next section.

Electron Transit Time

A number of objectionable effects arise in a triode oscillator when the time taken by an electron to travel from cathode to anode is comparable with the period of oscillation. In the first place the electron stream in a triode is density-modulated as it passes the grid so that, if the alternating grid potential has time to change appreciably during the transit of a group of electrons, a net flow of current is induced in the grid in phase with the grid potential, and grid circuit losses become excessive. In other words, the grid input conductance increases appreciably ; it is found, for a given valve, to be proportional to the square of the frequency. Further, the transconductance of the oscillator is affected, the phase angle between anode current and grid potential changing as the transit time increases, resulting in reduced efficiency. This effect can be overcome to some extent by adjustment of the feedback to the grid circuit, but the range of operation of the oscillator is limited.

The transit time effect is the most difficult problem to deal with. For a short electron transit time the electrode spacings must be very small. Since the electrode capacities must also be kept down, the whole structure requires to be physically small, and the power-handling capacity of the oscillator is therefore limited. We cannot then expect large power outputs from microwave triodes.

2. Microwave Triode Oscillators

The successful development of triodes which will operate down to centimetre wavelengths has been due largely to the introduction of the metal-glass disc-seal technique for the support of the electrodes. This allows a simple parallel plane electrode structure to be used, and the electrode leads then become circular discs projecting straight through the glass envelope. The new construction allows the valve to be adapted very simply to suitable oscillatory circuits ; it overcomes neatly the problem of electrode lead inductances, since the leads now become continuous with the walls of the external circuit ; and it permits an electrode structure of reasonable physical dimensions, whilst maintaining small interelectrode capacities as well as small electrode spacings.

A typical triode, the GL446 or 2C40 " lighthouse " valve, manufactured originally by General Electric Company Schenectady,[3] is illustrated in Fig. 228. The electrode discs

[3] E. D. McArthur, " Disk-seal tubes," *Electronics*, Vol. 18, February, 1945, pp. 98-102.

have spacings of a few thousandths of an inch and capacities
of one or two picofarads. The base of the pillar on which the
cathode is mounted is separated from the shell covering the

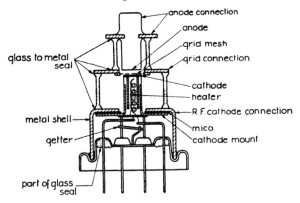

Figure 228.—Essential features of the lighthouse triode. For simplicity
the octal base and lower glass seal are not shown.

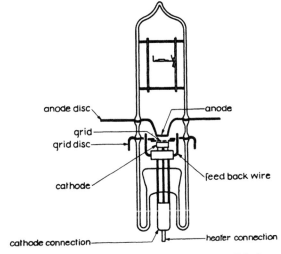

Figure 229.—Diagram of CV90 triode showing the parallel plane electrode
structure, feedback wires and electrode leads.

valve base by a thin ring of mica which provides DC insulation
and forms a by-pass condenser of sufficient capacity to ensure
reasonably good RF continuity at frequencies greater than

about 1,000 megacycles per second. The cathode and heater leads come out through pins in a normal octal base. The overall height of the valve is 2 inches.

Fig. 229 shows a similar type of triode, the CV90, made by the General Electric Company, Wembley,[4] and designed for direct insertion in a coaxial line circuit. The heater and cathode connections are in the form of a coaxial lead. Feedback capacity between anode and cathode is provided inside the valve by the introduction of a pair of wires connected to the base of the cathode and projecting through the grid plane to the neighbourhood of the anode. In this way the normal interelectrode capacity between anode and cathode is increased.

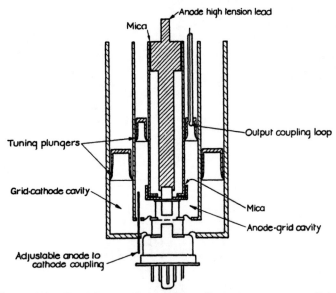

Figure 230.—Coaxial transmission line oscillator incorporating a lighthouse triode.

Coaxial Line Oscillator

Fig. 230 gives a sectional view of a coaxial line oscillator incorporating a lighthouse triode. The external circuit con-

    [4] J. Bell, M. R. Gavin, E. G. James, G. W. Warren, "Triodes for very short waves—oscillators," *J. Instn. Elect. Engrs.*, Vol. 93, Part IIIA, 1946, pp. 833-46; M. R. Gavin, "Triode oscillators for ultra-short wavelengths," *Wireless Engr.*, Vol. 16, 1939, pp. 287-95.

sists of two concentric lines, one within the other, the outer one completing the cathode-grid circuit and the inner, the grid-anode circuit. The disc-seal construction of the triode allows the electrode supports within the valve to become continuous with the walls of the external cavities. Contact is usually made by springy fingers pressing against the disc-seal around the whole circumference. Very good metallic contact is essential between all adjoining parts carrying radio frequency current if circuit losses and leakage of power are to be avoided.

The electrical length of each coaxial line is somewhat less than a quarter (or sometimes three-quarter) wavelength, and is controlled by a movable short-circuiting plunger which serves to vary the frequency of the oscillator. Each short-circuiting plunger is slotted to form a double ring of springy fingers which ensure good sliding contact with the cavity walls. Thus each line is equivalent to an inductive reactance which, with the terminating capacity formed by the corresponding pair of electrodes of the valve, completes one of the oscillatory circuits. The coaxial lines and the grid are at earth potential. DC insulation between the cathode and the shell (which is at the grid potential) is provided inside the valve, and a similar by-pass condenser is inserted at the anode contact ; a quarter wave RF choke is also provided in the anode high tension lead, which comes out through the centre of the oscillator. For the maintenance of oscillations some feedback is necessary between the anode-grid and cathode-grid circuits. For this purpose the anode to cathode capacity of the valve is too small, but frequently the leakage of power through the slots between the spring fingers which contact the grid seal is sufficient to sustain oscillations. If necessary, additional feedback is easily arranged by the insertion of a metal rod through the two coaxial cavities, as shown in the figure. The amount of feedback is then controlled by the depth of penetration of the rod. Power output from the oscillator is taken through a coaxial line terminated by a coupling loop of suitable dimensions which is inserted into the anode circuit, usually through the short-circuiting plunger because the magnetic field, with which the loop is coupled, is strongest there.

Cavity Oscillator

The essential features of the coaxial line oscillator just described are shown in Fig. 231 (*a*). Each of the oscillatory

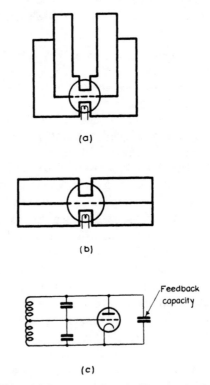

(a)

(b)

(c)

Figure 231.—Essential features and equivalent circuit of the coaxial line and cavity oscillators. (*a*) Coaxial line oscillator. (*b*) Cavity oscillator. (*c*) Equivalent circuit.

circuits is totally enclosed by metal walls and may be regarded as a special type of cavity resonator. It is thus a simple step to proceed to other more convenient cavity forms which carry out the same function. If the two coaxial line cavities are thought of as being straightened out parallel to the planes of the electrodes, instead of bending around at right angles, the construction outlined in Fig. 231 (b) is obtained. Each cavity is a flat cylinder with a central post which, for the anode-grid cavity, is formed by the anode pillar of the valve and in the case of the cathode-grid cavity, is the cathode mount and valve shell. The electric field in each cavity is substantially vertical, and strongest in the centre between the valve electrodes. The magnetic lines of force form closed circles in horizontal planes, with their centres on the vertical axis through the centre of the cavity. For purposes of design the cavities may be regarded as sections of radial transmission lines, short-circuited at the outer edges, and presenting inductive reactances between the corresponding valve electrodes.[5] An equivalent circuit which applies to both the coaxial line and cavity oscillators is shown in Fig. 231 (c). The natural resonant frequencies of the cavity circuits are not equal to the oscillation frequency, but lie on opposite sides of it, the anode-grid circuit having the higher resonant frequency.

A lighthouse oscillator using resonant cavities of the type just described is illustrated in Fig. 232.[5] The shell of the valve is earthed by a split-finger contact which forms part of the central pillar of the cathode cavity, and a special socket encloses the valve base to eliminate the residual power leakage which takes place through the internal by-pass from shell to cathode. A grid disc makes contact with the grid seal by spring fingers, and the grid is insulated from earth by a thin ring of mica which by-passes the grid disc to the cavity wall. The grid lead is brought out across the grid-cathode cavity through a small tube which, being close to the boundary wall, has little effect on the cavity inductance. The anode insulation incorporates a radial transmission line type resonant choke and by-pass, filled with trolitul to reduce the physical dimensions ; in effect, it presents a very low impedance at the insulating gap in the cavity wall and prevents leakage of RF power out of the anode cavity. Some feedback takes place from the

[5] R. N. Bracewell, " A microwave triode oscillator," C.S.I.R. Radiophysics Laboratory, Report RP 261, September, 1945.

anode cavity to the cathode cavity through the grid contact
fingers and by-pass, and additional feedback is obtained by
means of a probe passing from one cavity into the other.
Output from the oscillator is taken by means of a loop inserted

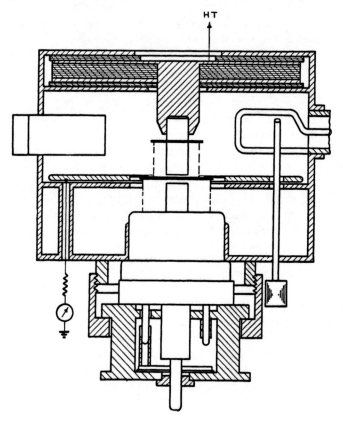

Figure 232.—Lighthouse triode cavity oscillator.

in the anode cavity ; coupling is a maximum when the plane
of the loop contains the axis of the oscillator. The coupling
loop and feedback rod are shown in the same plane in Fig. 232,
but would of course be displaced in the actual oscillator.
Tuning of the cavities is achieved by screwing metal plungers
in through the walls ; insertion of plugs reduces the cavity

inductance and so increases the oscillator frequency. Changes in the anode-grid cavity inductance cause much greater variations in the output frequency than equivalent changes for the cathode-grid cavity.

Plate XV (a) is a photograph of the oscillator just described showing the anode-grid cavity with the end-plate removed; the anode end of the lighthouse valve is visible in the centre. A spare valve and the special socket are shown on the right. This oscillator was made to operate at a wavelength of 25 centimetres, and the cavities were each 3 inches in diameter. A power output of 500 milliwatts was obtained, with an efficiency of 13 per cent.

At a wavelength of 10 centimetres the lighthouse triode can be made to produce about 100 milliwatts, and the wavelength limit is 7 or 8 centimetres. The CV90 triode is capable of slightly higher power outputs but has about the same frequency limit. For wavelengths less than 10 centimetres klystrons are invariably used.

3. The Klystron

In the klystron the finite electron transit time, instead of being a limitation as in the triode oscillator, is an essential factor in the operation of the valve, and the process of velocity modulation, by which the klystron functions, provides us with a means of producing oscillations at frequencies very much higher than can be attained by triodes.

The velocity modulation principle was suggested as early as 1935, by two German scientists,[6] as a means of producing very high frequency oscillations. Some further theoretical work on the subject was published in 1938,[7] but it was not until the following year, with the independent publications

[6] A. Arsenjewa-Heil and O. Heil, " *Eine neue Methode zur Erzeugung kurzer, ungedämpfter, elektromagnetischer Wellen grosser Intensität,*" Zeits. f. Physik, Vol. 95, July, 1935, pp. 752-62.

[7] E. Brüche and A. Recknagel, " *Über die Phasenfokussierung bei der Elektronenbewegung in schnellveränderlichen elektrischen Feldern,*" Zeits. f. Physik, Vol. 108, March, 1938, pp. 459-82.

of Hahn and Metcalf and of R. H. and S. F. Varian,[8] that the potentialities of the new type of microwave oscillator were fully realised. The name "klystron" was invented by the Varians to describe their oscillator which employed velocity modulation and also introduced special types of cavity resonators for the two tuned circuits associated with the valve. More recently a very useful modification of the Varians' double resonator klystron has been developed. This oscillator, designed originally by Sutton of the Admiralty Signal Establishment, is now usually called a reflex klystron. It requires only one cavity resonator, and although less efficient and having a smaller power handling capacity than its predecessor, is simpler to make and to operate. It is now invariably used in preference to the double resonator klystron as the local oscillator in centimetre wave radar receivers. However, the double resonator klystron is still of considerable interest to us, firstly as a logical introduction to the operation of the reflex klystron, and secondly as a microwave oscillator suitable for applications where an output power is required of magnitude greater than that possible with the reflex klystron. Consequently the double resonator klystron will be used here as the model for an introduction to the klystron principles and theory, and some typical double resonator valves will be described to illustrate the methods of design now commonly employed. With this discussion as an introduction, the operation of the reflex klystron and its application as a local oscillator in the radar receiver will be considered in detail in subsequent sections.

Velocity Modulation

The essential features of a double resonator klystron are shown in Fig. 233 (a). A parallel beam of electrons is accelerated from a cathode K to the first grid A by a direct potential difference V_0. If the emission velocities of electrons are neglected, it can be assumed that all electrons cross grid A

[8] W. C. Hahn and G. F. Metcalf, "Velocity-modulated tubes," Proc. I.R.E., Vol. 27, February, 1939, pp. 106-16 ; and R. H. Varian and S. F. Varian, "A high frequency oscillator and amplifier," Jour. Appl. Phys., Vol. 10, May, 1939, pp. 321-27.

(a) Internal view of a lighthouse triode cavity oscillator of the type illustrated in section in Fig. 232. A spare valve and the special valve socket are also shown.

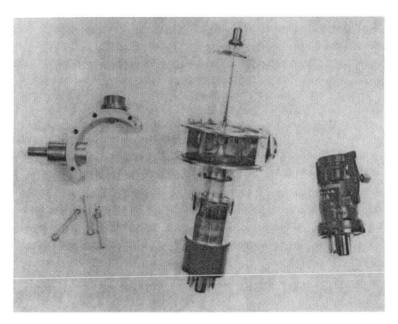

(b) Typical reflex klystrons. *Left*: the CV35 reflex klystron with the cavity opened up to show the internal construction (see p. 381). *Right*: the all-metal 723A reflex klystron showing the tuning mechanism and output probe (see p. 383).

PLATE XV.

in a uniform beam with a velocity v_0 given by $\frac{1}{2}mv_0{}^2 = eV_0$, where m is the mass and e the charge of an electron.

Suppose that between the grids B and A there exists a small alternating potential difference $V = V_1 \sin \omega t$, which is represented by the sine wave in Fig. 233 (b), and that these grids are sufficiently close together for the time of transit between them of an electron with velocity v_0 to be small compared with the period of the alternating potential. An electron passing from A to B at a time t', when B has its maximum negative potential relative to A, will be retarded, losing kinetic energy. An electron crossing AB a quarter of a cycle later, at the time t'', will emerge from grid B with unchanged velocity v_0, while an electron passing AB at time t''' will be accelerated by the

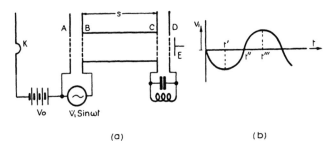

(a)   (b)

Figure 233.—Schematic representation of the klystron. (a) Essential components of the klystron. (b) RF potential of grid B relative to grid A.

alternating field, and so on. Thus the electron beam emerging from grid B, although still of virtually constant charge density, possesses a small velocity modulation superimposed on the steady velocity v_0.

The region between the grids B and C is assumed to be a field-free space. As the electron beam travels through this region the relatively fast electrons begin to catch up on the slower ones ahead of them, whilst the slowest electrons lag behind those of average speed. In this way the electrons tend to form themselves into groups or bunches, and a density-modulated beam is derived from the original velocity-modulated beam. It will be noticed that the electrons at the centre of each group will be those which crossed the grids A, B at a time such as t'', when the field there was passing through zero from the direction which retarded electrons to the direction which accelerated them.

If the beam is allowed to drift far enough in the field-free space, the fast electrons will actually overtake the slower ones originally ahead of them, and the bunches will begin to spread out again. Thus the electron groups will be most concentrated at a definite position down the tube BC, the position depending on the initial velocity v_0 of the electrons and on the amplitude of the alternating potential difference V_1. Suppose that the pair of closely spaced grids C and D is situated at this position, and that there is connected between them an oscillating circuit tuned to the angular frequency ω of the input signal V. If the phase of oscillations in this circuit is such that, at the moment when the centre of an electron group passes between C and D, the maximum retarding field exists across these grids, then energy will be given up by this electron bunch to the oscillatory circuit. By the time the next electron group has arrived at CD the phase of the oscillations will again be correct for the retardation of the electrons, so that the density-modulated electron beam continuously reinforces the oscillations.

The electrons emerging from the grid D with a small residual kinetic energy are finally collected on an auxiliary electrode E.

Since the electron stream crossing the first grids A and B is not density-modulated, it is clear that the power required to velocity-modulate the beam is very small. If the alternating potential $V_1 \sin \omega t$ is supplied from a separate source, the system acts as an amplifier. If a part of the energy in the oscillatory circuit connected to C and D is fed back in the correct phase to the grids A and B, the system becomes an oscillator.

Resonant Cavities for Klystrons

The most suitable form for each oscillatory circuit of the klystron is a cavity resonator, of which the grids form part of the boundary walls. A simple shape such as a sphere or cylinder is not satisfactory since the time of transit of an electron across the cavity requires to be small compared with the period of oscillation. The most suitable shapes are a toroidal or doughnut-shaped cavity or a cylinder with an inner post, as illustrated in cross section in Fig. 234 (a) and (b). The thin lines with arrows indicate the direction and relative intensity of the electric fields which develop inside the cavities when they are oscillating. In each of the cavities represented the grids are close together, and the electric field intensity is

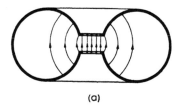

(a)

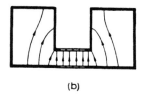

(b)

Figure 234.—Resonant cavities suitable for klystrons. The cavities have circular symmetry about the vertical axis. The electric fields are represented by thin lines carrying arrows. (a) Toroidal cavity. (b) Cylindrical cavity with inner post.

strongest in the narrow gap region where the electrons pass across the cavity ; the magnetic lines of force form closed circles in planes perpendicular to that of the section shown, with centres on the vertical axis of the cavity.

Fig. 235 (a) suggests a way in which the cavities may be regarded as being built up from a simple parallel plate condenser and loop inductance. An equivalent circuit of the cavities is represented in Fig. 235 (b). Klystron cavities are usually made of copper, for which the conductivity is large and the resistive losses in the walls correspondingly small.

Such cavities have large unloaded Q-factors and correspondingly large shunt resistances R.[9] The shapes of the actual cavities employed may vary slightly from those indicated in Fig. 234, but their features are essentially the same as for the cavities just discussed.

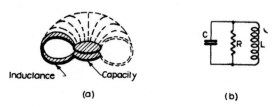

Inductance   Capacity

(a)   (b)

Figure 235.—Equivalent circuits for klystron cavities. (a) Development of a toroidal cavity from a condenser and loop. (b) The conventional LRC circuit to which a klystron cavity is roughly equivalent.

Tuning of klystron cavities may be carried out in either of two ways, the choice of which is dependent on the mechanical construction of the valve.

In the first method the effective capacity of the cavity is varied by altering the gap spacing. When this method is adopted, a portion of the cavity wall is made flexible so that the grids may be pulled further apart or pushed closer together with the aid of screw adjustments. This scheme is often used when the whole of the cavity forms part of the vacuum system. Reduction in the grid spacing increases the gap capacity and lowers the resonator frequency.

Secondly, the inductance of the resonator may be changed by inserting metal plugs into the cavity through the side walls where the magnetic field is relatively strong. This method is usually adopted when part of the resonator is external to the vacuum system. Insertion of plugs into the cavity reduces the effective inductance and increases the resonant frequency.

Essential Construction of the Klystron Oscillator

A schematic view of a klystron oscillator with its cavity resonators is shown in Fig. 236. The parallel electron beam is

[9] See Chapter VI, section 7, for the definition and calculation of R_{sh}.

produced by an electron gun with indirectly heated cathode. For convenience the resonator system is earthed, and the cathode is maintained at a negative potential with respect to earth. The first resonator, termed the buncher, produces velocity modulation of the electron stream. Beyond the enclosed drift space the second resonator, called the catcher, extracts energy from the now density-modulated beam.

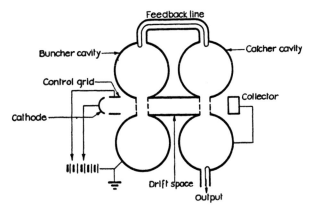

Figure 236.—Diagram of a klystron oscillator with cavity resonators.

Finally a collector at or near earth potential removes the electrons. The collector is so constructed as to dissipate readily the residual electron energy. High frequency power is fed into or abstracted from either cavity resonator by a loop terminating a suitable coaxial feeder, and coupled to the magnetic field inside the cavity. The degree of coupling may be adjusted by varying the size of the loop or by rotating it ; with a given loop the coupling is a maximum when the plane of the loop contains the axis of the cavity, in which position maximum coupling with the magnetic field is obtained. Feedback of a small amount of radio frequency energy from the catcher to the buncher is achieved by coupling loops joined by a short length of coaxial line, the amplitude of the feedback being controlled by adjustment of the loops, and the phase to a limited extent, by the length of the line.

4. Conditions for Oscillation

As has been mentioned during the qualitative discussion on the operation of the klystron, there are two main requirements for efficient oscillation. Firstly the catcher grids must be situated at the position along the drift-tube where the electron bunches are of the best form for the transfer of energy to the catcher circuit. Secondly, the central electrons in a bunch must pass the catcher grids when the radio frequency field between these grids has its maximum retarding value. A quantitative investigation of these requirements is of considerable assistance to an understanding of the operation of the klystron. We shall therefore examine the theory briefly.

A simple mathematical analysis of electron bunching and power transfer to the catcher, which applies under certain limiting conditions, was given by Webster in 1939 for the case of the double resonator klystron.[10] During the war this theory has been amplified and extended to cover much more general conditions. The principal contributors to the field have been Fremlin, Gent, Petrie, Wallis, Tomlin, and Barford.[11] In addition some work which follows the same lines has been published by Black and Morton, and by Harrison.[12]

We shall content ourselves here with a simple analysis since this forms a satisfactory basis for the more exact theories, and is sufficient to provide a general understanding of the functioning of the klystron. To begin with, the following assumptions will be made.

Electrons leave the cathode with zero velocity and travel in a parallel beam through the buncher grids, the loss of electrons to the grids being neglected. The amplitude of the alternating potential difference between the buncher grids

---

[10] D. L. Webster, "Cathode-ray bunching," *Jour. Appl. Phys.*, Vol. 10, July, 1939, pp. 501-8.

[11] J. H. Fremlin, A. W. Gent, D. P. R. Petrie, P. J. Wallis, and S. G. Tomlin, "Principles of velocity modulation," *J. Instn. Elect. Engrs.*, Vol. 93, Part IIIA, 1946, pp. 875-917; and N. C. Barford, "Theory of velocity modulation systems," Electric and Musical Industries Limited, Report RF/124, December, 1942.

[12] L. J. Black and P. L. Morton, "Current and power in velocity modulation tubes," *Proc. I.R.E.*, Vol. 32, August, 1944, pp. 477-82 ; and A. E. Harrison, "Graphical methods for analysis of velocity-modulation bunching," *Proc. I.R.E.*, Vol. 33, January, 1945, pp. 20-32.

is very much less than the direct accelerating potential acting between the cathode and the first buncher grid.[13] This implies that the amplitude of the velocity modulation impressed on the electron stream is small compared with the average velocity of the beam. The buncher grids are sufficiently close together for the time of transit of electrons between them to be negligible compared with the period of the alternating potential difference present there. The same condition applies to the catcher grids. De-bunching effects caused by the mutual repulsion of electrons in the beam will be neglected in the preliminary analysis.

The effect on the theoretical results of departure of conditions from those assumed above will be considered briefly later.

Current Distribution in the Drift Space

The first step in the analysis is the determination of the form of the electron bunching.

Features of the klystron relevant to the present discussion have been shown in Fig. 233 (a). Electrons are accelerated from the cathode to the first buncher grid A by a direct potential difference V_0, so that they cross this grid with a velocity v_0 given by

$$v_0 = \sqrt{\frac{2e}{m}\, V_0} \tag{1}$$

where m is the mass and e the charge of an electron. At the time t_1, an alternating potential difference $V_1 \sin \omega t_1$ is assumed to exist between the buncher grids B and A. Since the transit time between the grids is supposed negligible, electrons will emerge from the grid B, at the time t_1, with a velocity v given by

$$v = \sqrt{\frac{2e}{m}\, (V_0 + V_1 \sin \omega t_1)}. \tag{2}$$

[13] The phrase " radio frequency potential difference between a pair of cavity grids " will be understood to mean the line integral of the radio frequency electric field existing at the time between the grids (see Chapter VI).

When $V_0 \gg V_1$, this relation becomes

$$v = \sqrt{\frac{2e}{m} V_0} \left(1 + \tfrac{1}{2} \frac{V_1}{V_0} \sin \omega t_1 \right)$$

$$= v_0 + v_1 \sin \omega t_1 \tag{3}$$

where v_1 is equal to $\tfrac{1}{2} v_0 \, V_1/V_0$, and is the amplitude of the velocity modulation. If the length of the drift space is s, then the time t_2 of arrival at the catcher of an electron leaving the buncher at the time t_1, is given by

$$t_2 = t_1 + \frac{s}{v_0 + v_1 \sin \omega t_1}. \tag{4}$$

Now $V_0 \gg V_1$, and thus $v_0 \gg v_1$, so that the above equation may be written

$$t_2 \doteq t_1 + \frac{s}{v_0} - \frac{s v_1}{v_0^2} \sin \omega t_1. \tag{5}$$

The relation between t_2 and t_1 is shown graphically in Fig. 237 for two values of the velocity v_1, s and v_0 being fixed. Curve *I* corresponds to a case of incomplete bunching at the catcher, while curve *II* illustrates the case where some fast electrons have overtaken slow electrons which left the buncher ahead of them.

Suppose that i_0 is the beam current between cathode and buncher, and that i_1 and i_2 represent instantaneous values of the currents at the buncher and catcher respectively, i_1 being measured at the time t_1, and i_2 at the time t_2. The initial assumptions imply that $i_1 = i_0 = $ a constant. Now if the electrons passing the catcher in a small interval of time δt_2 are supposed to have departed from the buncher during a corresponding interval δt_1, then

$$i_2 \, \delta t_2 = i_1 \, \delta t_1$$

that is

$$i_2 = i_1 \frac{\delta t_1}{\delta t_2} = i_0 \frac{\delta t_1}{\delta t_2}.$$

and, in the limit,

$$i_2 = i_0 \frac{dt_1}{dt_2}. \tag{6}$$

In some cases, such as that indicated by the broken lines to curve *II* in Fig. 237, electrons arriving at the catcher at a particular time $t_2^{(a)}$, will have left the buncher at three

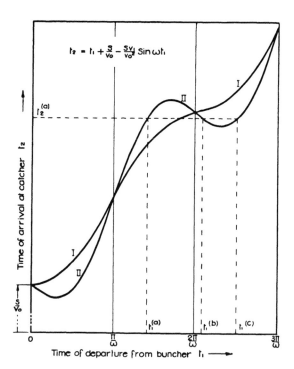

Figure 237.—Relation between time of arrival of electrons at the catcher and time of departure from the buncher, for two values of the amplitude of velocity modulation.

separate instants, $t_1^{(a)}$, $t_1^{(b)}$ and $t_1^{(c)}$. To cover such cases equation (6) should be written

$$i_2 = i_0 \, \Sigma \, \frac{dt_1}{dt_2}, \qquad (7)$$

where the summation includes values of dt_1/dt_2 for all the times t_1 corresponding to the particular time t_2. The values of dt_1/dt_2 may be obtained from the slope of the appropriate t_2 versus t_1 curve. Or, differentiating (5) with respect to t_1 and substituting in (7), we obtain for the beam current at the catcher

$$i_2 = i_0 \ \Sigma \left(\frac{1}{1 - \dfrac{s v_1 \omega}{v_0^2} \cos \omega t_1} \right) \tag{8}$$

or

$$i_2 = i_0 \ \Sigma \left(\frac{1}{1 - r \cos \omega t_1} \right) \tag{9}$$

where

$$r = \frac{s v_1 \omega}{v_0^2}. \tag{10}$$

The factor r given by (10) is usually referred to as the bunching parameter. It may be expressed in the alternative form

$$r = \tfrac{1}{2} \omega \tau_0 \frac{V_1}{V_0} \tag{11}$$

where $\tau_0 = s/v_0 =$ the transit time of an unmodulated electron in the drift space.

According to equation (9), i_2 may have a negative value for values of r greater than unity. This merely implies that electrons arriving at the catcher at successive instants left the buncher in the reverse order. Thus for the calculation of the beam current through the catcher grids the magnitude only of the factor $(1 - r \cos \omega t_1)$ should be taken, and equation (9) should be written

$$i_2 = i_0 \ \Sigma \left| \frac{1}{1 - r \cos \omega t_1} \right|. \tag{12}$$

Similarly equation (7) should be

$$i_2 = i_0 \ \Sigma \left| \frac{dt_1}{dt_2} \right|. \tag{13}$$

The form of the beam current i_2 is shown in Fig. 238 for three values of the bunching parameter r. The curves are most easily obtained by the use of equation (13) and the slope of the appropriate t_2 versus t_1 curve. We observe that for $r = 0\cdot5$ only a small degree of bunching has developed. For

$r = 1$ the current waveform has a single infinite peak, which is interpreted as corresponding to the arrival simultaneously, at the catcher, of electrons departing from the buncher at successive instants. For $r > 1$ a double-peaked wave is obtained, which spreads out with increasing r.

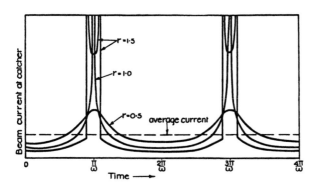

Figure 238.—Beam current at the catcher as a function of time, for three values of the bunching parameter r.

For constant v_1 and v_2, the current waveforms of Fig. 238 show the way in which electron bunching develops along the drift tube, increases in r corresponding to increases in the distance s.

To determine the best waveform for the transfer of energy to the catcher circuit we expand the expression (12) for i_2 as a Fourier series. In this way we obtain (for the derivation, see for example, Black and Morton[12]).

$$i_2 = i_0 \left[1 + 2 \sum_{n=1}^{\infty} J_n(nr) \cos n\theta_2 \right] \qquad (14)$$

where

$$\theta_2 = \omega t_2 - \frac{s\omega}{v_0} \qquad (15)$$

and $J_n(nr)$ is the Bessel function of the first kind of order n.

The current waveform is obviously rich in harmonics, making klystron frequency multipliers practicable.[14] Here we shall consider power transfer at the fundamental frequency only. The fundamental component of radio frequency current is

$$i_2 = 2i_0 J_1(r) \cos \theta_2. \tag{16}$$

Thus i_2 is a maximum when $J_1(r)$ is a maximum and $\cos \theta_2 = 1$. The maximum value of $J_1(r)$ occurs when $r = 1.84$, for which value of the bunching parameter the double-peaked current wave is well developed. From equations (15) and (5) it can be seen that the condition $\cos \theta_2 = 1$ corresponds to $\omega t_1 = 2n\pi$; this means that electrons at the centre of a bunch are those which left the buncher when the radio frequency field there was passing through zero from the negative to the positive half cycle, a fact that has already been observed in the qualitative discussion.

Power Transfer to the Catcher

Suppose that, at the time t_2, the alternating potential difference between the catcher grids due to oscillations in the catcher circuit is given by $V_2 \cos (\theta_2 - \phi)$, where ϕ is some phase angle and V_2 is positive in the direction to accelerate electrons. The transfer of energy from the electron beam to the catcher circuit during an interval of time dt_2 is then

$$dW = -i_2 V_2 \cos (\theta_2 - \phi) \, dt_2 \tag{17}$$

where i_2 is given by equation (16).[15]

When this expression is integrated over a complete cycle of the alternating potential, the net power transfer to the catcher is found to be

$$P_2 = -i_0 V_2 J_1(r) \cos \phi. \tag{18}$$

The power supplied to the electron beam is $P_0 = i_0 V_0$, so the efficiency of power conversion is

$$\eta = -\frac{V_2}{V_0} J_1(r) \cos \phi. \tag{19}$$

[14] A. E. Harrison, *Klystron Tubes*, McGraw-Hill, New York, 1947.

[15] Only the fundamental frequency term need be included since for all the other current components the average power transfer is zero.

The conditions for this expression to be a maximum are firstly that $\cos \phi = -1$. This means that when the centre of an electron bunch passes the catcher ($\theta_2 = 2n\pi$), the potential difference there is $-V_2$, that is, the maximum retarding value. This requirement has already been noted in the qualitative discussion. Secondly, for maximum efficiency $J_1(r)$ is a maximum and $r = 1\cdot84$, the corresponding value of the Bessel function then being $0\cdot58$. Thirdly, V_2/V_0 must be a maximum. Clearly V_2 cannot be greater than V_0, for in such a case most electrons would be thrown back by the field between the catcher grids. Thus the maximum allowable value for V_2 is approximately equal to V_0. The maximum theoretical efficiency of the ideal klystron oscillator is then

$$\eta_{max} \doteqdot 0\cdot58. \qquad (20)$$

It should be remarked that, in the above calculation of efficiency, power losses in the buncher have been neglected. Under the conditions we have assumed, the power required to velocity-modulate the electron beam is very small.

The relation (20) gives the maximum efficiency of power conversion. The actual output efficiency will of course depend on the output loading of the catcher, and the resistive losses in the catcher walls. The output efficiency has been discussed by Barford.[11]

Phase Relationship between Oscillations in Buncher and Catcher

We have seen that for maximum power transfer the central electrons in a bunch should pass the catcher grids when the radio frequency field across the latter has its maximum retarding value. These electrons crossed the buncher when the field there was passing through zero from the negative to the positive half-cycle. The time taken for these average speed electrons to travel from buncher to catcher is s/v_0. Thus the phase angle ψ of the catcher field relative to the phase angle of the buncher field should be given by

$$\psi = 2\pi n - \frac{\pi}{2} - \frac{s\omega}{v_0} \qquad (21)$$

where n is a positive integer. This phase angle can be adjusted by varying the accelerating potential V_0 which controls the velocity v_0, and, to some extent, by adjustment of the feedback line.

Recapitulation

To sum up, the conditions for most efficient power conversion in the double resonator klystron just considered are:

(a) The waveform of the electron current at the catcher must be that corrosponding to a value of the bunching parameter r of 1·84, where $r = s\omega v_1/v_0^2$.

(b) The phase difference ψ between oscillations in the catcher and buncher must satisfy the relation $\psi = 2\pi n - \pi/2 - s\omega/v_0$.

(c) The amplitude V_2 of the radio frequency potential difference between the catcher grids must be such as to take the greatest amount of energy from the electrons passing the grids without reversing their motion. That is, $V_2 \doteqdot V_0$.

Second Order Effects

In practical cases the conditions assumed in the simple theory are not always approached, and the results stated above are modified accordingly.

Firstly, we cannot always assume that the amplitude V_1 of the radio frequency potential difference between the buncher grids is very small compared with the direct accelerating potential V_0. In this case the approximate equations (3) and (5) are no longer valid. We now find that the curves of Fig. 237 are distorted, and the waveform of the electron beam current is not symmetrical (see Harrison[12]). The result of this effect is that the optimum value of the bunching parameter [which of course is no longer accurately represented by equation (10)] is different from the previously determined value, and further, the electrons at the effective centre of a bunch are no longer those which left the buncher when the radio frequency field there was zero. Consequently the phase condition for buncher and catcher oscillations is modified accordingly.

In the second place, in some cases the transit time of electrons between the buncher grids and also between the catcher grids cannot be assumed to be negligible compared with the oscillatory period. As far as the effect at the buncher is concerned, a finite transit time causes a decrease in the amplitude and a variation in the phase of the velocity modulation produced by a given buncher field, and the power absorbed by the buncher field is increased. At the catcher the effect of a finite electron transit time is to modify the waveform of the current induced in the catcher circuit so that it will no longer be exactly as represented in Fig. 238. The result is a decrease in the maximum efficiency, although the optimum value of the bunching parameter is not affected (Black and Morton[12]).

Thirdly, the repulsive forces acting between individual electrons produce de-bunching effects which become appreciable in certain cases. One such case is the high power oscillator, where relatively dense electron beams are necessary and the electron repulsion forces are correspondingly large. In fact, the de-bunching effect sets the most serious limitation to the production of very high powers from klystrons.

However, the modifications just mentioned are sufficiently small effects in practical klystrons for the theoretical analysis originally presented by Webster to remain a useful basis for design work. In practice, output efficiencies of about 20 per cent are realised with double resonator klystrons.

Advantages and Limitations of the Klystron

Two difficulties encountered in the microwave triode oscillator are overcome in the klystron. Firstly, in the triode, transit time effects cause losses in the grid circuit since the electrons are density-modulated as they cross the grid. In the klystron the electron beam is of uniform density as it passes the buncher, which corresponds to the triode grid circuit, so that the net loss of power in the buncher circuit is very small. In the second place, power dissipation becomes a serious problem in the triode, where the anode, of necessity small, collects the electron stream. In the klystron a separate and suitably designed electrode is used to collect the electrons and dissipate their residual energy.

As far as power output is concerned the chief limiting factor in the klystron is the de-bunching effect already mentioned. With regard to frequency limitations of the klystron, the first difficulty which arises is that of physical size. For wavelengths shorter than about one centimetre the cavity resonators required are too small for effective handling. In fact, the klystron in its present form has about the same frequency limit and the same cause of failure as most of the other microwave apparatus now in common use.

Another effect which may be mentioned here is the production of noise in a klystron oscillator. This noise is appreciable over a band of frequencies centred on the frequency of oscillation.[16]

5. Typical Double Resonator Klystrons

Some of the basic features of a klystron made by the Sperry Gyroscope Company[14] are shown in Fig. 239. Except for the glass seal-in of the leads to the electron gun and the coupling loop seals, the valve is entirely of metal construction, the cavity walls themselves enclosing part of the vacuum system. The electron gun consists of a cup-shaped cathode with non-inductive heater, and a control grid which focuses the electrons into a beam. The grids of both the buncher and the catcher resonators are approximately 0·03 inches apart, and tuning of the cavities is achieved by varying the grid spacings, the ends of the cavities being made flexible for this purpose. The tuning adjustments are made by three pairs of tuning struts, and a fine control of one pair of struts is obtained with the aid of a wedge operated by a graduated knob. The output and

[16] Local oscillator noise is discussed further in Chapter XII.

feedback loops are sealed into the cavities. The collecting electrode is a metal cylinder with heat-radiating fins.

Klystrons of the type just described have been made to operate over a wavelength band from 9 to 11 centimetres

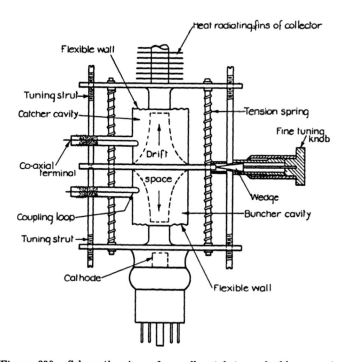

Figure 239.—Schematic view of an all-metal type double resonator klystron, showing two of the coupling loops and part of the tuning mechanism. There are actually three pairs of tuning struts placed symmetrically round the valve, one pair being controlled by the fine tuning knob.

with a CW power output of 20 watts, the accelerating potential being 2500 to 3000 volts.

A klystron of the type in which part of each cavity is external to the glass envelope is illustrated in Fig. 240. The particular valve shown was designed by Electric and Musical Industries

Limited for use as a relatively high power pulsed oscillator.[17] The cavity walls are brought out through the envelope in disc seals and the external parts clamped on to them. The buncher and catcher cavities have a common diaphragm, and the coupling between the two resonators is provided by slots in

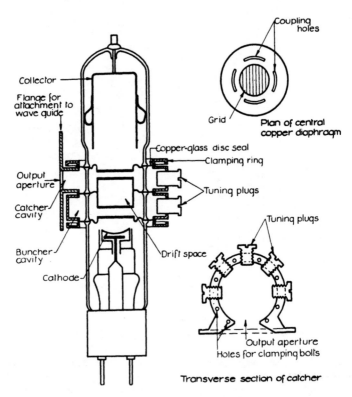

Figure 240.—High power double resonator klystron, with part of each cavity external to the glass envelope.

this diaphragm. This arrangement for feedback is preferable in the present case to one involving coupling loops and a coaxial line, since the valve operates at radio frequency powers of sufficient intensity to give trouble due to spark-over. For

[17] " EMI klystron PK150," Electric and Musical Industries Limited, Research Laboratories, Report RF/137, June, 1943.

the same reason the output from the catcher is taken through an aperture in the external wall of the cavity instead of by a loop, the catcher cavity being clamped directly on to the end of a waveguide feeder. Tuning of the cavities is achieved with metal plugs which reduce the cavity inductance, and hence increase the resonant frequency as they are screwed in through the outer cavity walls. The valve is designed to give pulsed power at a wavelength of 10 centimetres. The resonator and collector are earthed, the cathode being pulsed at a negative potential of 12·5 kilovolts. The klystron gives a peak power output of about 20 kilowatts, the allowable mean dissipation with air-blast cooling being 80 watts.

6. The Reflex Klystron

In the reflex klystron a single cavity resonator is used both as buncher and catcher. The basic construction is shown in Fig. 241 (a). Beyond the resonator is placed an electrode, called the reflector or repeller, maintained at a steady potential somewhat negative with respect to the cathode. Thus electrons emerging from the resonator encounter a DC retarding field which is strong enough to bring them to rest and reflect them back again towards the cavity. If the resonator is oscillating, the electron beam enters the reflector field in the forward direction with velocity modulation. The electrons with velocities above the average will penetrate further in the retarding field, and will take longer to return to the resonator grids, than those with lesser energies. Thus the reflected beam becomes density-modulated. The formation of bunches is illustrated in Fig. 241 (b) and (c). In (b) the radio frequency potential of grid B relative to grid A is plotted against the time t. An electron crossing the grids in the forward direction at the time t', when the RF field is a positive maximum, follows the path represented by curve I in (c) ; an electron crossing a quarter of a cycle later, at the time t'', has a shorter drift time in the reflector field as shown by the curve II, and an electron entering the reflector field at time t''' follows the still shorter path III. Thus electrons passing the resonator grids during the half period from t' to t''' tend to return to the resonator in a bunch. On the other hand electrons crossing in the subsequent half cycle will be spread out.

It will be noted that, whereas in the double resonator klystron the slowest electrons have the longest transit time in

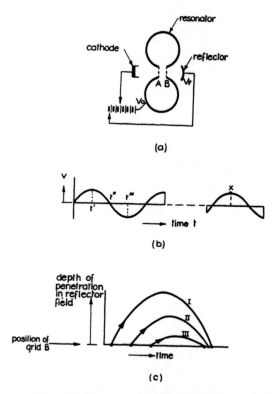

Figure 241.—The reflex klystron. (a) Essential features of the reflex klystron. (b) RF potential of grid B relative to grid A. (c) Electron motion in the reflecting field.

the drift space, in the case of the reflex klystron the fastest electrons have the longest transit time in the corresponding reflecting field. Thus in the former case the bunches form about electrons crossing the buncher when the field there is passing through zero from the direction which retards to that which accelerates electrons ; but in the latter case the bunches form about electrons crossing the resonator when the field there is changing from accelerating to retarding.

The degree of bunching of the beam on return to the resonator is obviously dependent on the accelerating potential V_0, which determines the initial speed of the electrons ; on the amplitude of the radio frequency field across the resonator grids, which is governed by the loading of the oscillator ; and on the strength of the retarding field, determined by the valve geometry and by the potential difference $(V_0 - V_F)$, where V_F is the reflector potential measured with respect to the cathode.

The electron bunches returning to the resonator will give up their energy to it if they meet there the maximum retarding field. Since the electrons are now passing the resonator in the reverse direction, this means that, at the moment when the centre of an electron bunch returns, the potential of grid B with respect to grid A should be a positive maximum. This corresponds to a peak such as the point marked X in Fig. 241 (b). Electrons at the centre of a bunch start out from the resonator at a time such as t'', so that for maximum power transfer to the cavity the number of cycles of oscillation which should occur during the transit time, in the retarding field, of electrons destined to become the centre of a bunch, is given by

$$N = (n + \tfrac{3}{4}) \text{ cycles}, \, n = 0, 1, 2......\qquad (22)$$

This phase condition makes an extra requirement which the potentials V_0 and V_F must satisfy, in order that oscillations be maintained in the klystron. The various values of n correspond to various possible modes of operation of the oscillator.

After their second transit through the resonator the electrons are finally collected on the outer walls of the cavity, which must therefore be designed to dissipate the residual energy involved.

Obvious practical advantages of the reflex klystron over the double resonator klystron are firstly, comparative ease of construction and secondly, simplicity of tuning. The initial tuning adjustment of the double resonator klystron is a complicated operation since two high Q cavities are involved.

Disadvantages of the reflex klystron are firstly, as we shall
see in the next section, that it is less efficient than the double
resonator oscillator. Secondly, in the single resonator klystron
the electrons end up on the cavity itself, instead of on an auxiliary electrode which can be more suitably designed for heat
dissipation. Thus the double resonator klystron can handle
higher powers than the reflex valve.

However, the reflex klystron is eminently suitable for use
as a local oscillator in the superheterodyne receiver, and is
most widely used as such in microwave radar.

7. Theory of Operation

The theoretical analysis of bunching and power transfer in
the reflex klystron is carried out along lines very similar to
those followed in the double resonator klystron theory.[18]
Again a simple theory only is presented here, with the adoption
of initial assumptions similar to those set out for the case of
the double resonator oscillator. We suppose that V_0 is the
direct accelerating potential difference between cathode and
resonator and that $V_1 \sin \omega t_1$ is the alternating potential
difference between the resonator grids at the time t_1, V_1 being
very much less than V_0. In the double resonator klystron
analysis we assumed for simplicity that the electron transit
time across the buncher grids was negligible, and that an electron crossing the gap at the time t_1 experienced the full potential
change $V_1 \sin \omega t_1$. Following general practice we shall, in the
present analysis, make some allowance for a small but finite
electron transit time across the resonator grids, with the introduction of a gap efficiency factor or modulation coefficient β,
defined by supposing the kinetic energy of an electron which
passes the centre of the resonator at the time t_1 to be changed
by an amount $\beta e V_1 \sin \omega t_1$. The coefficient β is less than unity
and depends on the transit time and on the configuration xf
the resonator grids, decreasing with increasing transit time.
We shall, however, neglect the time taken for an electron to
traverse the resonator gap in comparison with its time of
transit in the direct reflecting field.

Thus, at the time t_1 electrons may be supposed to enter the retarding field with a velocity v given by

$$v = \sqrt{\frac{2e}{m}(V_0 + \beta V_1 \sin \omega t_1)} \qquad (23)$$

$$\doteq \sqrt{\frac{2e}{m} V_0} \left(1 + \tfrac{1}{2}\beta\frac{V_1}{V_0} \sin \omega t_1\right). \qquad (24)$$

Let us assume for a start that the reflecting field between the resonator and the reflector electrode is uniform, and of magnitude E_0. Then an electron entering this field with velocity v will return to the resonator again after a transit time τ, given by

$$\tau = \frac{2mv}{eE_0} \qquad (25)$$

which, using equation (24), may be written

$$\tau = \tau_0\left(1 + \tfrac{1}{2}\beta\frac{V_1}{V_0} \sin \omega t_1\right) \qquad (26)$$

where τ_0 is the transit time in the reflecting field of an electron crossing the resonator grids when the alternating field there is zero.

Let t_2 be the time of return to the centre of the resonator gap of an electron which passes this point in the forward direction at the time t_1, so that $t_2 - t_1 = \tau$. Suppose i_0 denotes the beam current from the cathode, and i_1 and i_2 are the instantaneous values of the currents passing the centre of the cavity grids on the forward and return journeys respectively, i_1 being measured at the time t_1, and i_2 at the time t_2.

As in the previous analysis we have

$$i_2 = i_0 \ \Sigma \ \left|\frac{dt_1}{dt_2}\right|$$

$$= i_0 \ \Sigma \ \left|\frac{1}{1 + \dfrac{d\tau}{dt_1}}\right|. \qquad (27)$$

Differentiating (26) and substituting in (27),

$$i_2 = i_0 \; \Sigma \left| \frac{1}{1 + (\tfrac{1}{2}\beta\omega\tau_0 V_1/V_0) \cos \; \omega t_1} \right| \qquad (28)$$

or

$$i_2 = i_0 \; \Sigma \left| \frac{1}{1 + x \cos \omega t_1} \right| \qquad (29)$$

where

$$x = \tfrac{1}{2}\beta\omega\tau_0 V_1/V_0 \qquad (30)$$

and x is the bunching parameter.

A comparison of equations (29) and (30) with equations (9) and (11) of the analysis for the double resonator klystron shows that the form of the density-modulated beam which develops in the two types of valves is the same, and that the bunching parameters, x and r respectively, are identical except for the coefficient β which was omitted for simplicity in the derivation of the factor r. This identity is exact only for the case of the uniform reflecting field which we have been considering. It can be shown that for a reflecting field of general form the bunching parameter involves a factor g, called the reflector gain factor, which is constant for a given reflector field and a given mode (see Beck[18]). Thus in general the form of the electron current returning to the resonator is given by equation (29) when

$$x = g \, \beta\omega\tau_0 \, V_1/V_0, \qquad (31)$$

with $g = \tfrac{1}{2}$ for the special case of a uniform reflecting field.

As before, the expression for i_2 may be expanded as a Fourier series, leading to

$$i_2 = i_0 \left[1 - 2 \sum_{n=1}^{\infty} J_n(nx) \cos n\theta_2 \right] \qquad (32)$$

where

$$\theta_2 = \omega t_2 - \omega\tau_0. \qquad (33)$$

[18] The reflex klystron theory is due chiefly to Fremlin, Gent, Petrie, Wallis and Tomlin,[11] Barford[11], and A. H. Beck, "Fundamental limitations and optimum conditions in reflex klystron oscillators," Admiralty Signal Establishment Extension, Bristol, Report BR.S.445/45/XQ2, June, 1945. A useful qualitative discussion has been published by J. R. Pierce, "Reflex oscillators," *Proc. I.R.E.*, Vol. 33, February, 1945, pp. 112-18.

The fundamental component of the current returning to the resonator is therefore

$$i_2 = - 2i_0 J_1(x) \cos \theta_2. \tag{34}$$

This expression is a maximum when $\cos \theta_2 = - 1$, which corresponds to $\omega t_1 = (2n + 1)\pi$, so that, as we have already observed, electrons at the centre of a returning bunch are those which passed through the resonator on their first trip when the alternating field there was passing through zero from the positive to the negative half cycle.

If the radio frequency potential difference between the resonator grids at the time t_2 is taken as $V_1 \cos (\theta_2 - \phi)$, then the rate of transfer of energy from the returning electron beam to the resonator at the time t_2 is

$$P_{t_2} = - 2i_0 J_1(x) \cos \theta_2 . \beta V_1 \cos (\theta_2 - \phi). \tag{35}$$

Averaged over a complete cycle, the net power transfer to the resonator is

$$P = - \beta i_0 V_1 J_1(x) \cos \phi. \tag{36}$$

When the phase angle ϕ is correctly adjusted, this expression becomes

$$P = \beta i_0 V_1 J_1(x). \tag{37}$$

As we have already seen [equation (22)], the phase condition may be expressed in the form

$$\omega \tau_0 = 2\pi(n + \tfrac{3}{4}) \tag{38}$$

n being the mode number.

Neglecting the power required for bunching, the efficiency of conversion of direct current power into radio frequency power is

$$\eta = \beta \frac{V_1}{V_0} J_1(x) \tag{39}$$

which may be written

$$\eta = \frac{1}{g\omega\tau_0} \cdot x J_1(x). \tag{40}$$

As V_1 varies, this expression is a maximum when the bunching parameter $x = 2{\cdot}40$. Thus the condition for maximum power transfer is not the same as that for maximum radio frequency

current ($x = 1\cdot84$).   It will be observed that for good power
conversion efficiency we require a small reflector gain factor g,
and a small value of $\omega\tau_0$, which means a small mode number n.
Our theory is strictly valid only for very small ratios V_1/V_0,
but it has been shown that, after allowance is made for the
power used in bunching the beam, the theory may be successfully applied to practical cases where V_1 may be an appreciable
fraction of V_0 (see Beck[18]).   Of course a practical upper limit
is set to the magnitude of V_1 by the requirement that no
electrons be thrown back in traversing the resonator.   Thus
we must have $2\beta V_1 < V_0$.

A comparison of the expressions for conversion efficiency
for the double resonator and reflex klystrons shows that the
efficiency of the latter oscillator cannot be made as high as
that of its predecessor, where the relative amplitudes of the
radio frequency fields at buncher and catcher are controllable.

Turning to the question of output power from the reflex
klystron we note that the amplitude of the radio frequency
potential difference V_1 between the resonator grids varies
with the loading of the resonator, and the loading also controls
the distribution of power between the output circuit and the
shunt resistance of the cavity itself.   If the total parallel
impedance of the cavity at resonance when loaded with the
output circuit is R, then the condition for steady oscillation
in a given mode is

$$\frac{V_1^2}{2R} = P = \beta i_0 V_1 J_1(x). \qquad (41)$$

The significance of this relation may be readily understood by
reference to curves of the type shown in Fig. 242.[19]   In
Fig. 242 (a) the amplitude of the fundamental component of
radio frequency current $i_2 = 2i_0 J_1(x)$ is plotted against V_1, each
curve corresponding to a particular mode number n, which
fixes the factor $g\omega\tau_0$ in the bunching parameter x.   Each
curve has a maximum when $x = 1\cdot84$; the maxima are equal
for the different modes, but occur for smaller values of V_1 as
the mode number and hence the drift time τ_0 increases.   In
Fig. 242 (b) the radio frequency power transferred to the

[19] The ensuing discussion is based largely on the treatment by Pierce.[18]

resonator, that is $P = \frac{1}{2}\beta V_1 \cdot i_2$, is plotted against the same variable V_1. The maxima now occur for each mode number when $x = 2\cdot4$, and increase in magnitude with increasing V_1. The broken curves A, B and C on the same graph represent power absorbed in resonator and load, $V_1^2/2R$, for three

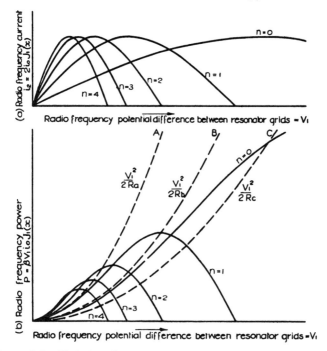

Figure 242.—Variation of current and power in the reflex klystron with the magnitude of the radio frequency potential difference between the resonator grids.

different values of R corresponding to three different degrees of coupling of the output circuit to the resonator. For a given circuit impedance R, steady oscillations will occur in a particular mode at the value of V_1 for which the appropriate power load curve intersects the power production curve corresponding to this mode. In the diagram the power load curve A intersects the power transfer curves for $n = 2$ or more; it can be seen that, for the particular loading corresponding to $R = R_a$, most power is obtained in the mode $n = 2$, although the coupling is a little greater than optimum since curve A intersects the curve for $n = 2$ at a point a little to the left of the maximum. The coupling is close to optimum for the modes $n = 3$ and 4, but the power output is smaller since their maxima are smaller.

The load curve A lies entirely above the curves for $n = 0$ and $n = 1$, implying that the loading of the resonator is too great for oscillation to occur for these two modes. If however the output coupling is reduced so that R is increased to R_b, the corresponding power load curve B cuts the power production curve for $n = 1$ at its peak, and maximum power is obtained for this mode. A still greater reduction of the resonator loading, with $R = R_c$ say, would allow of oscillations in the mode with smallest drift time and greatest power transfer, provided the shunt resistance of the cavity itself is large enough.

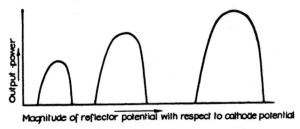

Figure 243.—Reflex klystron characteristic : variation of output power with reflector potential.

However, depending on the magnitude of this shunt resistance, we may find that for this mode most of the power is absorbed in the cavity itself, and the net power reaching the output circuit is less than in the case of the mode $n = 1$, with the greater cavity loading corresponding to $R = R_b$. Thus the best mode and resonator loading for a particular klystron depend upon the valve and cavity design.

In practice, the characteristics of typical reflex klystrons bear out the inferences we have just made. An example is given in Fig. 243, where the output power from an oscillator with fixed loading is plotted against the reflector potential, the latter becoming more negative with respect to cathode potential towards the right. An increasingly negative reflector potential corresponds to a decreasing drift time τ_0, and the peaks of the output curves occur at points for which the phase relation $\omega\tau_0 = 2\pi(n + \frac{3}{4})$ is satisfied; successive peaks correspond to successive modes, with n decreasing towards the right. We see that the available output power increases with decreasing mode number until the point is reached where the loading is too great for oscillations to occur.

8. Typical Reflex Klystrons

Plate XV (b) is a photograph of two typical reflex klystrons. The first, which is shown diagrammatically in Fig. 244 (a), is

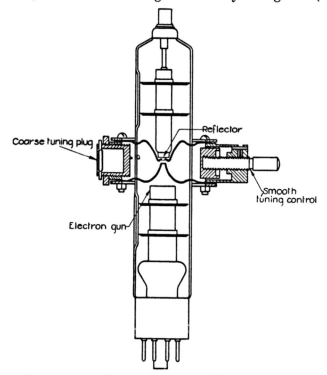

Figure 244 (a).—Construction of the CV35 reflex klystron.

the CV35, a reflex klystron of glass construction with partly external cavity, designed by Electric and Musical Industries Limited for operation at a wavelength of ten centimetres.

The reflector electrode is brought very close to the resonator gap in order to provide a strong reflecting field which, as we have seen, corresponds to a short electron drift time and relatively good efficiency, and the re-entrant portion of the cavity on this side is enlarged to accommodate the reflecting electrode. Grid structures at the cavity gap are found to be unnecessary,

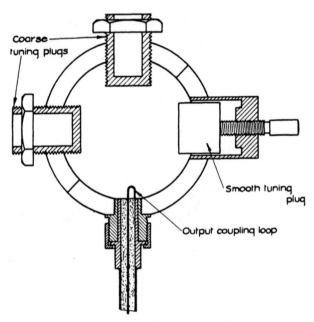

Figure 244 (*b*).—Horizontal section through the CV35 resonator showing the three tuning plugs and output coupling loop.

and two 2 millimetre diameter holes are provided for the passage of the electron beam. A section through the cavity at right angles to the valve axis is shown in Fig. 244 (*b*). The external cavity walls are made in two semi-circular pieces which fit round the valve and are clamped to the copper parts sealed into the glass envelope. Tuning is carried out by three plungers, one of which is spring-loaded and specially designed to give a smooth frequency variation. A coupling loop, which can be rotated in the cavity, takes the radio frequency power out. Since the magnetic lines of force in the resonator form closed circles in planes parallel to that of the section of Fig. 244 (b), it is clear that the coupling loop as shown in the figure is in the plane of minimum coupling to the cavity field, and that it should be rotated through 90 degrees to obtain maximum coupling.

A power output of 300 milliwatts at 10 centimetres wavelength is obtainable with this valve, the efficiency being about 3 per cent; an accelerating potential of 1500 to 2000 volts and a reflector potential of − 250 to − 300 volts (with respect

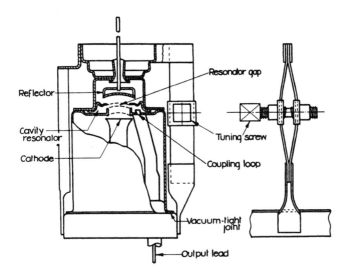

Figure 245.—Internal construction of the 723A reflex klystron.

to cathode potential) are required. The simplicity of construction of the CV35 will be readily appreciated. The technique has been applied to many types, including lower power versions which operate with quite small accelerating potentials. Special glass having good dielectric properties at microwavelengths is used in making these klystrons, in order to minimise power losses inside the cavity resonator.

Fig. 245 illustrates a reflex klystron of the all-metal type. It is the 723A, made by the Western Electric Company, for a wavelength of three centimetres. The resonator has become a rudimentary cavity near the top of the valve. Variation of the grid spacing is used to tune the cavity, and the tuning is achieved with the aid of two springy strips of metal mounted at the side of the valve and bowed outwards so that when squeezed together by the tuning screw, the distance between their ends is increased and the cavity grids are pulled further apart. The reflector electrode is at a relatively large distance from the resonator for a reason which will become clear in the next section. A small coupling loop takes energy from the resonator and feeds it through a concentric line built inside the valve to a probe projecting through the base. This probe is usually inserted directly into a waveguide feeder. The valve, being all-metal, is somewhat sensitive to temperature, and should be shielded from air currents to prevent frequency fluctuations. This valve is also shown in Plate XV (b).

Typical operating data for the valve are : accelerating potential, 300 volts ; reflector potential, − 20 to − 300 volts ; cathode current, 20 milliamperes ; power output, 80 milliwatts.

9. Frequency Stability and Electronic Tuning

In both the reflex and double resonator klystrons, small changes in the DC electrode potentials cause appreciable changes in the frequency as well as the power output of oscillation. In the reflex klystron in particular, variations in either the accelerating potential or reflector potential cause frequency variations. The effect is easy to visualise, for the resonator has a finite bandwidth and if, due to a change in potential, electron bunches return to the resonator after a drift time which is not exactly equivalent to $(n + \frac{3}{4})$ cycles, so that the resonator field is not exactly in its maximum retarding phase, frequency pulling will occur as well as a reduction in power transfer efficiency.

Typical curves relating output and frequency to reflector potential for a particular mode in a reflex klystron are shown in Fig. 246. The magnitude of the effect may be from $\frac{1}{2}$ to 1 megacycles per second for a one per cent change in reflector potential. On account of this frequency sensitivity a klystron used as a local oscillator requires a regulated power supply.

However there is a way in which this effect may be turned to advantage in the reflex klystron, and that is in electronic tuning of the oscillator over a small range by variation of the reflector potential. Electronic tuning has a very important application in automatic frequency control of the radar receiver (usually termed AFC), as has already been discussed in Chapter X.

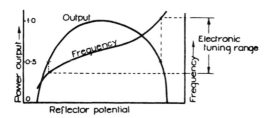

Figure 246.—Variation of power output and frequency with reflector potential in a reflex klystron.

For a particular mode the usable range of frequency for AFC purposes is usually taken to be that obtained for reflector potentials between the values at which the power output has decreased to one half the maximum power. In Fig. 246 the electronic tuning range calculated on this basis is indicated.

Klystrons intended for use with AFC, and for other purposes where electronic tuning is desirable, are specially designed to give as wide a range as possible. The design requirements for such valves may be deduced readily with the aid of the theoretical results already established, so we shall consider the problem briefly, following the treatments of Bleaney, Barford, and Finlay and Lasich.[20]

Suppose, at the time t, the radio frequency potential difference between the resonator grids is $V_1 \cos \omega t$. From the

[20] B. Bleaney, " Elementary theory of automatic frequency control of reflection oscillators," C.V.D. Research Group at Clarendon Laboratory, Oxford, Reports CL.Misc. 13 and 15 ; N. C. Barford, " The A.F.C. properties of reflection oscillators," Research Laboratories, Electric and Musical Industries Limited, Report RF/126, February, 1943 ; and E. A. Finlay and W. B. Lasich, " The design of a klystron for the L-band," C.S.I.R. Radio Research Laboratory, Melbourne, Report RP 238, February, 1945.

previous theory we know that the fundamental component of current returning through the resonator will be

$$i_2 = 2i_0 J_1(x) \cos(\omega t + \phi),$$

where ϕ is a phase angle which, for the condition of maximum retardation of the electron bunches, is zero. At the time t the power delivered to the resonator is

$$P_1 = \beta V_1 \cos \omega t.\ 2i_0 J_1(x) \cos(\omega t + \phi)$$

$$= 2\beta V_1 i_0 J_1(x) [\cos^2 \omega t \cos \phi - \cos \omega t \sin \omega t \sin \phi]. \quad (42)$$

Now the klystron cavity may be regarded as equivalent to a simple parallel LRC circuit. The impedance Z of such a circuit, measured at the angular frequency ω, which we shall suppose to differ by a small amount $\delta\omega$ from the resonant angular frequency ω_0 of the circuit, may be expressed in the form

$$\frac{1}{Z} = \frac{1}{R} + 2j\delta\omega C \quad (43)$$

where R is the resistive impedance at resonance, and C is the circuit capacity, effectively the capacity of the resonator gap.

At the time t the power dissipated in the resistive losses of the resonator plus the power supplied to the electromagnetic field is

$$P_2 = V_1^2 \left[\frac{\cos^2 \omega t}{R} - 2\delta\omega C \cos \omega t \sin \omega t \right]. \quad (44)$$

For steady oscillations, $P_1 = P_2$. Equating like terms we have

$$2\beta i_0 J_1(x) \cos \phi = V_1/R, \quad (45)$$

$$\beta i_0 J_1(x) \sin \phi = \delta\omega C V_1. \quad (46)$$

To find the total frequency range over which the klystron will oscillate without retuning of the resonator, we must determine the starting condition. This is obtained from equations (45) and (46) by supposing the amplitude of oscillations to approach zero, in which case $J_1(x)$ may be replaced by $\frac{1}{2}x$. Making this substitution, and using equation (31) which gives $x = \beta g \omega \tau_0 V_1 / V_0$, we have

$$\beta^2 i_0 (g \omega \tau_0 / V_0) \cos \phi = 1/R, \tag{47}$$

$$\beta^2 i_0 (g \omega \tau_0 / V_0) \sin \phi = 2 \delta \omega C. \tag{48}$$

These two equations give $\delta \omega$, the maximum departure from the resonance frequency ω_0 for which oscillations can still be maintained. Dividing (48) by (47) we have

$$\delta \omega = \frac{1}{2CR} \tan \phi = \frac{\omega_0}{2Q} \tan \phi. \tag{49}$$

Eliminating ϕ,

$$\delta \omega = \frac{\beta^2 i_0 g \omega \tau_0}{2 C V_0} \sqrt{1 - \frac{1}{(R \beta^2 i_0 g \omega \tau_0 / V_0)^2}}. \tag{50}$$

Thus for a wide frequency range $\delta \omega$, the reflex klystron cavity should have a low Q and a small gap capacity C, although the shunt resistance R should not be too small; in addition we require large values of the gap parameter β and the beam current i_0, and a small accelerating potential V_0. Also, a large reflector gain factor g and a long drift time τ_0 are desirable. The latter implies a high mode number, and klystrons designed especially for AFC use are made with a relatively long reflector field space. As we have already seen, the maximum power output decreases with increasing mode number, so that output efficiency has to be sacrificed to some extent for a wide electronic tuning range. Some of the requirements deduced above are conflicting because a large beam current i_0 means a large diameter beam, and hence large diameter resonator grids; this tends to increase the gap capacity C. For a low C the resonator gap should be large, but a wide gap and low accelerating potential V_0 lead to a small gap efficiency factor β. Thus suitable compromises between the various requirements must be made in the design of the AFC klystron.

The frequency range $\delta \omega$ derived above is that corresponding to a change in output power from maximum to zero. The useful frequency range is that obtained with an output power change from maximum to a specified fraction of the maximum, usually taken to be one half. This useful range may be derived by a method similar to the one adopted for determining $\delta\omega$, and the same general requirements for maximum electronic tuning range are thereby deduced.

As an example of the AFC range obtained in practical cases we may quote the characteristics of the 723A reflex klystron already described. This valve has a frequency range of 20 to 30 megacycles per second between the half power points, with a corresponding change in reflector potential of 20 to 30 volts.

A new method of tuning the klystron cavity, suitable for AFC purposes, has recently been introduced, particularly for the very high frequency reflex klystrons. It is known as electronically controlled thermal tuning. The valve is constructed according to the 723A pattern, and the resonator is connected mechanically to a metal structure made up partly of a material such as stainless steel which has a high coefficient of expansion. This member can be heated by electron bombardment from a suitably placed auxiliary cathode, the tuner cathode, the current being controlled by the potential of a tuner grid. Expansion of the heated member causes a flexing of the resonator wall and a corresponding variation in the gap spacing. The system can be made to have a time constant sufficiently small for AFC applications to be possible.

CHAPTER XII

FREQUENCY CONVERTERS

THE well-known superheterodyne principle of reception is to combine the received signal with a locally generated oscillation so as to produce a third signal at a frequency usually referred to as the intermediate frequency. At the wavelengths used in radar, the generation of the local oscillation and the combining or mixing operation take place in separate units. The local oscillator has already been discussed (Chapter XI), and this chapter will be concerned with the mixer, which consists essentially of a non-linear element and associated radio frequency circuits. The term frequency converter is used to denote the combination comprising the local oscillator and mixer.

The first radar sets were designed for wavelengths in the metre region, where multi-grid mixer tubes of the type used at wavelengths greater than 10 metres had poor signal to noise ratios. Triode and diode mixers with lumped element circuits were therefore the usual practice. When the invention of the cavity magnetron enabled centimetre wavelengths to be exploited, the most important receiver advance was the successful development of mixing crystals for use with coaxial or waveguide circuits. This type of mixer is universally employed at wavelengths less than 10 centimetres, and the latest units are able to compete successfully with special tube mixers at any wavelength less than about 25 centimetres.

1. Frequency Conversion Theory

Elementary Mechanism

To produce frequency conversion it is essential for the mixer to contain an element which has a non-linear relation between current and applied voltage :

$$i = f(v)$$

where i and v refer to the variations from the steady current

and voltage values respectively. If v is the sum of a signal
of frequency ω_1 and a local oscillation of frequency ω_2, it
is clear that an element for which $f(v)$ is a linear function
will produce current variations containing only the original
frequency components ω_1 and ω_2. However, for a non-linear
function such as

$$i = a_0 + a_1 v + a_2 v^2$$

there will be generated, in addition to these frequencies,
current components having combination frequencies of value
$2\omega_1$, $2\omega_2$, $(\omega_1 - \omega_2)$ and $(\omega_1 + \omega_2)$. The difference frequency $(\omega_1 - \omega_2)$ is normally chosen as the intermediate
frequency and may be selected from the mixer by means of a
suitable filter. The amplitude of this component is proportional to the product of the signal and local oscillation
amplitudes, and whereas the former is normally very small,
a large local oscillation is always employed.

The essential mechanism of frequency conversion is similar
for other non-linear functions except that additional combination frequencies are produced. The complete theory of mixer
operation involves an analysis of the action of the mixing
element in association with the radio frequency and intermediate frequency circuits. Thus the impedances of the
various circuit elements to the frequency components in the
mixer and their interaction on one another are important
factors. Some general results of the theory are given later
in this section.

Performance Parameters

The performance of a mixer may be expressed in terms of
the two parameters, conversion gain and noise temperature.
The former is a measure of the efficiency with which the signal
frequency energy is converted to intermediate frequency
energy.   At long wavelengths it has been the custom to express
the conversion gain as a voltage ratio but at microwavelengths
a power ratio is more satisfactory.   Accordingly, the conversion
gain G is defined as the ratio of the output power at intermediate frequency to the input power to the mixer at the signal

frequency. The conversion gain may be greater than unity if the mixer contains a source of amplification. Mixers composed only of passive elements, however, have a maximum conversion gain of unity and in practice the value is always less.[1] A common practice is to express the conversion gain in decibels.

The ultimate performance of a receiver is limited by its total noise output. If the mixer is replaced by an equivalent resistor of resistance equal to the output impedance of the mixer at the intermediate frequency, the receiver noise output will be due partly to shot effect and partition noise in the tubes (primarily the first stage), and partly to the thermal agitation of the electrons in the equivalent resistor. The Johnson noise, as the latter component is commonly called, is equal to kTB where k is Boltzmann's constant, T is the absolute temperature of the resistor and B is the effective noise bandwidth of the receiver. It is found that in practice mixers generate more noise than the equivalent resistor at room temperature. The ratio of the available output noise power of the mixer to kTB is called the noise temperature T_N of the mixer. This term is used since it is numerically equal to the number of times the absolute temperature of the equivalent resistor would have to be increased to give the same noise output as the mixer at room temperature.

An additional source of noise in the mixer, which must be considered at very short wavelengths, arises in the local oscillator. This noise has a frequency distribution which is a maximum at the local oscillator frequency and falls off on either side owing to the selectivity of the resonant circuit of the local oscillator. The magnitude of this noise at signal frequency, therefore, depends on the separation of the signal and oscillator frequencies, that is on the intermediate frequency. It is not usually significant until wavelengths as short as a few centimetres are reached, when the intermediate frequency

---

[1] The word " gain " is used here in accordance with common usage and in the same sense as in the term amplifier gain. The reciprocal of the conversion gain as defined above is often called the conversion loss. It should be realised, that the words " gain " and " loss " used in this sense do not signify that the value of the quantity is zero in the case where the output power equals the input power. As stated above the value is then unity.

becomes a small percentage of the operating frequency. It may be reduced by increasing the intermediate frequency or practically eliminated by the use of recently developed types of balanced mixer described in section 5.

General Theoretical Results

The analysis of the operation of various types of mixer will not be reproduced here, as it has been adequately discussed in the literature. Instead, some general results of the theory for common types of mixer together with references, are given below.

The operation of triodes as mixers is now reasonably well understood. At wavelengths longer than a value which depends on the design of the tube a conversion gain greater than unity is possible. A full discussion of their performance may be found in a paper by Malter.[2]

The theoretical treatment of diode and crystal mixers is complicated by the large interaction between the input and output circuits. This is due to a reverse conversion process in which intermediate frequency energy is reconverted by the local oscillator to signal frequency energy. An analysis of diode operation has been given by James and Houldin and by Herold.[3] The conversion gain is usually much less than unity, which is in part due to circuit losses. However, even with zero circuit loss and with large local oscillator drive, the conversion efficiency depends on the impedance of the radio frequency circuits to the image frequency.[4] Herold, Bush and Ferris[5] have shown that the conversion gain even for zero circuit losses, cannot be increased above $\frac{1}{2}$ ($-$ 3 decibels) if the image frequency conductance of the input circuit is equal to the signal frequency conductance. A gain

[2] L. Malter, " General superheterodyne considerations at ultra-high frequencies," *Proc. I.R.E.*, Vol. 31, October, 1943, pp. 567-75.
[3] E. C. James and J. E. Houldin, " Diode frequency changers," *Wireless Eng.*, Vol. 20, January, 1943, pp. 15-27 ; and E. W. Herold " Frequency mixing in diodes," *Proc. I.R.E.*, Vol. 31, October, 1943, pp. 575-82.
[4] Image frequency is equal to the signal frequency plus or minus twice the intermediate frequency.
[5] E. W. Herold, R. R. Bush and W. R. Ferris, " Conversion loss of diode mixers having image frequency impedance," *Proc. I.R.E.*, Vol. 33, September, 1944, pp. 603-9.

of unity is possible only when the image frequency impedance is zero or infinite.

The theory of crystal mixer operation is very similar to the diode theory, the difference arising from the finite reverse conductivity of the crystal as compared with the essentially zero reverse conductivity of the diode. This leads to decreased conversion gain and increased local oscillator power consumption.

2. Tube Mixers

Triodes and Diodes

At wavelengths in the region of a metre, the use of triode mixers has been common, since a conversion gain greater than unity may be obtained. Fig. 247 is a typical circuit for

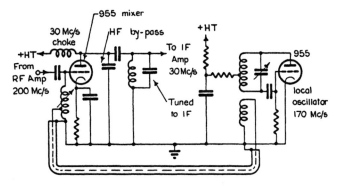

Figure 247.—Frequency converter for use at 1½ metres using a 955 acorn triode mixer. Energy from the type 955 local oscillator is injected into the grid circuit of the mixer.

operation at 200 megacycles per second. The mixer is a 955 acorn tube, and grid injection of the local oscillator energy is employed. One or more stages of radio frequency amplification are used at this wavelength to improve the signal to noise ratio (see Chapter XIII).

It becomes increasingly difficult to design such a receiver as the wavelength is reduced. The gain of the radio frequency amplifier falls to a point where immediate conversion gives better results. Ultimately, the gain of the triode mixer becomes so low that diode mixers can be designed to give superior results. The wavelength region in which these effects occur depends upon the tubes available, and the form of the radio

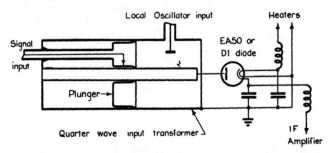

Figure 248.—Diode mixer suitable for operation at 50 centimetres. The tuned circuit is a resonant, quarter wave, coaxial line.

frequency circuits which may be used with them. When the acorn type tube represented the most advanced triode design, immediate conversion by a diode mixer was preferable at wavelengths of about 50 centimetres. Fig. 248 is the circuit

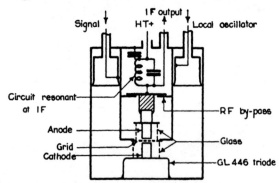

Figure 249.—A GL446 disc seal triode used as a microwave mixer. The construction of this type of tube permits operation at wavelengths as short as 10 centimetres.

of a 50 centimetre mixer using a type D1 or EA50 television diode.

Recently, disc seal triodes such as the CV90 and the GL446 "lighthouse"[6] have been developed for use with cavity

[6] E. D. McArthur, "Disc-seal tubes," *Electronics*, pp. 98-102, February, 1945.

resonator circuits as described in Chapter XI.   The application of this type of triode has reduced the wavelength at which gains greater than unity may be obtained from radio frequency amplifiers and mixers.   A typical arrangement using a GL446 tube is shown in Fig. 249.   This type of mixer may be used between a wavelength of about 10 centimetres and the wavelength at which the cavity type circuits become too bulky. The performance of crystal mixers is superior, however, at wavelengths less than 25 centimetres.

Other Mixer Tubes

A special type of tube which gives good results as a mixer is the beam deflection tube.   In this device the input signals are arranged to deflect an electron beam so that more or less of the beam current is deflected past an obstructing post on to a collector electrode.   Separate deflecting circuits are provided for the signal and local oscillator inputs, and it is possible to control the shape of the output-current versus deflecting-voltage characteristic by variation of the geometry of the obstruction.   The latest tubes of this type developed by the Radio Corporation of America have given good performance down to 10 centimetres.   However, the crystal mixer is still somewhat superior in this region, and as yet the beam deflection mixer has not found application in radar design.

Tubes employing the velocity modulation principle can be used as mixers and experiments have been made with various types.   They have proved to be much noisier than crystals, however, and the method is not employed in existing microwave practice.

3. Microwave Crystal Rectifiers

Construction

Crystal rectifiers developed for centimetre radar are constructed in cartridge form as shown in Fig. 250.   The rectifying contact is provided by a sharpened tungsten wire which bears on a piece of polished silicon.   Both are mounted in a small ceramic tube, the tungsten whisker being connected to a metal cap at one end, and the silicon to a metal pin at the other (in American crystals these connections are reversed).[7] A screw adjustment enables the pressure of the tungsten wire on the silicon to be varied during manufacture. When a satisfactory rectifying action has been achieved, the cartridge is impregnated with a filler through a small hole in the ceramic tube. This prevents the entrance of moisture and ensures a robust unit by increasing the mechanical stability of the contact.

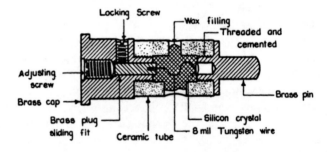

Figure 250.—Sectional view of the tungsten-silicon crystal cartridge developed in Britain for microwave operation. The drawing is 2¼ times full size.

Rectification by Crystals

The fundamental theory of crystal rectification has been the subject of much research which will not be discussed here.[8] To a great extent however, the successful performance of mixer crystals is the result of the accumulation of a vast amount of empirical data on the preparation and treatment of the silicon and the polishing and etching of the contact surface. It has been found for example that the presence of minute

[7] B. Bleaney, J. W. Ryde and T. H. Kinman, "Crystal valves," *J. Instn. Elect. Engrs.*, Vol. 93, Part III A, 1946, pp. 847-54; J. H. Scaff and R. S. Ohl, "Development of silicon crystal rectifiers for microwave radar receivers," *Bell Syst. Tech. J.*, Vol. 26, 1947, pp. 1-30; W. E. Stephens, "Crystal rectifiers," *Electronics*, Vol. 19, 1946, pp. 112-9.

[8] S. J. Angello, "Semi-conductor rectifiers," *Elect. Engng.*, Vol. 68, 1949, pp. 865-72.

amounts of such impurities as aluminium and boron in the silicon have a marked effect on the results. Experiments on the type and amount of impurity to be added have resulted in improved performance.

Equivalent Circuit

Fig. 251 is the equivalent circuit usually used to represent the rectifying contact. C is the capacity of the boundary layer and is in parallel with a resistance R representing the

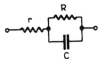

Figure 251.—The equivalent circuit of the rectifying contact of a crystal rectifier. The resistance R representing the contact is large for voltages in one direction and small for those in the other.

contact itself. The resistance R is large in one direction and small for sufficiently large voltages in the other. In series with both is the bulk resistance r of the silicon.

The capacity C, which may be of the order of 1 picofarad, is one of the limiting factors in using these units at higher and higher frequencies, since it shunts the resistance of the contact in the backward direction. Various shapes of cats-whisker tip are used (conical, wedge shaped, etc.), but the reduction of C by reduction of contact area is complicated by the necessity for low current densities to prevent the crystal being "burnt-out" too easily by leakage power from the aerial duplexing switch.

The resistance r represents a loss of power by heating, and, for good power conversion efficiency, should be small.

Welded Contact Crystals

Recent research[9] on crystals using germanium as the semi-conductor showed that unusual properties could be obtained by welding the whisker to the germanium at the point of

[9] H. Q. North, "Properties of welded contact germanium rectifiers," *J. Appl. Phys.*, Vol. 17, 1946, pp. 912-23.

contact. The welding was produced by the passage of a large, momentary DC current. This resulted in a value of r of about 3 ohms, which is about one tenth of the value for normal silicon crystals.

Measurements show that the conversion efficiency is very high but unfortunately the crystals are found to be much noisier than the usual tungsten-silicon type.

Crystal Burn-out

As stated in the chapter on aerial duplexing, crystals may be " burnt-out " by excessive TR leakage power. This does not necessarily mean that the crystal ceases to function entirely, although this might happen with high enough leakage levels. However, if leakage exceeds the limits discussed in Chapter IX, both conversion gain and noise temperature may be impaired, and the receiver signal to noise ratio then decreases.

Since "spike" energy in the TR leakage pulse is now believed to be the main factor causing impairment, crystals are subjected after manufacture to a burn-out proof test, before acceptance measurements of conversion loss and noise temperature are made. The proof test consists in exposing them to short DC pulses of the order of duration of the spike, and of energy 0·3 to 2·0 ergs depending on the application for which the crystals are intended. They are then classified as medium or high burn-out types respectively.

Apart from the effects of TR leakage power, crystals may also be damaged by improper handling. Static charges accumulated by an operator in a dry atmosphere must be discharged before the crystal is inserted in a mixer. Connecting leads must be shielded against induced voltages from neighbouring electrical apparatus and when not in the mixer, the crystal cartridge should be wrapped in metal foil, if likely to be exposed to intense radio frequency fields.

Performance Figures

It is usual to rate a crystal cartridge intended for a particular wavelength as having a certain conversion gain and noise temperature at a specified rectified DC current. This is the performance which would be obtained in a mixer with zero circuit losses and having matched input and output impedances. The circuit losses of properly designed microwave mixers are negligible, so that the crystal performance is the performance of the complete mixer. Both the conversion gain and noise temperature of a crystal are functions of the rectified DC current through the crystal resulting from the injected local oscillator power. Increasing the latter increases the conversion gain, but makes the crystal noisier so that the noise temperature increases. There is, therefore, an optimum value of rectified current for which the overall noise factor of the receiver is a minimum. This value usually varies from 0·3 to 1 milliampere depending on the individual crystal and the noise factor of the intermediate frequency amplifiers. The local oscillator power input corresponding to this current range is about 1 milliwatt.

The following table gives the present rejection limits of conversion gain and noise temperature for various crystals. The performance of the average crystal is somewhat better.

TABLE 7—CRYSTAL PERFORMANCE

Wavelength		Conversion gain G (decibels)	Noise Temperature T_N (times)
10 centimetres		−6·5	2
3 centimetres	high burn-out	−8·0	2·7
	medium burn-out	−6·5	2·7
about 1 centimetre		−8·5	2·5

4. Design Factors

The design of a crystal mixer involves several factors, some of which are fundamental to all mixers, whereas others are imposed by the requirements of particular radar sets.

Input Impedance Matching

For maximum conversion gain all the signal energy must be transferred to the crystal. This requires that the input impedance to the mixer at the signal frequency be matched

to the characteristic impedance of the input feeder, which may
be coaxial or waveguide. The input impedance is a function
not only of the crystal itself but also of the geometry of its
surroundings, and must be specified for a particular pair
of terminals in the mixer. It is found to vary considerably
from crystal to crystal.

The process of design is then to measure, in a particular
mixer, the complex impedance of a large number of crystals
and to select an " average crystal " from among them. By
adjustments of the mixer dimensions or by the addition of a
suitable matching transformer, the average impedance of the
mixer with this crystal may then be matched to the input
transmission line. Provided the scatter of impedances about
the average is not too large (say, a voltage standing wave
ratio on the feeder of 0·5 or greater), the reflection loss with
the extreme crystals in the batch will be small (less than
0·5 decibels) and no tuning adjustments need be provided. The
unit is then known as a preplumbed or fixed-tuned mixer and
is good practice in radar design. This is partly because tuning
controls are subject to misadjustment, and partly because the
leakage power from a TR switch is dependent on the impedance
presented to it by the crystal during the transmitter pulse.
Hence tuning operations may lead to wide variations of leakage
power and possible crystal burn-out. Sometimes, however,
the impedance scatter may be too great to be ignored and a
tunable mixer is necessary. The latter type of mixer is often
desirable for use in laboratory measuring apparatus. In
particular cases, it has been found possible to use the tuning
control on the TR cavity to assist in adjusting the crystal
impedances to the matched condition.

During the transmitter pulse, leakage power from the TR
switch changes the crystal impedance from its low level value.
This causes the leakage power to vary from the value available
to a matched load, and it is desirable to design the mixer so
that the change of input impedance is such as to cause a
decrease in leakage power. In this way, the alteration of
crystal impedance with power level gives a degree of self-
protection. In most cases, the required conditions may be

achieved simply by proper choice of the length of transmission line between TR switch and crystal. This will not affect the impedance presented to the switch at low signal levels, since the crystal in the latter case matches the impedance of the line.

Local Oscillator Injection

The local oscillator power is usually coupled into the mixer by means of an adjustable probe, loop or window. This coupling must not be too tight, otherwise some of the signal energy in the mixer will be lost by transfer to the local oscillator circuits. It has been shown[10] that the fraction of the incoming signal power which is lost in these circuits is equal to the utilised fraction of the local oscillator power available at the mixer. Hence if 5 per cent loss of signal power can be tolerated, the coupling between the mixer and local oscillator must be weak enough to inject only 5 per cent of the power available at the terminals of this coupling. Since the injected power must be about 1 milliwatt, the available local oscillator power at the terminals must be not less than 20 milliwatts.

Owing to the necessarily weak coupling, the resulting load on the local oscillator is largely reactive so that its output may vary greatly with tuning, with consequent difficulty of adjustment. An improvement in this respect results if some of the local oscillator power is fed into a dummy load, or if attentuation is inserted between the oscillator and the mixer coupling terminals. Both these methods require increased oscillator power to maintain the same available power at the mixer. A better solution has recently been obtained, with the development of the " magic tee " type mixer described in section 5.

Radio Frequency Filtering

The difference or intermediate frequency is selected from the mixer by means of a filter which must prevent leakage of radio frequency energy. A simple by-pass capacity is some-

[10] R. V. Pound, *Microwave Mixers* (Massachusetts Institute of Technology, Radiation Laboratory Series, Vol. 16), McGraw-Hill, New York, 1948. pp. 136-40.

times used, but more effective are filters making use of sections of transmission line which are a quarter wavelength long at the radio frequency. Since the connection between the mixer and the intermediate frequency amplifiers is normally coaxial in form, the quarter-wave filters may be arranged in either of the two ways shown in Fig. 252. In each case the short circuited sections ABC are resonant at the radio frequency and introduce an impedance which is nominally infinite to this frequency between the points AC in series with the coaxial line. These sections have a negligible series impedance to the intermediate frequency. Hence, over a band centred on the resonant frequency, the leakage of radio frequency energy is prevented to an extent which depends on the number of filter

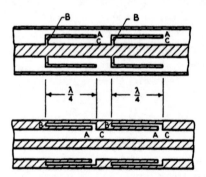

Figure 252.—Two methods of constructing coaxial, quarter wave filter sections. This type of filter is often used to prevent leakage of RF energy from a mixer into the IF circuits.

sections used. Normally one section is sufficient. The design of a radio frequency filter must take into account the output capacity to the intermediate frequency. It is desirable to keep this capacity to a minimum when wide band amplifiers are required.

Special Applications

Modern radar converters are usually required to perform other functions in addition to frequency conversion of the received echo signal. It is common in microwave radar equipments to use automatic frequency control (AFC) circuits to maintain correct tuning of the local oscillator relative to the magnetron frequency. The discriminator circuit of the AFC unit may be operated by the intermediate frequency output from the signal mixer during the transmitter pulse. This output however, contains the characteristic " spike " generated by gas discharge TR switches (see Chapter IX, section 5) and the frequency components in this spike tend to confuse the discriminator operation. It has become the practice, therefore, to provide a second mixing chamber and crystal coupled to a common local oscillator. A very small fraction of the transmitter power is transferred from the main feeder to this AFC mixer and after conversion to the intermediate frequency, operates the discriminator. The modulation envelope of this pulse is merely that of the transmitter pulse, so that the unwanted " spike " frequencies are absent.

When the radar set is required to operate with beacons (see Chapter XVII), the signal received from the latter in response to the transmitter pulse is often on a frequency different from that of the radar. A second local oscillator coupled to the signal crystal is then provided, and is tuned to the correct frequency relative to the beacon frequency.

With the additional facilities described above, the complete radar frequency converter unit comprises a signal frequency mixer with echo and beacon local oscillators and an AFC mixer.

5. Typical Crystal Converters

Adjustable Mixer

Fig. 253 shows an early design of radar mixer in which the crystal is plugged into a quarter-wave coaxial cavity tuned by the capacity between an adjustable plate and the central conductor. The signal is coupled into the cavity by means of a probe, the insertion of which provides a second matching adjustment. Another adjustable probe, loosely coupled, injects the local oscillator power. The mixer is often mounted directly on the intermediate frequency preamplifier chassis, so that a short connection is possible to the mixer output

terminal on the bottom of the unit. This output is by-passed
to radio frequency energy by means of the capacity shown.
Connection between the probes and the rest of the radio fre-

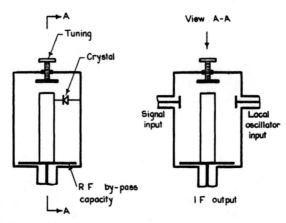

Figure 253.—An early type of adjustable crystal mixer for microwave
operation. The variable capacity tunes the cavity to resonance.

quency system is by means of flexible feeder. Although
superseded in radar practice by other types, this form of
mixer is often convenient for laboratory experimental work.

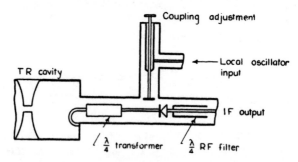

Figure 254.—Coaxial crystal mixer for direct connection to the TR
switch cavity. All adjustments, except the local oscillator coupling,
are fixed.

Coaxial Type Mixers

Fig. 254 is an example of a coaxial-type mixer which is mounted
rigidly on the TR switch cavity. The signal energy, coupled

out of the TR cavity by means of a loop, is matched to the
" average " crystal impedance by means of a quarter-wave
transformer and no tuning adjustments are provided.   Other
types of fixed impedance transformer might of course be
used.   The local oscillator power is injected by means of an
adjustable probe and the output, after passing the quarter-
wave radio frequency filter, is fed to the intermediate frequency
amplifier by a length of flexible coaxial cable.   Access to the
crystal cartridge is obtained by unscrewing the filter unit.

Waveguide Mixers

A waveguide crystal converter for use in the 3 centimetre
wavelength region is shown in Fig. 255.   In this case the

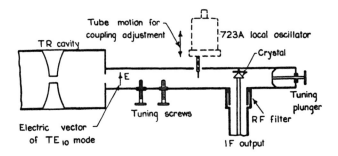

Figure 255.—Crystal converter for 3 centimetre operation using the
TE_{10} mode in rectangular waveguide.   The local oscillator is a low-
voltage, reflex klystron.

crystal is mounted perpendicular to the axis of the guide.
The cap of the crystal is earthed and the pin plugs into the
centre conductor of the coaxial intermediate frequency output
feed, which is filtered as before by a quarter-wave section.
A waveguide plunger behind the crystal and a pair of matching
screws provide tuning adjustments, if the scatter of crystal
admittances is too great for preplumbing.   Other standard
forms of fixed or variable matching transformer may be used.
The local oscillator is a 723A low voltage klystron, which has
been described in Chapter XI.   This tube is mounted in a
socket attached to the mixer so that the small output aerial

projects into the waveguide through a hole. Sometimes a quarter wave filter is incorporated in this aperture to prevent leakage. A screw mechanism enables the tube to be moved up or down so that the insertion of the aerial is varied. This provides a control of oscillator coupling. A waveguide mixer of this type is coupled directly to the TR cavity by means of a window as shown.

Fig 256 illustrates a method of local oscillator injection which gives improved loading of the oscillator. The oscillator

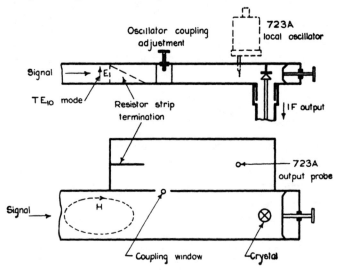

Figure 256.—Waveguide converter with an added resistive load for the local oscillator. This improves the stability of operation during tuning adjustments.

probe feeds an auxiliary waveguide which is terminated in its characteristic impedance. A portion of the power flowing down this auxiliary guide is coupled into the mixing chamber through a window and the coupling is adjusted by the insertion of a screw as indicated.

AFC and Beacon Facilities

As previously described, some frequency converters are required to operate on either echoes or beacon signals and to provide an output suitable for automatic frequency control. The converter described in the previous paragraph (Fig. 256) may be developed to include these facilities by the addition of a beacon local oscillator and an AFC mixer, as shown in Fig. 257. A third crystal and the cavity shown coupled

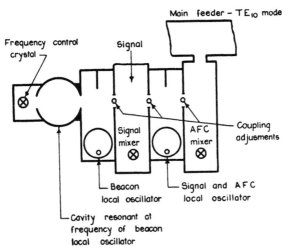

Figure 257.—Waveguide crystal converter having a separate AFC mixer and a beacon local oscillator, the frequency of which is controlled by the preset resonant cavity.

to the beacon oscillator represent a method of stabilising the frequency of the latter relative to the beacon frequency. The resonant frequency of the cavity is accurately preset to the required oscillator frequency. In conjunction with suitable circuits, the crystal output may then be used to supply a correcting voltage to the electronic tuner of the oscillator whenever its frequency differs from resonant frequency of the cavity.

Balanced Mixers

Various forms of balanced mixer using two crystals have been devised. The one which has received the most development uses the magic tee junction described in Chapter VII.

Fig. 258 shows the arrangement. There is no mutual coupling between the E and H branches of the junction. Hence, if the echo signal is fed into the E branch and the local oscillator output into the H branch there will be no loss of signal energy

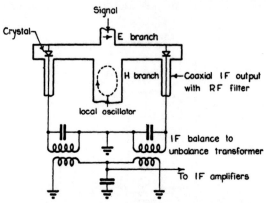

Figure 258.—A balanced crystal mixer using a " magic-tee " waveguide junction, the four branches of which are shown in the one plane for convenience.

to the oscillator circuits. Also all the available power from the local oscillator is utilised, one half going to each crystal. As the latter represents a fairly well matched load, the operation of the oscillator is stable with tuning.

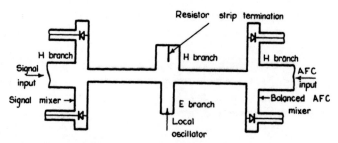

Figure 259.—A waveguide unit which has balanced signal and AFC mixers and uses three " magic-tee " junctions.

Another excellent feature is that noise originating in the local oscillator may be practically eliminated in the output. Signals from the E branch excite in-phase currents in the two crystals, whereas noise from the H branch excites out-of-phase currents. The intermediate frequency outputs from the crystals are combined in the push-pull transformer shown. By winding the latter in the proper sense, the signal currents may be made to add and the noise currents to cancel. This feature is especially important at very short wavelengths as discussed in section 1, p. 391. It has been found that small differences in the characteristics of the two crystals do not affect the cancellation of the noise to any significant extent.

Fig. 259 shows the way in which this circuit is developed to include an AFC mixer. Cross coupling between the signal and AFC mixers is inherently small.

6. Measurement of Performance

Conversion Gain

Measurement of conversion gain may be made in a straight-forward fashion by noting the amount of power from a radio frequency signal generator which must be injected into the mixer to give a certain receiver output. If the frequency converter is then replaced by an intermediate frequency

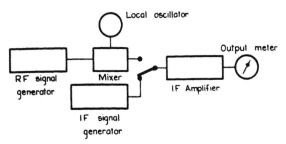

Figure 260.—Apparatus for the measurement of the conversion gain and the noise temperature of a mixer. The RF signal generator is used only as a matched load when noise temperature measurements are made.

signal generator, and the power output of the latter adjusted to give the same receiver output reading, the conversion gain is given by the ratio of the intermediate to the radio frequency power. This measurement is illustrated diagrammatically in Fig. 260. In practice the method requires great care in matching both the input and output impedance of individual crystals.

Simplified methods of measuring the conversion gain have been devised but cannot be described in detail here.[11]   One of these depends on the fact that the gain correlates quite closely with the rectified direct current produced by a CW radio frequency oscillator matched to the mixer.   In the second method, the oscillator is modulated at a low frequency and the amplitude of this frequency component across a suitable load in the mixer output is measured.   In both methods, the local oscillator is of course switched off.

Noise Temperature

The noise temperature of a mixer may be measured as follows (see Fig. 260).   With the mixer operating under normal conditions, the receiver gain is adjusted to give a convenient noise output reading.   The mixer is then replaced by an intermediate frequency signal generator modified to have the same output impedance as the mixer, so that its noise output when switched off is equal to kTB.   It is then switched on and the output increased until the original receiver noise output reading is duplicated.   The power delivered by the generator is then equal to the excess noise power of the mixer above kTB.

$$\text{Hence } T_N = \frac{\text{noise output of mixer}}{kTB}$$

$$= \frac{kTB + \text{excess noise}}{kTB}$$

$$= \frac{1 + \text{power delivered by generator}}{kTB}.$$

For this measurement the radio frequency signal generator in

[11] H. C. Torrey and C. A. Whitmer, *Crystal Rectifiers* (Massachusetts Institute of Technology, Radiation Laboratory Series, Vol. 15), McGraw-Hill, New York, 1948, pp. 198-232.

Fig. 260 is switched off and merely acts as a load matched to the mixer input.

Alternatively, if the noise factors of the intermediate frequency amplifier (N_{IF}) and of the overall receiver (N_R) are known, and the conversion gain (G) is measured, T_N may be calculated from the known relation between these quantities.[11]

$$N_R = \frac{T_N + NF_{IF} - 1}{G}.$$

Mixer Input and Output Impedance

A knowledge of these parameters is important both in design and the measurement of mixer performance.

The input or radio frequency impedance of a mixer may be measured by standard microwave impedance methods. These techniques usually involve the use of some form of standing wave measurement and have already been described in Chapter VII.

A measurement of the output impedance of a mixer at the intermediate frequency may be made by means of a suitable impedance bridge with the local oscillator switched on. Alternatively, sufficient external resistance may be inserted in series with the mixer output to reduce the rectified direct current to half its value when the output is short-circuited. This value of resistance is then equal to the output impedance of the mixer.

CHAPTER XIII

AMPLIFIERS

THE amplification of the signals received by a radar system is divided between amplifiers operating at signal, intermediate and video frequencies. The general considerations involved in the design of these units have been outlined in Chapter X. In the present chapter, each type of amplifier is discussed in more detail, following introductory sections on certain fundamental aspects.

1. Sensitivity

As explained in Chapter X, the ultimate performance of an ideal receiver in which the valves generate no noise is determined by the thermal radiation absorbed by the aerial. When

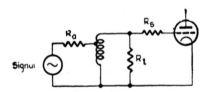

Figure 261.—The equivalent circuit of the input stage of a receiver. The sources of noise are the aerial resistance R_a, and the valve noise represented by fictitious resistors R_t and R_s.

the latter is in radiative equilibrium with its surroundings, the noise power at the receiver input is the Johnson noise in a resistor at ambient temperature whose resistance is equal to the radiation resistance ; the noise factor of such a receiver is unity.

The noise factor of a practical receiver is greater than unity because of noise generated in the valves. It is usual to represent the various sources of noise in a valve by the fictitious resistors shown in Fig. 261, which is the equivalent circuit of the input stage of a receiver. The resistor R_t is equal in value to the input resistance to the grid due to electronic loading. By assigning an effective temperature greater than room temperature to this resistor, its thermal noise may be taken to represent the induced grid noise of the valve.[1] The resistor R_s in series with the grid and at room temperature is equivalent to all other noise sources in the valve, such as shot effect, electrode partition noise, etc.[2]

For all practical purposes the overall noise factor of the receiver is determined by the noise generated in the input stages because the gain of these stages raises the signal and noise powers to levels where subsequent noise contributions are insignificant. For optimum noise factor, the input circuit should be mismatched to the generator,[3] but unless the noise factor is very low, the advantage to be gained is small. Thus at centimetre wavelengths a match is sufficient, but at metre wavelengths a special adjustment is usually called for. The same considerations apply in the case of an input direct to a crystal mixer, but here the first stage of the intermediate frequency amplifier must also be designed for best noise-factor.

The overall noise factor N_{rec} of the receiver[3] is given by

$$N_{rec} = N_1 + \frac{N_2 - 1}{G_1} \qquad (1)$$

where N_1 is the noise factor of the input stage, G_1 is the power gain of the input stage and N_2 is the noise factor of the rest of the receiver. This expression shows that the noise factor of the receiver approaches that of the first stage if the gain of the latter is sufficiently large.

The noise generated by a valve when used as a signal-frequency amplifier is less than that of the same valve used as a mixer.[4] Thus, as shown by equation (1), the noise factor of a receiver may be reduced by using one or more stages of signal-

[1] D. O. North and W. R. Ferris, " Fluctuations induced in vacuum tube grids at high frequencies," *Proc. I.R.E.*, Vol. 29, February, 1941, pp. 49-50.

[2] W. A. Harris, " Fluctuations in space-charge limited currents at moderately high frequencies, Part V," *R.C.A. Rev.*, Vol. 5, April, 1941, pp. 505-24.

[3] H. T. Friis, " Noise figures of radio receivers," *Proc. I.R E.*, Vol. 32, July, 1944, pp. 419-22.

[4] L. Malter, " General superheterodyne considerations at ultra-high frequencies," *Proc. I.R.E.*, Vol. 31, October, 1943, pp. 567-75.

frequency amplification provided the gain of each amplifying stage is large which will be so if the signal frequency is low enough. If the signal frequency is so low that the circuits of the receiver have to be " loaded down " with external shunt resistance in order to achieve the desired bandwidth, the gain of the valve as an amplifier is higher than its gain as a mixer.[4] As the signal frequency is increased, however, the external loading necessary to give the desired bandwidth decreases and becomes zero at a certain frequency owing to the increase in electronic loading. Above this frequency, the gain of an amplifying stage falls faster than the gain of a mixer stage because the anode load impedance of the amplifier decreases with increasing signal frequency, whereas that of the mixer remains constant since it operates at intermediate frequency. Ultimately the gain of the amplifying stage is reduced to the point where, despite the lower noise resistance of the valve, a better receiver noise factor is possible by immediate conversion. The frequency at which this happens depends on the valves which are available.

2. The Product of Bandwidth and Stage Amplification

The amplifying valves in a radar receiver operate under conditions in which the amplification of each stage is limited by the required bandwidth. This restriction is equally true for television or any other application where the circuits are required to have comparable speeds of response. Although developed primarily for radar purposes, the techniques described in this chapter consequently apply to a much wider field.

Wheeler[5] has shown that the theoretical limit of stage amplification for a given bandwidth is

$$A = \frac{g_m}{\pi BC} \tag{2}$$

where g_m is the transconductance of the valve, B is the bandwidth, and C is a capacitance which is a function of the anode

[5] H. A. Wheeler, "Wide band amplifiers for television," *Proc. I.R.E.*, Vol. 27, July, 1939, pp. 429-37.

capacitance of the valve, the input capacitance to the following stage and wiring capacitances. The limit of amplification for a given valve may be approached quite closely by using suitable interstage coupling networks. Less complex coupling circuits can achieve at most half the ultimate performance, but are often used for simplicity or convenience. For any given valve and type of coupling network the product of amplification and bandwidth is a constant ; it is also a constant fraction of the theoretical limit. The designer has then to choose his valves and circuits to realize a stage performance which will satisfy the overall amplification and bandwidth requirements.

Sections 4 and 5 deal with the design aspects and relative merits of various circuits for intermediate- and video-frequency amplifiers. Similar considerations apply to the amplification and bandwidth characteristics of signal-frequency amplifiers. However, because of electronic loading, the bandwidth of a signal-frequency amplifier nearly always exceeds the requirement and is not an important factor in the design.

3. Signal Frequency Amplifiers

The limitations imposed by electronic loading on the highest frequency at which a valve may be operated as a signal-frequency amplifier have been discussed in section 1. This limit is also considerably influenced by the physical form of the valve. In the microwave region the radio frequency circuit with which the valve is associated is usually a section of transmission line or a resonant cavity. It is essential that the valve should be constructed so that the connection to this circuit involves as little discontinuity as possible. This requirement leads ultimately to designs in which the circuit and valve can be distinguished electrically only because the former is the region where electromagnetic energy is stored and the latter the region where the fields and electrons interact.

Pentodes such as the 954 acorn and the 6AK5 miniature type have often been used as signal-frequency amplifiers at frequencies as high as 200 megacycles per second. It is well known, however, that, owing to the absence of screen partition noise, the value of the noise resistance R_s is much less for triodes than for pentodes. For example, Burgess,[6] who has experimentally confirmed the theoretical expression for R_s, gives values for a high-slope triode and a pentode of 310 ohms and 880 ohms respectively. It was difficult in the past to take advantage of the better noise characteristics of triodes because their large grid-anode capacitance made some form of stabilisation necessary. Neutralisation was sometimes employed for this purpose but was found to be rather critical at frequencies of 200 megacycles per second. A recent method of ensuring the stability of triodes is to use a grounded-grid circuit in which the grid is connected to ground and is designed specially to act as a shield between cathode and anode.[7] The input signal is applied between the cathode and ground and the load impedance connected between anode and ground. The noise resistance R_s under these conditions is only slightly greater than it is for the same valve connected as a conventional amplifier. The input impedance of a grounded-grid amplifier is inherently low (see below) but it may be of the same order as or even greater than the grid input impedance of an ordinary amplifier at higher frequencies due to the severe effects of electronic loading on the latter. In general, grounded-grid triode amplifiers are to be preferred to grounded-cathode pentodes above 100 megacycles per second but the latter are used below this frequency because electronic loading is no longer excessive.

When the frequency is higher than about 500 megacycles per second, the construction of valves such as the 954 and the 6AK5 is inadequate owing to lead inductance and other sources of stray reactance. To overcome these difficulties, disc-seal triodes such as the GL446,[8] and the CV90 were developed.

[6] R. E. Burgess, Signal/noise characteristics of triode input circuits," *Wireless Engineer*, Vol. 22, February, 1945, pp. 56-61.

[7] M. Dishal, "Theoretical gain and signal to noise ratio of the grounded grid amplifier at ultra-high frequencies," *Proc. I.R.E.*, Vol. 32, May, 1944, pp. 276-84; and M. C. Jones, "Grounded grid radio frequency voltage amplifiers," *Proc. I.R.E.*, Vol. 32, July, 1944, pp. 423-9.

[8] E. D. McArthur, "Disc-seal tubes," *Electronics*, Vol. 18, February, 1945, pp. 98-102.

These valves, which have been described in Chapter XI have disc connections to the electrodes and are intended for use with coaxial or other form of cavity resonator. Close spacing of the electrodes ensures that transit-time effects are as small as possible. The grid of the disc-seal triode (Fig. 262) completely shields the grid-cathode cavity from the grid-anode cavity so that the operation is of the grounded-grid type.

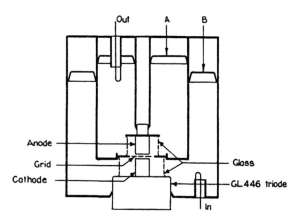

Figure 262.—Signal frequency amplifier using a disc seal triode (GL446) in a grounded-grid circuit. A and B are tuning plungers. This diagram indicates the RF arrangements only and shows no provision for the introduction of DC potentials.

With this type of valve it is possible to design amplifiers with substantial gain at frequencies as high as 3,000 megacycles per second.

Characteristics of the Grounded-grid Amplifier

Fig. 263 shows a single-stage grounded-grid amplifier ; the amplification ratio is

$$\frac{V_2}{V_1} = \frac{R_L}{\rho + R_L} (\mu + 1) \tag{3}$$

where μ is the amplification factor and ρ is the anode resistance, or $(\mu + 1)/\mu$ times that for a grounded cathode amplifier. Thus both types of circuit have essentially the same amplification ratio for large values of μ.

The input resistance R_t is low and is given by

$$R_i \doteqdot \frac{\rho + R_L}{\mu + 1} \tag{4}$$

so that the voltage step-up obtainable in the input transformer is correspondingly small.   The input resistance given by the formula is due to anode-circuit impedance " reflected " into the input circuit.   The formula neglects circuit losses and transit-time conductance.   At frequencies where the latter assumption is untrue the input resistance becomes the parallel combination

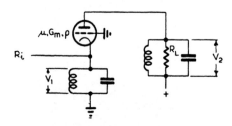

Figure 263.—A single stage grounded-grid amplifier.   The resistance between cathode and ground is R_i (equation 4) and the input resistance of the stage is R_i transformed to any desired value by the input circuit.

of R_i, and R_t the resistance due to transit-time loading.   In practice, valves may have sufficient cathode-lead inductance to change this input resistance.   As a result, although theoretical figures are a valuable guide, it is necessary in most receivers to design the input circuit experimentally for the most favourable performance.   This occurs when the input impedance is slightly mismatched to the generator.

For $\mu \gg 1$, the optimum load impedance (Jones,[7] p. 429) is given by

$$R_L = \rho \sqrt{1 + R_t g_m} . \tag{5}$$

If the succeeding valve is a pentode, a π matching network as shown in Fig. 264 (a) is suitable for transforming the grid input resistance R_g of the pentode to a value approaching the optimum.   When R_g is low it is desirable to make C_2/C_1 as large as possible ; however an upper limit is imposed by the required bandwidth.   Tuning adjustments may be made by a trimming condenser in parallel with C_2, or a large variable condenser in series with L. When a grounded-grid stage is followed by another of the same type, the transformation ratio is large and the circuit of Fig. 264 (b) has been found very suitable. The inductance L_2 is small and may be a short metal strip or a piece of thick wire.

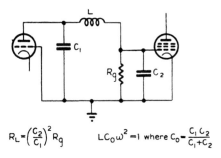

$$R_L = \left(\frac{C_2}{C_1}\right)^2 R_g \qquad LC_0\omega^2 = 1 \text{ where } C_0 = \frac{C_1 C_2}{C_1 + C_2}$$

Figure 264 (a).—A circuit suitable for transforming the grid input resistance of a pentode to a value R_L which is a suitable load impedance for the grounded-grid stage.

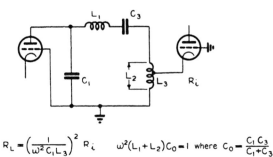

$$R_L = \left(\frac{1}{\omega^2 C_1 L_3}\right)^2 R_i \qquad \omega^2(L_1 + L_2)C_0 = 1 \text{ where } C_0 = \frac{C_1 C_3}{C_1 + C_3}$$

Figure 264 (b).—An impedance transforming circuit which is preferred when the following stage is also a grounded-grid triode.

The noise factor (Jones,[7] p. 428) of a grounded-grid amplifier (Fig. 265) for the matched input condition is

$$N = \sqrt{2 + \frac{R_1}{R_0 A_1^2} + \frac{R_2}{R_0 A_1^2 A_2^2}} \qquad (6)$$

where R_1 is the equivalent noise resistance of the first valve,

R_2 is the equivalent noise resistance of the second valve, R_0 is the generator impedance, A_1 is the amplification of the input circuit ($= V_1/V_0$) and A_2 is the amplification of the first stage ($= V_2/V_1$).

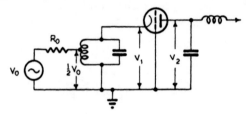

Figure 265.—Grounded-grid input stage. The voltages V_0, V_1, V_2 are used in deriving equation (6) which is based on the matched input condition.

Other analyses of grounded-grid amplifiers have been published[9] which include the effect of transit-time loading on noise factor and power gain. The reader is referred to these papers for a more detailed treatment.

A circuit recently developed in America uses two triodes and enables the low noise properties of triodes to be realised

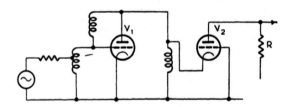

Figure 266.—A low noise input circuit which uses a neutralised triode followed by a grounded-grid triode.

without suffering the low input impedance of a grounded-grid circuit. Fig. 266 shows the arrangement. The triode V_1 is operated with its cathode grounded and is neutralised. However, stability is ensured primarily by using the input imped-

[9] J. Foster, "Grounded-grid amplifier valves for very short waves," *J. Instn. Elect. Engrs.*, Vol. 93, Part IIIA, 1946, pp. 868-47; S. G. Tomlin, "Operation of the earthed grid amplifier at very high frequencies with large transit angle," Standard Telephones and Cables Valve Laboratory Report G32, September, 1941.

ance to V_2 as the load impedance of V_1 and, since V_2 is a grounded grid triode, this input impedance is very low. The current amplification of V_2 is approximately unity so that the amplification of the two stages together is $g_m R$, which is the amplification the first valve would have with a load resistance R. The combination may be regarded as the equivalent of a single grounded-cathode triode with zero grid-anode capacity, and a transconductance and noise resistance equal to those of V_1.

Table 8 shows typical noise factors for amplifiers operating between 30 and 200 megacycles per second. Numerical values are given for various types of valves and circuits. Although given here to illustrate the performance of signal-frequency amplifiers these noise factors apply equally well to the performance of intermediate-frequency amplifiers operating at the stated frequencies.

TABLE 8—TYPICAL NOISE FACTORS FOR AMPLIFIERS OPERATING BETWEEN 30 AND 200 MEGACYCLES PER SECOND

Type of circuit	First stage	Second stage	Noise factor of amplifier (decibels above kTB)		
			30 Mc/s	70 Mc/s	200 Mc/s
Pentodes	6AK5	6AK5	3	4·5	12
	6AC7	6AC7	5·5	8	
	954	954			13
Grounded-grid triodes	6J4	6J4			8
	RL37	RL37		4	9
Grounded-grid triode and pentode	RL37	954			10
	GL446	954			10
Neutralised triode and grounded-grid triode	6J4	6J4	2		6
	6J6	6J6	2		
	6AK5	6J4	2·5		
	6AK5	6J6	2·5		

Using a GL446 disc-seal triode, a single-stage amplifier may be designed to have a noise factor of about $11\frac{1}{2}$ decibels and a power gain of 10 times at 3000 megacycles per second. As indicated by equation (1) the overall noise factor of a receiver using a signal-frequency amplifier of this type would be at least $11\frac{1}{2}$ decibels, but at present it is possible to obtain lower values with receivers in which the first stage is a crystal converter. Crystal converters are therefore preferred in the region of 3000 megacycles per second, but disc-seal amplifiers of the type described become of value at frequencies below about 2000 megacycles per second.

4. Intermediate-Frequency Amplifiers

Gain and Bandwidth

The bandwidth of an intermediate-frequency amplifier is normally defined as the frequency difference between points on the response curve 3 decibels below the maximum. This definition is convenient for most calculations involving tuned circuits and is also useful when considering noise phenomena. The noise-power formula kTB (Chapter X) gives the noise power transmitted by an ideal band-pass filter which has uniform response over the band B and is zero elsewhere. In practical band-pass devices, the bandwidth used in this formula is very closely equal to the 3-decibel bandwidth. The only exception to this is a band-pass filter consisting of one single-tuned stage. The 3-decibel bandwidth also provides useful information about transient response, but in this connection the bandwidth defined with respect to the 15-decibel points is also significant (see p. 432).

The simplest design of intermediate-frequency amplifier employs sufficient cascaded stages to give the required amplification and bandwidth, each of the stages having a single resonant circuit tuned to the operating frequency. This design is always preferred for production purposes ; alignment is straightforward and tolerances on components have the least effect on the overall performance. The product, amplification times bandwidth for a single resonant circuit, is $g_m/2\pi C$ which is half the theoretical limit given in section 2. When several stages of the same amplification are cascaded, the overall bandwidth is reduced and the fractional reduction is given by $(2^{\frac{1}{n}} - 1)^{\frac{1}{2}}$ where n is the total number of tuned circuits.[10] This factor is plotted as curve A in Fig. 270. For an eight-stage amplifier the product of amplification times bandwidth is reduced 10·5 decibels, so that for the amplifier to have an overall bandwidth of 3 megacycles per second, each stage must have a bandwidth of 10 megacycles per second.

The product of amplification times bandwidth for each isochronous stage is constant for a given valve, but the number of stages n for a given overall amplification may be chosen at will. If a small number of high-amplification stages is used, each stage will have a relatively small bandwidth. As n is increased, the necessary amplification per stage becomes less so that the bandwidth of each stage may be increased. The resultant increase in overall bandwidth, however, is opposed by the reduction factor described above. Ultimately this factor predominates and the overall bandwidth decreases with further increase of n. Hence at a certain value of n there is a maximum overall bandwidth for a given overall amplification. With 6AC7 valves and a required overall amplification of 10^5, the maximum bandwidth occurs with 23 stages and is 6 megacycles per second.

In practice, amplifiers with isochronous stages composed of a single resonant circuit have been used for bandwidths of 2 megacycles per second or less. This covers the majority of radar requirements. The use of more elaborate schemes is not justified because of the added production complexity and the difficulty of servicing. Isochronous circuits have the unique advantage that an amplifier using them may be aligned on any convenient signal. Each circuit is tuned for maximum output and it is seldom important if the centre of the pass band is slightly different from the nominal intermediate frequency.

[10] G. E. Valley and H. Wallman (Eds.), *Vacuum Tube Amplifiers* (Massachusetts Institute of Technology, Radiation Laboratory Series, Vol. 18), McGraw-Hill, New York, 1948, pp. 166-92.

When greater bandwidths are needed, stagger tuning has proved quite satisfactory.[11]   Stagger tuning may be applied to a group of r stages, where r is a number greater than one. Groups containing 2, 3 and 4 stages are called pairs, triples and quadruples respectively and the general case is referred to as an "r-uple." The complete amplifier contains one or more

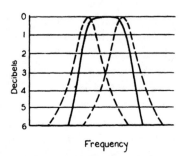

Figure 267.—The component resonance curves and the resulting overall response in an exact staggered pair, $r = 2$.

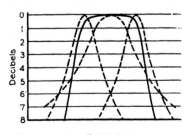

Figure 268.—The component resonance curves and the resulting overall response in an exact staggered triple, $r = 3$.

r-uples as required. The tuned circuits in each r-uple are resonated at selected frequencies within the pass band and by adjusting the bandwidths and frequencies to appropriate values the overall response curve may be given a flat top. This is

[11] A. B. Thomas, "Stagger-tuned intermediate frequency amplifier design," *J. Instn. Engrs., Aust.*, Vol. 22, No. 6, 1950, pp. 141-8; D. Weighton, "Performance of coupled and staggered circuits in wide band amplifiers," *Wireless Engr.*, Vol. 21, 1944, pp. 468-77; G. E. Valley and H. Wallman.[10]

illustrated for a staggered pair and a triple in Figs. 267 and 268. The method of design which yields a flat top is termed exact staggering. An exact staggered r-uple has the same overall bandwidth times mean stage amplification as one single-tuned stage, where the term mean stage amplification is used because, in general, each stage in the group has a different amplification. Stagger tuning therefore eliminates the reduction in bandwidth which occurs when isochronous circuits are cascaded.

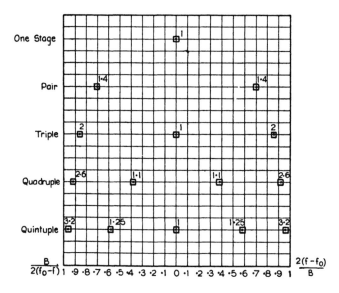

Figure 269.—Design chart for staggered r-uples whose centre frequency is f_0 and overall bandwidth B. The horizontal scale is a function of the frequency and a value of 1 represents the 3 db. points of the r-uple. The number attached to each point is the Q of that circuit, expressed in terms of f_0/B.

Large values of r in a staggered amplifier lead to critical adjustments because high Q circuits are necessary near the edges of the pass band. A small value of r is therefore preferred provided the performance requirements can be satisfied ; for example, four staggered pairs are superior to two quadruples. Design data are given in Fig. 269 for r equal to 2, 3, 4, 5. Because a staggered amplifier has the same mean stage ampli-

fication times overall bandwidth product as a single tuned
stage the value of this product is one half the theoretical limit.
Fig. 270 shows the fractional reduction in bandwidth when
r-uples are cascaded for r equal to 2 and 3.

The effect of stagger tuning may be illustrated by the case
of a 6-stage isochronous amplifier with a bandwidth of 2 megacycles per second, in which conversion to two staggered triples

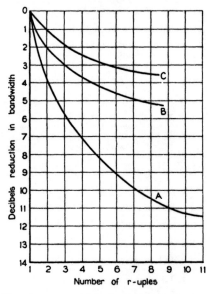

Figure 270.—The degree to which the 3 db. bandwidth of an amplifier
of n r-uples is less than the bandwidth of identical constituent r-uples.
Curve A is for n single stages, curve B for n exactly staggered pairs and
curve C for n exactly staggered triples.

increased the bandwidth to 5 megacycles per second without
altering the overall gain.   Each set of three stages was converted to a triple by leaving one stage without alteration,
doubling the load resistors of the other two and then staggering
their resonant frequencies so that their selectivity curves are
each 6 decibels down where they intersect at the mid-band
frequency.   The gain of each stage at the centre of the band is
thus just what it was originally but the overall bandwidth is
two and a half times as great.

It will be noted that compared with isochronous amplifiers stagger-tuned amplifiers use higher load resistances and that, moreover, the anode load is reactive at the resonant frequency of the grid circuit. For this reason with single-ended valves care must be exercised in the wiring layout to keep stray grid-anode capacity down to a sufficiently low figure ; if this is not done there will be interaction between anode and grid circuits and possibly oscillation. In some instances transit-time grid loading will limit the Q which can be obtained in a tuned circuit and hence prevent the use of those staggered r-uples requiring high Q circuits near the edges of the pass band.

Transient Response

In a radar receiver it is the transient response and not the steady-state frequency response which is of prime importance, but it has been the practice to specify the latter for convenience. When an intermediate-frequency amplifier is to reproduce a given pulse, the overall bandwidth B is frequently chosen as $1/\delta$ where δ is the pulse duration (see Chapter X). This gives optimum visibility of small signals above noise and may be used if a build-up time (from 10 to 90 per cent of final amplitude) of the same order as the pulse duration is acceptable. This design basis is quite satisfactory for isochronous amplifiers because response to a suddenly applied voltage is closely specified by bandwidth. In stagger-tuned amplifiers, on the other hand, the shape of the response curve must be taken into account. The increase in the product of amplification times bandwidth which results from staggering is obtained at the cost of a more rapid decrease of amplification outside the pass band. This increased steepness of the sides of the response curve is accompanied by considerable curvature of the phase characteristic. As the value of r in a staggered amplifier is increased the amplified pulse changes in shape and in particular, starts to show ringing.

The main aspects of transient response and their importance in a radar receiver may be summarized as follows :

(a) The build-up time sets a limit to the narrowest pulse the amplifier can handle and hence to the resolution of the radar system.

(b) The overshoot and subsequent ringing is unimportant at the start of the pulse but the time taken by the transient on the tail to decay to a small value determines the extent of pulse widening on strong signals.

(c) The time to reach 50 per cent of the final amplitude is of interest when high-accuracy range measurements are required.

(d) The time of rise to some small fixed percentage of the final amplitude is of importance when range is measured from the point where the pulse emerges from the noise.

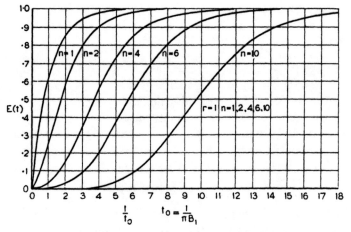

Figure 271.—The RF envelope $E(t)$ of the output from an amplifier tuned to f_0, when an RF signal of frequency f_0 is suddenly applied to the input. The amplifier consists of n single tuned stages ($r = 1$) each of bandwidth B.

An analysis of a single circuit and a coupled pair by Gent[12] shows that the exact mathematical treatment essential for a true picture of transient response is very laborious. Using some simplifying approximations, Eaglesfield[13] makes a

[12] A. W. Gent, "The transient response of RF and IF 'circuits to a wave packet," Standard Telephones and Cables Valve Laboratory, Report G70, October, 1942.

[13] C. C. Eaglesfield, "Carrier frequency amplifiers. The unit step response of amplifiers with single and double circuits," *Wireless Engineer*, Vol. 22, November, 1945, pp. 523-32.

comparison between double circuits and staggered single circuits for television amplifiers. It is concluded that when a limit is imposed on the allowable overshoot, single and double circuits give very similar performances.

An alternative approach[14] is based on the assumption that the response of the equivalent low-pass filter to a DC pulse is identical with the response of the band pass filter to a radio frequency pulse. Figs. 271, 272 and 273 from the paper by

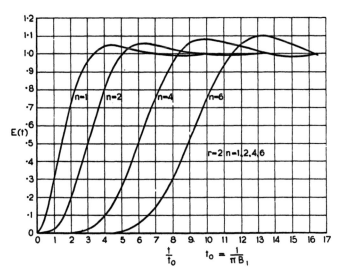

Figure 272.—The RF envelope $E(t)$ of the output from an amplifier tuned to f_0, when an RF signal of frequency f_0 is suddenly applied to the input. The amplifier consists of n exactly staggered pairs ($r = 2$) each of bandwidth B.

Twiss are calculated from this video analogue and apply to an IF amplifier of n staggered r-uples. They are plotted to a time scale t/t_0 in which $t_0 = 1/\pi B_1$, where B_1 is the bandwidth of one r-uple. This makes it possible to display the transient response of the video analogue in terms of the parameters r and n alone. The curves illustrate the deterioration

[14] G. E. Valley and H. Wallman (Eds.), *Vacuum Tube Amplifiers* (M.I.T. Radiation Laboratory Series, Vol. 18), McGraw-Hill, New York, 1948, pp. 274-300. R. Q. Twiss, T.R.E. Report T. 1834, April 1945.

of waveform resulting from the decrease in the overall band-width as the number of cascaded r-uples is increased. The relation between the build-up time and the overall bandwidth is plotted in Fig. 274 which is derived by combining the results of Fig. 270 with those of Figs. 271, 272 and 273. This graph shows that the build-up time is inversely proportional to the overall bandwidth and is almost independent of r and n.

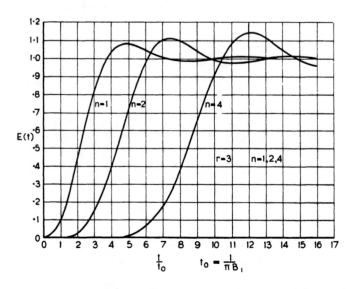

Figure 273.—The RF envelope $E(t)$ of the output from an amplifier tuned to f_0, when an RF signal of frequency f_0 is suddenly applied to the input. The amplifier consists of n exactly staggered triples ($r = 3$) each of bandwidth B.

Experimental verification of the theoretical information on transient response was undertaken by the author to test the approximation on which this information is based. The performance of a 70-megacycle amplifier with 0·1 microsecond pulses was deduced from experiments in which the frequency was scaled down by a factor of 100.

Pulses which could be varied in duration were generated at a frequency of 0·7 megacycles per second and applied to the circuit under test. A typical waveform for the input pulse

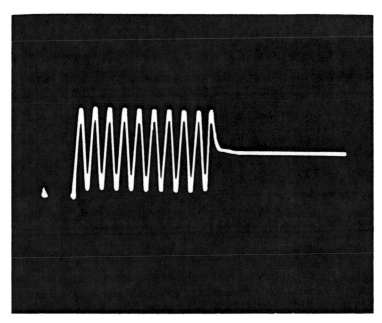

(a) The waveform of the RF pulse which was applied to stagger tuned amplifiers to confirm the theoretical treatment of transient response (see Plates XVII and XVIII).

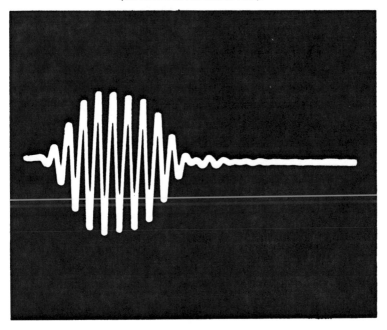

(b) The response of the staggered quadruple of Plate XVIII (a) to a 7 cycle pulse.

PLATE XVI.

see p. 431]

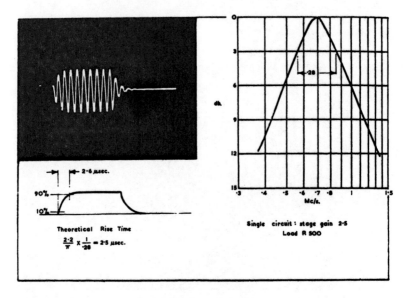

(a) Single tuned stage.

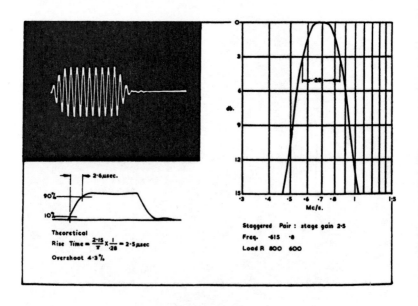

(b) Staggered pair.

PLATE XVII.—Experimental transient response results.

see p. 431]

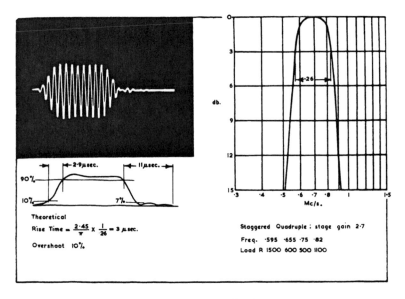

(a) Staggered quadruple.

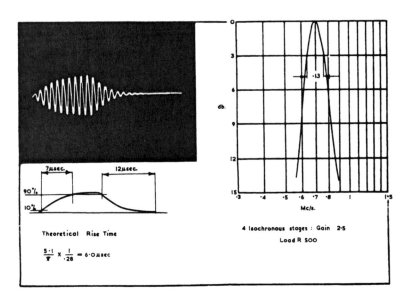

(b) Four isochronous stages.

PLATE XVIII.—Experimental transient response results.

see p. 431]

is shown in Plate XVI (a). There is an initial transient due to resistance in the shock excited circuit of the generator and also a delay in fully brightening the trace.

Plates XVII (a), (b), XVIII (a), (b), give the results of the measurements together with theoretical values for the time of rise and the overshoot ; the agreement is very good. In most cases a pulse duration of greater than 7 cycles per second (corresponding to 0·1 microseconds at 70 megacycles per

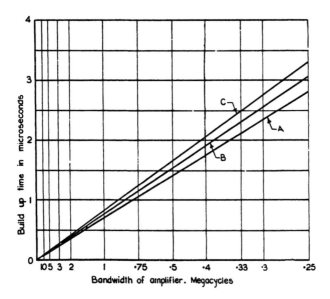

Figure 274.—The relationship between build-up time (10% to 90%) and overall bandwidth (3 db.) in an amplifier for, one single stage (curve A), 4 staggered pairs (curve B), and 4 staggered triples (curve C). Frequency is plotted to a reciprocal scale to illustrate the inverse relationship and the diagram shows also the small influence of the method of tuning.

second) was used in order to separate the initial and final transients. Plate XVI (b) shows the response to a 7-cycle pulse.

It has been deduced empirically that the duration T of the transient from 5 per cent of final amplitude to a point where the ringing has decayed to 1 per cent of the steady state value is specified quite closely by $T = 4/B_T$ where B_T is the bandwidth referred to the 15-decibel points of the frequency response curve. This is illustrated in Plate XVIII (*a*) and (*b*). For the same stage gain the duration of a strong pulse is not significantly reduced by stagger tuning although there is an improvement in rate of rise on the leading edge.

Feedback Amplifiers

Inverse feedback may be applied to the stages of an intermediate-frequency amplifier as shown in Fig. 275. The theoretical performance of a feedback amplifier is the same as that of a coupled-circuit amplifier[15] or a stagger-tuned amplifier

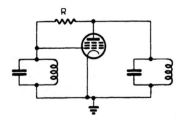

Figure 275.—The basic circuit of a single-stage IF amplifier with negative feedback.

However, the lining-up procedure is more complicated and the bandwidth of the amplifier changes if its gain is controlled by variation of grid bias.

Noise Factor

The figures given in Table 8, section 3, for the noise factors of signal-frequency amplifiers are also applicable to intermediate-frequency amplifiers operating at the stated frequencies. If a valve mixer is used, the noise generated in the intermediate-frequency amplifier will normally be overshadowed by mixer noise. In centimetre-wave radar sets, however, reduction of the noise factor of the intermediate-frequency amplifier is important because it is preceded only

[15] G. E. Valley and H. Wallman (Eds.), *Vacuum Tube Amplifiers* (Massachusetts Institute of Technology, Radiation Laboratory Series, Vol. 18), McGraw-Hill, New York, 1948, pp. 232-69.

by a crystal mixer (see equation (1) in section 1). In this case the use of a triode input stage is preferable as is shown by the figures in Table 8. The best performance at present is given by the neutralised triode plus grounded-grid amplifier described in section 3.

Gain Control

Gain control of radar receivers is usually effected by altering the transconductance of variable-slope valves. While straightforward, this method may be undesirable because a change of cathode current simultaneously alters both the conductive and susceptive components of valve input admittance. Of these effects the latter is of greater consequence because valve capacitance and strays usually comprise the total tuning capacity. The former becomes important when damping of the tuned circuit by grid cathode conductance is an appreciable part of the total damping, as in amplifiers of high amplification and narrow bandwidth or in staggered r-uples with large values of r. Variations of input admittance with gain-control setting may be minimised by the use of cathode degeneration.[16] An unbypassed cathode resistor of 30 ohms effectively stabilises the input capacity of a 6AC7 and has been widely used ; the change in conductance which occurs when this value of resistance is employed is not excessive for most circuits. Another method of minimising change of input admittance is by simultaneous variation of the bias on the control and suppressor grids.[17] The control grid is operated at 10 to 30 per cent of the total bias depending on the characteristics of the valve. A disadvantage of this scheme is that a large voltage, 50 volts or more, is needed for adequate control. In addition, the input conductance is high and the stabilisation is not as good as when cathode degeneration is employed.

Two useful forms of automatic gain control are swept gain and instantaneous automatic gain control (IAGC). In the

[16] R. L. Freeman, " The use of feedback to compensate for vacuum tube input capacitance variations with grid bias," *Proc. I.R.E.*, Vol. 26, November, 1938, pp. 1360-6 ; and F. L. Smith *Radiotron Designers Handbook*, Amalgamated Wireless Valve Company, Sydney, 3rd edition, 1940, p. 96.
[17] C. Lockhart, " Gain control of radio frequency amplifiers," *Electronics, Television and Short Wave World*, August, 1940, pp. 365-8.

former a variation of amplification is superimposed on the normal fixed value set by the gain control knob, so that the receiver gain is low at the time of occurrence of the transmitted pulse and increases with time thereafter, i.e. with increase of range. Strong nearby echoes and weak distant ones then appear to be of comparable strength. This prevents short-range echoes from overloading the amplifier when sufficient sensitivity is employed to detect distant targets. The change in the aerial coverage diagram due to the variation of sensitivity with range should not be overlooked in aerial designs for sets which use swept gain.

IAGC is a development from the automatic volume control of broadcast receiver practice. Because there is no continuous carrier to operate an AVC system of conventional type, each echo pulse is made to provide its own control. This requires a system which will come into operation in a time short compared with the pulse duration and return the receiver to full sensitivity equally rapidly following each received pulse. The negative output from a diode detector is applied to the control grids of some of the stages through a cathode follower, the low driving impedance of which is necessary if a sufficiently short time constant is to be realised. The circuit time constants are adjusted so that a rectangular input pulse emerges from the receiver with an amplitude which rapidly diminishes as the IAGC comes into action. This form of echo distortion is very suitable for breaking up clutter such as that produced by nearby ground, buildings, etc. but is only fully effective with an IF amplifier of generous bandwidth.

Stability

Due attention must always be given to the prevention of regeneration and instability. Regeneration can be detrimental long before oscillation occurs, and is apparent as a reduction in bandwidth and asymmetrical distortion of the frequency response characteristics. To ensure stability the attenuation between output and input circuits must exceed the overall amplification by a considerable margin. This is readily achieved by suitable disposition and shielding of connecting leads, and by adequate filtering of the detector output circuit ; at this point care is needed to filter out the IF component while retaining adequate video bandwidth. Elaborate shielding of each amplifier stage has been found unnecessary as the stage gain is usually low and individual stages show little tendency to regeneration. Standard practice dictates that power-supply leads to each stage shall be filtered, heaters by-passed, and earth returns arranged to minimise circulating currents in the chassis. The external field of coils should be prevented from interlinking outgoing leads or the wiring of adjacent stages, while by-pass condensers are chosen to resonate in series with their own leads at the operating frequency.

A suitable division of overall gain between RF, IF and video sections is important. Excessive video gain is undesirable because the response to low frequencies may result in feedback or interference from power supply mains. In microwave radar sets, the radio frequency units are often mounted near the aerial in an RF " head." It is common practice to use an intermediate-frequency preamplifier in the latter to amplify the signal before transmission to the remainder of the receiver through lengths of feeder which are sometimes quite long. The preamplifier effectively divides the intermediate-frequency amplification between two units and reduces the difference between the input and output level of either amplifier.

The second detector in a radar set is usually a diode because of the high signal handling capacity of this type. However, it is not possible to design such a detector for high efficiency because the circuit impedances are determined by bandwidth considerations in both the intermediate and video frequency circuits. The efficiency is normally only 25 to 40 per cent[18] a fact which must not be overlooked when designing an amplifier for a given overall gain and peak output or when estimating the control voltages available for IAGC.

Typical IF Amplifiers

The valve types commonly used in the intermediate-frequency amplifiers of radar receivers are few in number—namely the

[18] O. H. Schade, " Analysis of rectifier operation," *Proc. I.R.E.*, Vol. 31, July, 1943, pp. 341-61.

6AK5, 6AG5, EF50 and 6AC7. To reduce the number of valve types in an equipment, the 6J6 double triode has been proposed for use in the intermediate-frequency amplifier.[19] When the 6J6 is used in this way signals are applied to one triode section, which is used as a cathode follower driving the second triode section connected as a grounded-grid amplifier. The gain and stability of such an amplifying stage compares favourably with those of a pentode circuit and the noise characteristics are superior.

A typical intermediate frequency amplifier using 6AK5 miniature pentodes is illustrated in Plate XIX (*a*) and (*b*).

5. Video Amplifiers

Video amplifiers for radar purposes are similar to those for television as regards both frequency response and transient performance. The product of amplification times bandwidth is also an important design factor, the same as in intermediate-frequency amplifiers. However, a requirement which is peculiar to radar video amplifiers is the necessity for avoiding paralysis of the amplifier following the overloading which occurs during the transmitted pulse or the reception of strong echoes. Another special aspect is concerned with methods of "limiting" signal amplitude, whereby weak echoes are amplified normally but strong ones are unable to rise above a fixed maximum.

Frequency and Transient Response

High stage amplification associated with satisfactory phase characteristics over the required frequency band is obtained by the well known method of adding reactive elements to the simple resistance-capacity coupling circuit. Although the steady-state frequency-response characteristics have been widely used when specifying video amplifier performance, the

[19] G. C. Sziklai and A. C. Schroeder, "Cathode-coupled wide-band amplifiers," *Proc. I.R.E.*, Vol. 33, October, 1945, pp. 701-9.

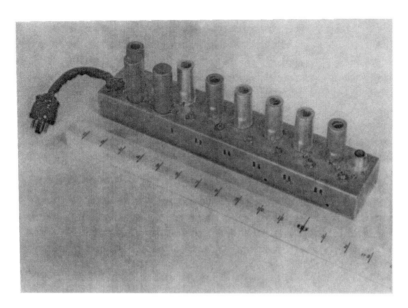

(a) Top view.

(b) Bottom view showing component layout.

PLATE XIX.—Typical IF amplifier using a 6AK5 miniature pentode.

see p. 436]

transient response is of greater concern to the radar designer.[20] However, because the input pulse is already distorted by the IF amplifier, the published theoretical treatments based upon response to a step function give a somewhat inadequate picture of video amplifier behaviour and compensating circuits which have a relatively sharp cut-off may prove to be acceptable. If limiting occurs in a low-level video stage, however, the original rectangular form of the pulse tends to be restored and the theoretical treatments are then more adequate.

Paralysis

Video circuits must be capable of recovering from strong signals quickly enough for weak ones to be detected immediately afterwards.[21] Following the disturbance of quiescent operating conditions by a strong pulse, an overshoot may be

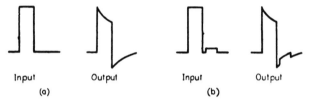

Input   Output   Input   Output

(a)   (b)

Figure 276.—(a) The way in which a small grid coupling condenser produces overshoot. (b) How the overshoot from a strong echo causes a weak adjacent echo to be lost.

produced even when high frequency compensation is satisfactory. The most common cause of overshoot is a coupling condenser of inadequate size which loses a significant amount of charge during the pulse [Fig. 276 (a)]. If the input consists of a strong signal followed by a weak one, the output will then be as in Fig. 276 (b). On a Class A display without limiting, the weaker echo would still be detectable, but on an intensity-modulated display the overshoot would be responsible for a

[20] A. V. Bedford and G. L. Fredendall, " Transient response of multi-stage video amplifiers," Proc. I.R.E., Vol. 27, April, 1939, pp. 277-84 ; A. V. Bedford and G. L. Fredendall, " Analysis, synthesis, and evaluation of the transient response of television apparatus," Proc. I.R.E., Vol. 30, October, 1942, pp. 440-57 ; and H. C. Kallman, R. E. Spencer and C. P. Singer, " Transient response," Proc. I.R.E., Vol. 33, March, 1945, pp. 169-95.
[21] G. E. Valley and H. Wallman, (Eds.), Vacuum Tube Amplifiers (Massachusetts Institute of Technology, Radiation Laboratory Series, Vol. 18), McGraw-Hill, New York, 1948, pp. 113-52.

blind region immediately following the strong signal. When this type of distortion takes place in an early stage of the amplifier, the subsequent amplification of the overshoot may easily be sufficient to cause cut-off in one of the video stages and any signals during this period will be completely lost.

In practice, it is quite feasible to use values of coupling time constant which will maintain the amplitude of a one microsecond pulse constant to one per cent or better. Screen and especially cathode by-pass condensers however cause much greater trouble. If a cathode resistor is not by-passed, overshoot does not occur but the resultant degeneration produces a loss of gain. However, it is often so difficult adequately to by-pass a cathode circuit that degeneration is deliberately permitted. Another method is to use a small cathode by-pass condenser to reduce degeneration but to choose its value so that the cathode time constant is approximately equal to that in the anode circuit. The time constant effects in the anode and cathode circuits then tend to cancel so that overshoot is eliminated.

Time constant calculations must be based on the worst possible cases. Strong clusters of echoes returning during the first 10 to 20 miles of the time base are often so close together that the effect resembles one long pulse lasting for 100 microseconds or more. The amplifier must recover from this signal rapidly enough to prevent the occurrence of paralysis with resulting blind spots.

When limiting of the signal level in an amplifier is used, the relative amplitude of a small overshoot produced in an early stage is increased many times. In particular, the transmitted pulse which is often strong enough to cause overloading in the IF amplifier may produce a serious overshoot unless the design is adequate. If time constants cannot be made long enough, the performance may be improved by using conventional methods of low frequency compensation.

Limiting

Protection of the screen of the cathode ray tube in an intensity-modulated display requires restriction or limiting of the modulating voltage so that strong echoes do not produce more than the permissible brilliance. The acceptable range of voltage is usually less than the cut-off bias of the cathode ray tube. Limiting is also important in preventing a strong positive signal from causing a flow of grid current in any of the video amplifier stages. If this is not done the amplifier becomes paralysed for a long period after the end of the signal. This effect depends on the fact that flow of grid current has the same effect as inserting a very much smaller grid resistor. The coupling condenser therefore charges more quickly than it would with the grid negative, so that an appreciable potential builds up across the condenser during the pulse. At the end of the pulse, the charge acquired by the condenser discharges through the grid resistor and the return to quiescent conditions is quite slow as shown in Fig. 277.

Figure 277.—The waveform produced when overshoot results from the flow of grid current and the return to quiescent conditions is by discharge through the grid resistor.

To prevent strong signals causing grid current, one of the stages in which the pulses applied to the grid are negative is allowed to be driven into the cut-off region by signals which are excessive. The peak output of such a stage is equal to the DC drop across the load resistor and is limited to a value which can be satisfactorily accepted by the subsequent stage or the cathode ray tube.

Cathode Followers

It is frequently necessary to divide the video amplifier into two units, one being on the receiver chassis and the other close to the cathode ray tube. The units are connected by a low impedance coaxial cable which unless of undue length is equivalent to a lumped capacity at video frequencies. Because of its low output impedance a cathode follower is nearly always used to drive such a cable and is entirely satisfactory provided certain precautions are taken.[22]   For positive input pulses, the output impedance of a cathode follower is always low.   If its value is R_0, the cable capacity C is charged with a time constant CR_0.   When the input pulse is negative, however, the cathode follower may become cut-off under certain conditions and the time constant will then increase to CR_c where R_c is the value of the cathode resistor.   If this time constant is excessive, distortion of the waveform will occur.   Each circuit must be analysed individually to ensure that the grid is not driven negative at a rate much in excess of that at which the cathode is able to follow with time constant CR_0.   Cathode followers which are required to handle negative pulses should be arranged to operate at low level.

[22] B. Y. Mills, "Transient response of cathode followers in video circuits," *Proc. I.R.E.*, Vol. 37, 1949, pp. 631-3; H. Goldberg, "Some considerations concerning the internal impedance of the cathode follower," *Proc. I.R.E.*, Vol. 33, 1945, pp. 778-82.

CHAPTER XIV

DISPLAY CIRCUITS

1. The Cathode Ray Tube

THE heart of radar display systems is the cathode ray tube and as explained in Chapter X there are two types in common use, those with electrostatic deflection and focusing systems and medium persistence screens, and those with magnetic deflection systems and long persistence screens. The former are used when a deflection modulation display such as Class A is required, the latter for intensity modulated displays such as the PPI, because magnetic deflection of the beam is much superior when increased screen brilliance with good focusing at large deflections is required.[1]

The properties of some typical tubes used in radar practice are given in Table 9.

TABLE 9—TYPICAL CATHODE RAY TUBES USED IN RADAR

Tube type	Screen size (inches)	Screen persistence	Deflection and focusing	Normal HT voltage & polarity (kilovolts)
1802 (5BP1)	5	Medium	Electrostatic	−2
ACR 1	5	Medium	Electrostatic	−4
VCR139A	2½	Medium	Electrostatic	−800 volts
VCR526	2½	Medium (daylight viewing)	Electrostatic	−1·3 and +2·5
5CP7	5	Long	Electrostatic	−2 and +2
5FP7	5	Long	Magnetic	+4 to +6
7BP7	7	Long	Magnetic	+4 to +6
12DP7	12	Long	Magnetic	+4 to +6
3DP1	3	Medium	Electrostatic (radial deflection)	−1·5

The last tube in this table has an unusual electrode system which is used to produce radial deflection. The construction

---

[1] H. Moss, "The electron gun of the cathode ray tube," *J. Brit. I.R.E.*, Vol. 5, 1945, pp. 10-25, and Vol. 6, 1946, pp. 99-129.

of the tube is similar to a normal electrostatic type except for
the addition of an axial electrode attached to the face of the
tube (Fig. 278) which with the conducting coating inside the
tube forms a coaxial deflecting system. The waveform to be
examined is applied to these electrodes and the beam is deflected along a radius drawn from the centre of the tube face.[2]
A circular time base is applied to the normal deflection plates.

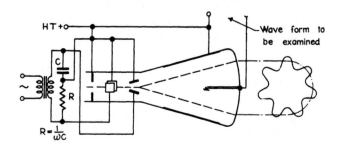

Figure 278.—Radial deflection cathode ray tube.

Information concerning the various types of screens has
been given in Chapter X and it is sufficient to note that no
special circuit techniques are required either for the skiatron
or the daylight viewing tube.

Deflection and Focusing Coils

The widespread use of coils for magnetic deflection and focusing
has led to standardised designs. The focusing coil is nearly
always of the type indicated in Fig. 279 (*a*) and consists of a
solenoid with a large number of turns (about 35,000) encased
in a magnetic shield, with a small air gap to concentrate the
field and produce a thin electron lens. For optimum focus the
coils require about 10 to 15 milliamperes at about 200 to 250
volts.[3] Permanent magnets have also been used successfully
for this purpose in lightweight sets.

[2] O. S. Puckle, *Time Bases*, Chapman and Hall, London, 1943, p. 74.

[3] T. Soller, M. A. Starr and G. E. Valley (Eds.), *Cathode Ray Tube Displays*
(Massachusetts Institute of Technology, Radiation Laboratory Series,
Vol. 22), McGraw-Hill, New York, 1948, pp. 93-110.

Three types of deflection coil are illustrated in Fig. 279
(b), (c) and (d). The square iron core coil of (b) is generally
of most value. Two windings on opposite sides of the core
are connected in series so that their fluxes are in opposition,

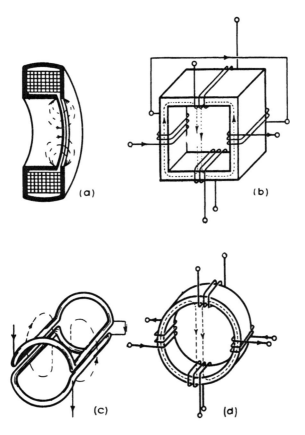

Figure 279.—Deflection and focusing coils for magnetic deflection cathode
ray tubes. (a) Focus coil. (b) Iron core deflection coil. (c) Air core
deflection coil. (d) Iron core deflection coil.

the result being a practically uniform field over the path of
the electron beam. Using a similar pair of windings on the
other two sides of the core, two independent deflections at
right angles are possible and this type of coil then corresponds

to the X and Y plates of an electrostatic cathode ray tube. Push pull operation can be obtained by using two windings on each leg. It is usual to enclose the coil in a conducting shield which by eliminating the external flux practically halves the inductance and makes a faster sweep possible (section 6). Plate XX shows this type of coil mounted in its shield. The air core coil of Fig. 279 (c) gives a deflection in one direction only. It is often used in PPI systems, where it is rotated mechanically in synchronism with the aerial. It has the disadvantage that if a conductor intercepts some of the return flux outside the coil serious distortion of a fast sweep may result. The toroidal iron core of Fig. 279 (d) has been used extensively in America. It has four equally spaced coils which may be inter-connected to produce either two deflections at right angles as with the square coil or a deflection in one direction as with the air core coil.

There is not a great deal of difference in the performance or efficiency of any of the coils described above, provided those using iron cores are shielded so as to avoid the production of useless flux.[4]

High Tension Power Supplies

Conventional high voltage transformers and diode rectifiers have been used extensively for producing the HT (2 to 6 kilovolts) required for cathode ray tubes, but recently the demand for good regulation and small size and weight has led to some research into improved systems. High frequency oscillator techniques have been found very effective for this purpose. If the transformer is designed so that the conduction period of the diode rectifier corresponds to the conduction period of the oscillator, good regulation can also be achieved by this method.[5]

[4] A. Woronow, "Iron-cored deflection coils for cathode-ray tubes," *J. Instn. Elect. Engrs.*, Vol. 93, Part IIIA, 1946, pp. 1564-74.
[5] O. H. Schade, "Radio frequency high voltage power supplies," *Proc. I.R.E.*, Vol. 31, 1943, pp. 158-63.

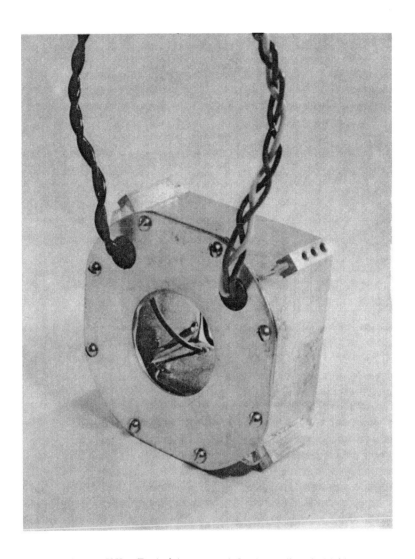

PLATE XX.—Typical iron core deflection coil and shield.

2. Transient Response of Basic Circuits

The performance of display circuits depends to a large extent on the transient response of the simple two element resistor-reactor networks shown in Fig. 280. The networks are com-

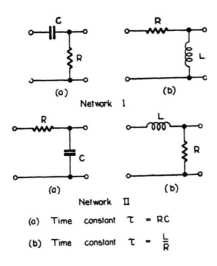

(a)   (b)

Network I

(a)   (b)

Network II

(a) Time constant τ = RC

(b) Time constant τ = $\frac{L}{R}$

Figure 280.—Basic networks for display circuit calculations.

posed of resistance-inductance and resistance-capacitance pairs and may be divided into two types each of which has a frequency characteristic and transient response determined only by its time constant τ. The transient response of each type of network has been collected in Table 10, for applied voltages which may be square, sawtooth or sinusoidal in form. These are the fundamental waveforms of a display system.

Considering the response to the first two waveforms, it is seen that networks of type I give an output proportional to the differential of the input voltage for large values of t/τ and they are therefore often known as differentiating circuits. Networks of type II on the other hand give an output proportional to the integral of the input waveform for small values of t/τ and are therefore often known as integrating circuits. When a sine wave is applied to the networks it is seen that, after the initial transient has subsided, the phase of the wave has been shifted by the same amount as for a con-

TABLE 10—TRANSIENT RESPONSE OF THE BASIC CIRCUITS TO COMMON WAVEFORMS

Applied Voltage V_1	Network	Transient Waveform V_2	Expression for Transient Response V_2		
			Exact	$\frac{t}{\tau}$ small	$\frac{t}{\tau}$ large
$V_1 = V_0$	Type I		$V_0\, e^{-\frac{t}{\tau}}$	$V_0(1-\frac{t}{\tau})$	0
	Type II		$V_0\left(1-e^{-\frac{t}{\tau}}\right)$	$V_0\frac{t}{\tau}$	V_0
$V = \alpha t$	Type I		$\alpha\tau\left(1-e^{-\frac{t}{\tau}}\right)$	αt	$\alpha\tau$
	Type II		$\alpha\left[t-\tau\left(1-e^{-\frac{t}{\tau}}\right)\right]$	$\frac{1}{2}\alpha\tau\left(\frac{t}{\tau}\right)^2$	$\alpha(t-\tau)$
$V = V_0 \sin\omega t$	Type I		$\frac{V_0\,\omega\tau}{\sqrt{1+(\omega\tau)^2}}\left[\sin(\omega t+\theta)-\frac{e^{-\frac{t}{\tau}}}{\sqrt{1+(\omega\tau)^2}}\right]$ where $\tan\theta=\frac{1}{\omega\tau}$	Not important	$\frac{V_0\,\omega\tau}{\sqrt{1+(\omega\tau)^2}}\sin(\omega t+\theta)$
	Type II		$\frac{V_0}{\sqrt{1+(\omega\tau)^2}}\left[\sin(\omega t-\phi)+\frac{\omega\tau\,e^{-\frac{t}{\tau}}}{\sqrt{1+(\omega\tau)^2}}\right]$ where $\tan\phi=\omega\tau$	Not important	$\frac{V_0}{\sqrt{1+(\omega\tau)^2}}\sin(\omega t-\phi)$

tinuous wave. For perfect differentiation or integration this phase shift would be 90 degrees but this is only approached in practice as a limiting condition.

3. Gate Generators

In a radar display the cycle of operations which is initiated by the transmitted pulse continues for a specified time interval and then all circuits are returned to a quiescent condition. The waveform required to control the electronic switching operation is usually a square pulse termed a "gate," and can be produced by a multivibrator with one stable and one unstable position of equilibrium.[6]

[6] O. S. Puckle, *Time Bases*, Chapman and Hall, London, 1943, pp. 50-9.

Two common types of multivibrator or "flip-flop" are
shown in Fig. 281 (a) and (b) together with typical waveforms
at important electrodes. Fig. 281 (a) is an anode-coupled and
Fig. 281 (b) a cathode-coupled flip-flop. A point of interest is
that the grid leaks are returned to the high tension supply in
order to obtain greater stability since dV_{g_1}/dt is then greater
at time T (see Fig. 281). Each type has its particular advantages ; for instance the anode-coupled flip-flop can easily be
arranged to give equal current drains through the two tubes

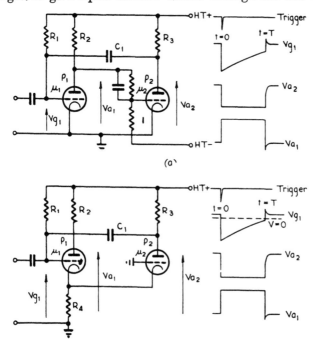

Figure 281.—Typical "flip-flop" circuits. (a) Anode coupled flip-flop.
(b) Cathode coupled flip-flop.

and thus to present a constant load to the power supply independent of duty cycle. It also requires a smaller drain for a
given transition time, and a smaller negative trigger pulse
for initiation of the cycle. The current drain of the cathode-
coupled circuit is normally higher and is also unequal in the
two tubes since R_3 is usually greater than R_2. A large negative
trigger pulse is required but the performance of the circuit is

not so susceptible to supply voltage fluctuations because of degeneration in the cathode resistor, and it suffers less from jitter of the gate length, probably for the same reason. Cathode-coupled flip-flops are often triggered by a positive pulse on the grid of the non-conducting tube.

Assuming ideal distortionless triodes, the gate periods for the two types are as follows :

(a) *Anode coupling* :

$$T = 2\cdot3\, C_1\!\left(R_1 + \frac{R_3\rho_2}{R_3 + \rho_2}\right)\!\log_{10}\frac{\left(2 + \dfrac{\rho_2}{R_3}\right)}{\left(1 + \dfrac{\rho_2}{R_3}\right)\!\left(1 + \dfrac{1}{\mu_1}\right)}. \quad (1)$$

(b) *Cathode coupling* :

$$T = 2\cdot3C\!\left[R_1 + \frac{R_3(\rho_2 + \mu_2 R_4)}{R_3 + \rho_2 + \mu_2 R_4}\right]\!\log_{10}\!\left\{\frac{2 + \dfrac{\rho_2 + \mu_2 R_4}{R_3}}{1 + \dfrac{\rho_2 + \mu_2 R_4}{R_3}} - \dfrac{R_4}{R_2 + R_4 + \rho_1}}{1 + \dfrac{1}{\mu_1} - \dfrac{1}{\mu_2\!\left(1 + \dfrac{R_3 + \rho_2}{\mu_2 R_4}\right)}}\right\}. (2)$$

where the symbols have the meanings indicated in Fig. 281. Stability is seen to be dependent mainly on the ratio

$$\frac{\rho_2 + \mu_2 R_4}{R_3},$$

where R_4 is zero for anode coupling. Since μ_2 is more nearly constant with changing voltages than ρ_2 the reason for greater stability with cathode coupling is clear. Accurate gate length calculations are often necessary if not much adjustment is available to take care of components near the limits of their tolerance. In this case it is usual to work directly from the tube characteristics rather than to use the above equations.

The tube most commonly employed in flip-flop circuits in American and recent Australian equipment is the 6SN7GT, consisting of two triodes electrically similar to the 6J5 and in the same GT envelope.

Other types of circuits often used as gate generators include single tube transitron circuits which are more compact if double triodes are not available, and phantastron circuits which are described in detail in section 6.   Both types suffer somewhat from a critical dependence on tube characteristics so that replacements often require to be specially selected.   Once working however, they are quite stable, particularly the phantastron which is used when the elimination of jitter is important.

4. Clamping Circuits

In the transfer of waveforms through transformer or capacitor coupling circuits the DC component of the wave is lost.   The effect is illustrated in Fig. 282 where a recurring negative gate is passed through the two coupling circuits shown.   It is

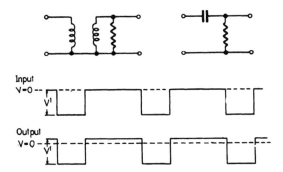

Input
V=0

Output
V=0

Figure 282.—Loss in DC level across coupling reactors.

seen that, referred to earth potential, the negative voltage swing has been reduced in amplitude and a positive swing has been added.   The magnitude of the effect is dependent on the duty cycle of the gate.   In general the insertion of the fixed DC level in such waveforms is important in display circuits, since most operations are controlled by instantaneous voltages and since waveforms are usually gated with a duty cycle which may vary considerably under different operating conditions.

There is a number of circuits which can be used for inserting a DC level.[7] They are called " clamps " and can be either " single-ended " or " double-ended." A single-ended clamp such as the diode of Fig. 283 (a) can restore in one direction only, its action being simply that of a peak rectifier which serves to fix the DC potential of a regularly recurring peak in

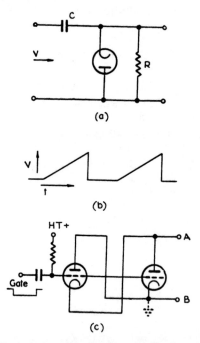

(a)

(b)

(c)

Figure 283.—Typical clamping circuits. (a) Single-ended diode clamp. (b) Typical sawtooth waveform which must be clamped. (c) Double-ended triode clamp.

the waveform. For example, the starting potential of the sawtooth wave of Fig. 283 (b), when applied through the capacitor C to the clamp of Fig. 283 (a), is fixed at zero even for wide variations of the duty cycle. If a negative sawtooth

[7] J. W. Sherwin, "Clamping tubes," Massachusetts Institute of Technology, Radiation Laboratory, Report 572, May, 1944; and B. Chance *et al.* (Eds.), *Waveforms* (Massachusetts Institute of Technology, Radiation Laboratory Series, Vol. 19), McGraw-Hill, New York, 1949, pp. 40-100.

were used with the above circuit, the finishing potential of the sawtooth wave would be fixed. To clamp the starting potential in this case it would be necessary to reverse the diode.

If it is desired to clamp the intermediate portion of a waveform, or if the waveform can change polarity, it becomes necessary to use a double-ended clamp. A typical example of this type is illustrated in Fig. 283 (c). While the tubes are conducting, current can flow in both directions through the clamp, so that point A is joined to the point B by a low impedance, usually of the order of 1000 ohms. When a negative gate is applied to the grids, both tubes are cut off and their impedance becomes effectively infinite, so that it is clearly possible to make a waveform start at a fixed potential and a fixed time merely by returning the point B to that potential, and applying a gate at the correct time. Design factors relating to the impedance of the waveform generator and the various time constants involved are sometimes important if the waveform is not zero when the clamp is being applied. However the restoration of DC level to a simple gated waveform of either polarity usually presents little difficulty since such waveforms are quiescent during the clamping period. With the waveform of Fig. 283 (b), the clamp would be released during the period of the sawtooth and reapplied when the voltage had returned to almost zero.

Many other forms of clamps have been used but will not be described here. In the next section the type of single-ended triode clamp commonly used in sawtooth generators is described. Naturally it may also be used for the insertion of a DC level into a recurrent waveform. Similarly the double-ended triode clamp of Fig. 283 (c) may be used as a sawtooth generator when an output of either polarity is desired.

5. Sawtooth Generators

To display a radar echo it is usually necessary to sweep the cathode ray tube spot across the tube face so that its displacement is always proportional to time. The production of such a sweep usually resolves itself into two parts, firstly the production of a linear sawtooth waveform, possibly with

correcting voltages added, and secondly the amplification of
the waveform. This requires a voltage amplifier in the case
of electrostatic deflection and a current amplifier in the case
of magnetic deflection. The generation of a linear sawtooth
is of particular interest since it is also the basis of a common
method of range measurement to be discussed more fully in
section 9. Some extremely useful circuits have been evolved
for the production of a truly linear sawtooth. It is instructive
to trace the development of these circuits starting with the
simple but fundamental case of charging a capacitor through
a large resistor.

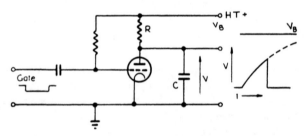

Figure 284.—The basic circuit of a sawtooth generator.

A typical circuit is shown in Fig. 284. The triode clamp
is normally conducting with both anode and grid only slightly
above earth potential owing to the small currents being passed.
A negative gate applied to the grid cuts off the anode current
and the voltage across capacitor C begins to rise exponentially
to the HT supply potential, thus,

$$V = V_B\left(1 - e^{-\frac{t}{RC}}\right) = V_B\left\{\frac{t}{RC} - \frac{1}{2!}\left(\frac{t}{RC}\right)^2 + \frac{1}{3!}\left(\frac{t}{RC}\right)^3 \ldots\right\}. \quad (3)$$

The rise of voltage is at first linear but distortion, represented
by the second and higher powers of t, increases rapidly. For a
given output voltage any degree of linearity may be achieved
by increasing the charging potential, but of course this is not
always desirable or even possible. For example, a 100 volt
sawtooth, linear to 0·1 per cent, would require a supply potential
of 50 kilovolts. The problem of increased linearity has therefore been attacked by other methods.

In order to keep the charging current flowing into C more uniform, and hence make the voltage rise more linear, an inductor is sometimes placed in series with the charging resistor R. This method is often very effective on fast sweeps but may be cumbersome if a slow sweep is desired. Theoretically

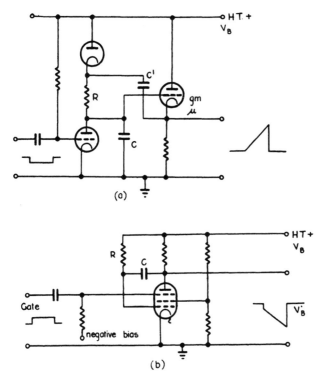

Figure 285.—Feedback circuits to give a more nearly linear sawtooth. (a) Cathode feedback. (b) Anode feedback (Miller effect).

any required degree of linearity may be achieved in this way by making the inductance sufficiently large.

By far the most common method, however, is to introduce some form of feedback,[8] and the two most generally used circuit

---

[8] B. Chance, "Some precision circuits used in waveform generation and time measurement," *Rev. Sci. Instrum.*. Vol. 17, 1946, pp. 396-415.

types are those of Fig. 285. In Fig. 285 (a), R is the charging resistor for C and also the grid resistor of a cathode follower. The sawtooth wave applied to the cathode-follower grid is transferred to the cathode load, which is coupled to the top of R, so that the voltage across R is kept nearly constant. Thus the current flowing into C is nearly constant and the voltage increase across it nearly linear. It can be shown that for circuits of this type the linearity is the same as if the capacitor C were charged through an ohmic resistance from a high potential supply. The magnitude of this equivalent supply voltage is $V_B/(1 - A)$, where A is the amplification of the cathode follower, so that linearity is improved by a factor of $1/(1 - A)$. In order to obtain good linearity the amplification is frequently increased by increasing the value of the cathode resistor and returning it to a negative voltage. The value of the coupling capacitor C' is also of importance since any change of voltage across it during the sawtooth is equivalent to a reduction in cathode-follower gain. The effect of the value of C' is apparent from the expression for the improvement factor which is given approximately by

$$\frac{1}{1 - A} = \frac{g_m R_K\left(1 + \dfrac{1}{\mu} + \dfrac{1}{g_m R_K}\right)}{1 + \dfrac{g_m R_K}{\mu} + \dfrac{g_m R_K C}{C'}\left(1 + \dfrac{1}{u} + \dfrac{1}{g_m R_K}\right)}. \quad (4)$$

The diode could be replaced by a resistor but this would be in parallel with the cathode load of V_3 and hence would reduce the amplification. A slight change of sawtooth magnitude would also occur with changing duty cycle. The type of circuit just described is often called a " bootstrap " and is one of the most convenient for obtaining a moderately linear sawtooth. One of its advantages is that the output is obtained at a low impedance level.

Fig. 285 (b) shows another form of feedback circuit. Whilst the last circuit utilised a cathode follower to increase the effective value of the charging resistance and thus to increase linearity, the present scheme uses the Miller effect to increase the effective value of the capacitor. The latter is connected between the grid and anode of a high gain pentode and the grid is returned to a high potential through the charging resistor R. During the quiescent period the anode current is reduced to zero by a negative potential on the suppressor grid. To initiate the sawtooth a positive gate is applied to the suppressor, raising its potential to zero and switching on the tube. The anode potential immediately falls by an amount approximately equal to the grid base and then begins to decrease exponentially. The degree of linearity can again be described by assigning an equivalent voltage towards which the anode is charging. In this case, the voltage is negative and equal to $-AV_B$ so that the linearity is improved by a factor of $(1 + A)$.

This type of circuit can also be arranged as a flip-flop giving, in addition to the linear sawtooth, a gate with a very precise and accurately controllable length which is often used for ranging purposes or producing jitter-free delayed sweeps.

In this form the circuit is termed a " phantastron," and a typical arrangement is shown in Fig. 286. The suppressor is negatively biased in the stable state by the voltage developed across the cathode resistor R_K by the screen current. A short positive pulse triggers the tube by temporarily removing this bias and allowing anode current to flow thus driving the grid negative so that the drop across R_K is reduced almost to zero. The bias on the suppressor at the end of the initiating pulse is therefore almost zero and the anode voltage falls exponentially as before. A typical anode characteristic and loadline is shown in Fig. 286 (b). It is clear that when point A is approached the anode current ceases to increase, feedback is removed, and the grid voltage rises rapidly. This causes an increase in screen current, and hence an increase in suppressor bias so that the anode current decreases until the quiescent condition is attained. The duration of this cycle of operations is dependent on the initial voltage of the anode, which is usually prevented from rising above a predetermined value by a diode connected as shown in the circuit diagram. Owing to the practically linear fall of anode voltage, the gate length is quite accurately proportional to the setting of the potentiometer *R*. In practice a high value of plate load (up to 5 megohms) is often required and it is then usual to connect the anode to the feedback capacitor via a cathode follower. The circuit is then able to recover much more rapidly.[9]

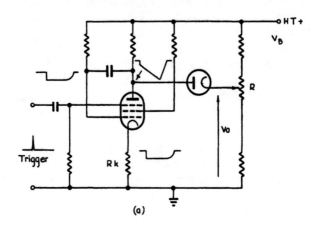

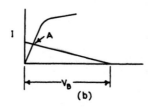

Figure 286.—The phantastron. (*a*) A typical circuit. (*b*) An anode current characteristic illustrating the operating conditions.

Feedback circuits reduce the curvature of the sawtooth wave but there still remains a residual effect due to the finite gain of the feedback system. Even this may be undesirable and it has been found possible in theory to eliminate all first order curvature by adding another feedback loop to the circuit of Fig. 285 (*a*) as shown in Fig. 287. The action of the additional

[9] F. C. Williams and N. F. Moody, "Ranging circuits, linear time base generators, and associated circuits," *J. Instn. Elect. Engrs* , Vol. 93, Part III A, 1946, pp. 1187-98.

resistor R_1 and capacitor C_1 is merely to feed a portion of the integrated output back to the grid of the cathode follower. This feedback voltage has the correct form to counteract the first order curvature of the sawtooth and by correct adjustment of C_1 and R_1 can be made to eliminate it entirely. If

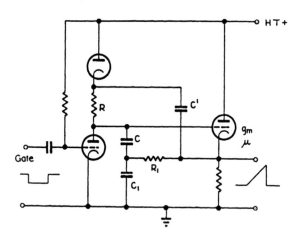

Figure 287.—Cancellation of first order curvature in a sawtooth.

C and C_1 are equal, as is usually the case, the requirement for cancellation is that :

$$R_1 = \frac{R}{4}\frac{(2A' - 1)}{(1 - A)} \tag{5}$$

where A' is the amplification of the cathode-follower alone, nelgecting the capacitor C'. Since the amplification A is involved, perfect cancellation cannot be relied upon in practice, but a reduction in curvature by a factor of five is quite feasible.[10]

6. Sweep Circuits

Electrostatic Deflection

In order to convert the linear sawtooth wave to a linear cathode ray tube sweep, an amplifier with push-pull output is norm-

[10] B. Chance, V. Hughes, E. F. MacNichol, D. Sayre and F. C. Williams (Eds.), *Waveforms* (Massachusetts Institute of Technology, Radiation Laboratory Series, Vol. 19), McGraw-Hill, New York, 1949, pp. 274-8.

ally required to eliminate trapezium distortion and defocusing of the trace. The ideal way of converting an unbalanced to a push-pull voltage is by means of the cathode
coupled or "long tail" amplifier of Fig. 288 (*a*).[11] This
circuit, in conjunction with one of the sawtooth generators
already described, is the usual form of deflection circuit. Its
advantages are freedom from power supply ripple and distortion, stability, and the ease with which it is possible to
apply a shift potential merely by returning one grid to a
variable potential. It is possible to reduce distortion and
increase stability still further by the introduction of cathode
feedback resistors as shown in Fig. 288 (*b*). For equal load

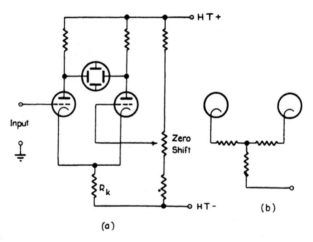

Figure 288.—Long-tail amplifier for electrostatic deflection cathode
ray tubes.

resistors, output balance is dependent on the ratio of the
common cathode resistance R_K to the resistance looking into
each tube from the cathode junction, including feedback
resistance if present. In order to obtain a good balance
without too large a voltage drop, the cathode resistor is often
replaced by a pentode.

11 O. S. Puckle, *Time Bases*, Chapman and Hall, London, 1943, pp. 119-25.

Magnetic Deflection

Any of the deflection coils described in section 1 give a deflection proportional to current with an accuracy sufficient for most practical requirements. The problem of producing a linear sweep therefore becomes that of producing in the coils a current which has a linear sawtooth waveform. This is often difficult, particularly when a fast sweep is required, because of the large inductance of the coil and the inevitable stray capacitance which shunts it.

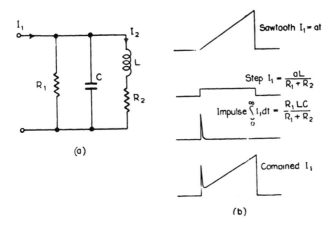

Figure 289.—The production of a linear current sawtooth in deflection coils.

An equivalent circuit of a deflection coil is given in Fig. 289 (a), where the impedance of the generator is given by R_1, the stray capacitance by C, and the inductance and resistance of the coil by L and R_2. The distributed capacitance of the coil may be considered to be included in C without great error. It can be shown[12] that to produce in the coil a current which has a perfectly linear sawtooth waveform, the applied current I_1 must consist of a linear sawtooth component, a step component, and an impulse as shown in Fig. 289 (b). The magnitude of the additional component waveforms is dependent

[12] T. Soller, M. A. Starr and G. E. Valley (Eds.), *Cathode Ray Tube Displays* (Massachusetts Institute of Technology, Radiation Laboratory Series, Vol. 22), McGraw-Hill, New York, 1948, pp. 356-83.

on the circuit constants of Fig. 289 (a) and the rate of sweep, a, as follows :

$$I_1 = a\left[t + \frac{L}{R_1 + R_2} + \frac{R_1 L C}{R_1 + R_2}\, \delta(t)\right] \qquad (6)$$

where $\delta(t)$ represents an impulse of unit magnitude. If the driving tube is a triode or pentode with the coil in its anode

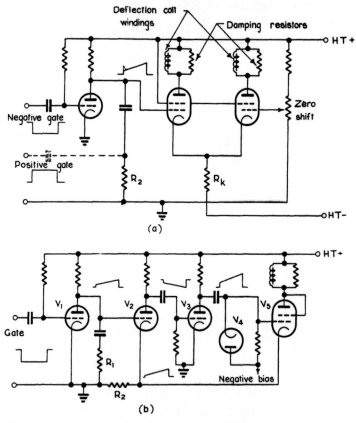

Figure 290.—Sweep circuits for magnetic deflection cathode ray tubes.
(a) Push-pull deflection utilising a long-tail amplifier. (b) A current feedback deflection circuit.

circuit, R_1 is large and the corrections small for most sweep velocities so that the impulse is often omitted.

The addition of the extra waveform components usually requires a slight modification to the sawtooth generators described in the last section. A simple deflecting circuit employing a long tail amplifier and useful for sweeps of duration greater than 50 to 100 microseconds is shown in Fig. 290 (a). The step is obtained by R_2 in series with C, and if required an impulse may be added by feeding a positive gate through a small capacity coupled to the top of R_2 (shown in dotted lines in the diagram). The linearity may be improved by any of the methods already discussed. Output tubes usually employed in this circuit are the 6V6, 6L6 or 807. When deflection coils which are not connected in push-pull are used, current feedback amplifiers are often employed to achieve good linearity,[13] a typical circuit being shown in Fig. 290 (b). The sawtooth amplifier consists of V_2, V_3 and V_5, current feedback being applied from V_5 to V_2 across the common cathode resistor R_2. The tube V_5 is normally biased well beyond cut-off so that there is initially no current through the deflecting coils and also no feedback voltage across R_2. The gain cf the amplifier is therefore extremely high and at the start of the sawtooth waveform the grid potential of V_5 rises rapidly until the tube begins to conduct. The voltage across R_2 then follows accurately the voltage at the grid of V_2, and the required current waveform is produced. An impulse of current may be added as in the previous circuit or by delaying the application of feedback with a low pass RC network in the cathode of V_2. The diode V_4 is a clamp to make the operation independent of the duty cycle.

For very fast sweeps a low impedance driving source is desirable. If R_1 in Fig. 289 (a) is of the order of a few hundred ohms and the sweep length less than about 100 microseconds, it is found with standard deflection coils that the impulsive and sawtooth components of the driving waveform become negligible, leaving only a simple step component to be applied. In this case the current increases exponentially with an initial rate of V/L, so that for the sweep to be linear the time constant

[13] L. Mautner, " A shipborne mechanical rotation PPI," Massachusetts Institute of Technology Radiation Laboratory, Report 62-8, June, 1943.

$L/(R_1 + R_2)$ must be much greater than the sweep time. This type of operation is made possible by using a cathode follower or, more efficiently, by using the circuit of Fig. 291 In this, a pentode or beam tetrode is biased by a negative grid voltage to plate current cut-off. The sweep is initiated by a positive gate which reduces the grid bias suddenly to zero. The resultant negative swing on the anode reduces its potential almost to zero (necessarily so since the anode current is initially zero) so that it is well below the knee of the anode

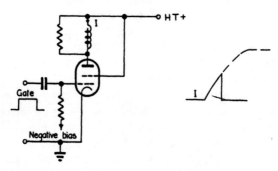

Figure 291.—A circuit for the production of fast sweeps on magnetic deflection cathode ray tubes.

characteristic curve. Under these conditions the effective anode resistance is of the same order as that obtainable with a cathode follower, but the voltage swing utilises the full HT supply. Using this type of circuit with two 807 driving tubes in parallel and an HT supply of 600 volts it has been possible to produce a sweep of approximately 1 inch per microsecond on a 7 inch tube. Faster sweeps are undoubtedly possible.

Recovery Time

In order to make the most efficient use of a sweep the recovery time is frequently required to be very short. This often leads to considerable trouble when using shielded iron core deflection coils. Eddy currents induced in the shields decay comparatively slowly when the current is cut off and may quite easily cause a deflection of the spot lasting until the next sweep commences. It can be shown that this deflection is always in the same direction as the sweep and proportional to it in magnitude so that if the sweep changes in magnitude the deflection is immediately apparent. A PPI display under these conditions exhibits an open centre as illustrated in

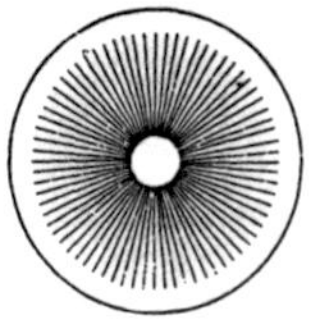

Figure 292.—Representation of a PPI display in which the time between sweeps is insufficient for deflection coil recovery.

Fig. 292. This effect often necessitates the use of unshielded or partly shielded coils, with consequent loss in efficiency on fast sweeps.

7. Pattern Stabilisation

The effect of mains voltage fluctuations on pattern size is nearly always of importance. In fixed installations where weight and complexity of the equipment are not too important it is possible to regulate all voltage supplies, and if feasible this is undoubtedly the most effective method. For light-weight portable sets, however, sufficiently good stabilisation can usually be achieved by careful design of compensating circuits. The factor of importance here is that the decrease of cathode ray tube deflection sensitivity with increasing gun voltage tends to be counterbalanced by the increase of sweep velocity with increasing supply voltage. If these two changes can be made equal and opposite the pattern will be a constant size.

In an electrostatic tube the problem is simple since deflection sensitivity is inversely proportional to the gun voltage and therefore to the mains voltage, so that it is merely necessary to produce a sweep directly proportional to mains voltage. The sawtooth generator usually satisfies this condition automatically, providing the sweep amplifier gain is independent of supply voltage.

Compensation for a magnetic tube is more difficult however since its deflection sensitivity is inversely proportional to the square root of the gun voltage. The requirement is therefore to produce a sweep velocity proportional to the square root of the mains voltage. It can be shown that the circuit of Fig. 293 performs this function, provides that $R_1/R_2 = V_1/V_2$.

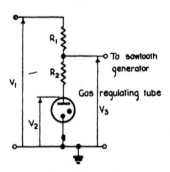

Figure 293.—A circuit for the stabilisation of the pattern in a magnetic deflection cathode ray tube.

The voltage V_3 then varies approximately as the square root of V_1 over a limited range.[13]

Compensation is thus usually possible for slow variations of mains potential within narrow limits. It is generally impossible to adjust time constants so that protection against transients is also achieved.

8. Blocking Oscillators

Blocking oscillators have proved to be extremely useful in many display applications particularly in ranging systems. A typical circuit is shown in Fig. 294 (*a*) and relevant waveforms in Fig. 294 (*b*). The circuit differs from that for a conventional oscillator principally in the extremely large coupling between anode and grid circuits, the respective coils being wound on a common iron core and therefore having a coefficient of coupling near unity.

In the example given, the number of turns on each winding is assumed to be equal, with the result that the voltage changes across the windings are approximately equal and opposite, and the peak grid current is practically equal to the peak anode current. If the grid coupling condenser C is small, a large negative voltage sufficient to stop further oscillation will be produced at the grid within a single cycle. Oscillation will recommence only when this negative voltage has decayed

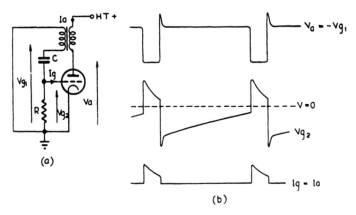

Figure 294.—Typical blocking oscillator circuit and waveforms.

through the grid leak R to the tube cut-off potential. The output therefore consists of a series of pulses with a repetition frequency determined mainly by the time constant RC, and a pulse length determined by the coil inductance and/or the grid coupling capacitance. Stray capacitance across the coil is normally not significant. The peak cathode current is usually very high and values up to 1 ampere are quite common with low power triodes such as the 6J5. The output impedance can therefore be made extremely small.

Blocking oscillators have found their main uses in stable relaxation oscillators, in triggered pulse generators, or in frequency dividing chains.[14] The use of hipersil cores and pulse transformer techniques has improved their performance im-

[14] R. Benjamin, "Blocking oscillators," *J. Instn. Elect. Engrs.*, Vol. 93, Part IIIA, 1946, pp. 1159-75.

mensely and the production of pulses of less than half a microsecond duration presents no difficulties.

9. Range Measurement

The measurement of range is essentially the measurement of
small time intervals. The time delay of received echoes is
one microsecond per 164 yards of range so that the time intervals to be measured extend up to about 2500 microseconds,
whilst accuracy is sometimes possible to about 0·06 microseconds or 10 yards. The basic method of measurement is
to compare the time interval between the transmitter pulse
and the reception of the echo with a time scale obtained from
an oscillator of known frequency. This does not represent
the actual measuring procedure on many systems but it is
always the basis of their calibration.

It is necessary that the range oscillator and transmitter
be accurately synchronised and this may be done in two ways.
Either the transmitter is fired at a fixed part of the oscillator
waveform, or the oscillator is switched on suddenly when the
transmitter fires, and switched off again at the conclusion of
the time base. In each method it is necessary to define
accurately a given point on a sine wave, usually the crest of
the wave in the case of CW oscillators, and always the zero
point of the wave in the case of gated oscillators. This is done
by producing short pulses at the specified instant which can
be used in the cathode ray tube display as range markers
(usually after mixing with the video signal) or in the case of
CW oscillators, for actuating counting-down circuits controlling
the transmitter firing. By interposing a calibrated phase
shifting network between the original wave and the range
mark generator it is possible to measure intermediate ranges
with accuracy.[15] This is in fact the most common technique
in precision ranging equipments.

[15] C. A. Laws, "A precision-ranging equipment using a crystal oscillator
as a timing standard," *J. Instn. Elect. Engrs.*, Vol. 93, Part IIIA, 1946,
pp. 423-40.

CW Range Oscillators

For very accurate work, crystal controlled oscillators are the most common and usually require the oscillator to be running continuously and the transmitter to be fired in synchronism. Methods will be described subsequently for combining a crystal controlled oscillator and a rotary gap modulator (one of which employs a triggered crystal oscillator), but they suffer from some limitation of duty cycle so that a hard tube modulator is usually necessary. CW range oscillators are also used in the calibration of other types of ranging system so that the method is of fundamental importance.

The most convenient method of counting down to the pulse repetition frequency is by means of blocking oscillators which are extremely well suited for this type of operation. Not only do they produce low impedance, short duration pulses which may be used directly for transmitter synchronisation and for range marks if required, but they have a small phase error. It is phase error which is important in a ranging equipment since it represents an actual error in range ; frequency division to a predetermined sub-multiple of the crystal frequency on the other hand is not always essential. In this respect a single blocking oscillator, dividing over a frequency ratio of 200 : 1, has been used, the maximum phase error resulting in a range error of only $\pm$ 20 yards.

In addition to the phase shifter plus marker technique, an excellent method for accurate ranging is to use the sine wave crystal oscillator output to produce a circular sweep on a radial deflection cathode ray tube. The video signal is placed on the central electrode and increments of range are marked out round the tube face. A gate of a duration equal to the oscillator period may then be applied to the tube to examine any particular range. This is often employed in conjunction with a Class A display which covers the useful range of the set and is used for searching.

The circuit of a simple system[16] involving a circular sweep generator, a blocking oscillator for range marking and a

[16] T. Soller, M. A. Starr and G. E. Valley (Eds.), *Cathode Ray Tube Displays* (Massachusetts Institute of Technology, Radiation Laboratory Series, Vol. 22), McGraw-Hill, New York, 1948, pp. 296-302.

frequency dividing blocking oscillator for firing the transmitter is shown in Fig. 295. The crystal oscillator V_1 generates a circular sweep by means of the transformer T_1 in its anode circuit, one of the secondary windings being tuned 45 degrees below resonance and the other 45 degrees above resonance.

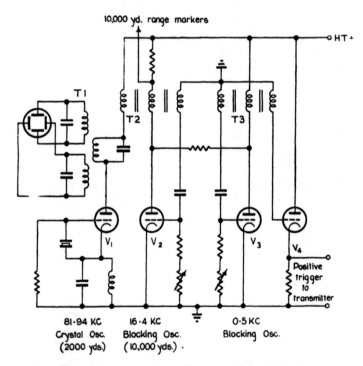

Figure 295.—A typical frequency dividing circuit for use with a crystal controlled range oscillator.

The anode current of V_1 consists of a series of pulses which is used to synchronise the blocking oscillator V_2, producing 10,000 yard range marks. The blocking oscillator V_3 is synchronised by the output of V_2 at approximately 500 cycles per second and feeds a transmitter trigger pulse through the cathode follower V_4. The accuracy with this type of circuit in which no special precautions have been taken as regards phase locking is about $\pm$ 20 yards.

If greater accuracy is required a coincidence circuit, of
which an example is shown in Fig. 296 is normally employed.
The series of pulses from the range oscillator, in this case
spaced 2,000 yards or 12 microseconds apart, is used to
synchronise a blocking oscillator at the pulse repetition frequency (1·5 kilocycles per second). The trailing edge of the
blocking oscillator pulse triggers a gate generator, producing a
positive gate of about 200 volts amplitude and a duration of

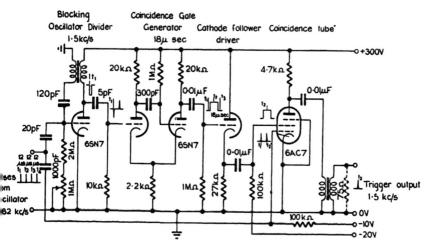

Figure 296.—A frequency dividing circuit with accurate phase lock for
a precision ranging equipment.

18 microseconds. The coincidence tube is so arranged, with
negative bias on its grid and screen, that an output is only
possible when an oscillator pulse and this gate coincide. The
pulse immediately following the one which triggers the blocking
oscillator divider therefore appears at the anode of the coincidence tube and is used for triggering the transmitter. The
only phase errors then are those which occur in the production
of the original pulses from the ranging oscillator, and in the
triggering of the transmitter. Both these will normally be
extremely small.

It is possible to employ a CW crystal oscillator when using
a rotary gap modulator, providing the duty cycle is sufficiently

low. The procedure is to interpose successive sweeps triggered by the transmitter and carrying radar signals with sweeps triggered by the range oscillator and carrying range marks alone. Owing to screen and eye persistence the two sweeps are seen simultaneously, and give exactly the same effect as if the respective signals had been mixed prior to placing on the screen.[15]

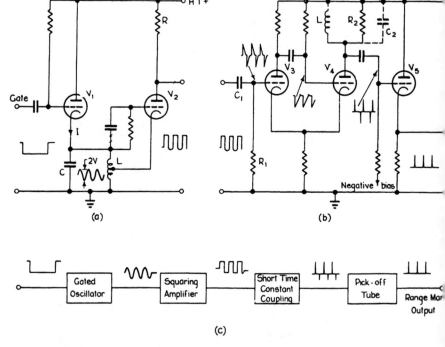

(a)   (b)

(c)

Figure 297.—Gated oscillator and range mark generator. (a) The gated oscillator. (b) The range mark generator. (c) A typical block diagram.

Gated Oscillators

When used with rotary gap modulators, or any modulator where the firing time cannot be accurately specified, the ranging oscillator is normally gated. A typical circuit for this purpose is shown in Fig. 297 (a) together with the important waveforms. The tube V_2 is connected across the tank circuit

LC to form a Hartley oscillator and V_1 is a clamp tube which is normally conducting, the current being 7 to 10 milliamperes in most applications. The clamp thus presents a low impedance across the tank circuit and prevents the build up of oscillations. A negative gate applied to V_1 cuts off this tube, removes the clamp and at the same time shocks the circuit into oscillation. By adjusting the anode load R and the tapping point on the coil it is possible to compensate exactly for dissipation in the tank circuit and to maintain a constant amplitude of oscillation given by

$$V = I\sqrt{\frac{L}{C}} \tag{7}$$

where V is the peak value of the oscillation and I is the interrupted cathode current of V_1. If V_2 is unbiased there is a very rapid change of anode current every time the grid crosses the zero axis so that corresponding times are accurately defined.

Circuits of this type are not often employed if the highest accuracy in ranging is required but are very satisfactory for search sets where the required 1 to 2 per cent accuracy can be achieved without any special precautions. In some systems which employ this method for precision ranging a constant check is kept on the frequency by means of a built-in crystal calibrating oscillator, and in this case, by careful design the accuracy can be made to approach that of the crystal oscillator itself.

As mentioned previously crystal oscillators have also been successfully gated but the circuit is quite complex at present and has not been used extensively. The problem here is not to initiate the oscillation—any suddenly applied voltage across the crystal will do this—or to maintain it at constant amplitude over the few cycles normally required since the Q is so high it cannot decay appreciably in this time. The difficulty arises in completely stopping the oscillation at the end of the gate and before the next sweep commences. To achieve this, it has been found necessary to apply suddenly a large amount of negative feedback thus producing across the crystal terminals a low impedance which rapidly damps the oscillation.[17]

There are many methods by which range marks can be produced from a gated sine wave but they can usually be summarised by the block diagram of Fig. 297 (c). As shown there the sine wave is passed through a high gain distorting amplifier which produces a square waveform output which crosses the time axis at points which coincide with the corresponding points of the sine wave. This output is then fed through a short time constant coupling circuit to produce a series of positive and negative pulses from which a " pick off " tube selects those of correct polarity, and possibly improves their shape.

One of the most useful forms of circuit which can be used to follow an oscillator like that of Fig. 297 (a) is shown in Fig. 297 (b). In this the tubes V_3 and V_4 form a cathode-coupled flip-flop which, in conjunction with V_2 of Fig. 297 (a), may be considered as the high gain distorting amplifier. The output of V_2 is fed through a fairly short time constant C_1R_1 to avoid difficulties of DC restoration, and is used to trigger the flip-flop. Pulses are produced by the circuit LR_2C_2, which is approximately critically damped and in which C_2 represents stray capacitance. The tops of the positive pulses are " picked off " by the cathode follower V_5 which is biased beyond cut-off, and appear as range marks across its cathode. It is generally possible to obtain range marks of about 20 to 30 volts amplitude and less than one microsecond duration by this method. If shorter duration, larger amplitude or lower output impedance is desired it is usual to replace the cathode follower by a blocking oscillator which is biased so that it can only fire when triggered by these pulses.

Resistance Capacity Ranging Circuits

Although the oscillator techniques already described represent the basic method of range measurement, many sets employ systems derived from the sawtooth generators of section 5. There are two variations of this type of system in common use,

[17] D. J. Mynall, "A pulsed crystal oscillator circuit for radar ranging," *J. Instn. Elect. Engrs.*, Vol. 93, Part IIIA, 1946, pp. 1207-14.

one being to start a sawtooth of known slope coincident with the transmitter pulse and measure its voltage at the instant the echo returns, the other to measure the starting potential

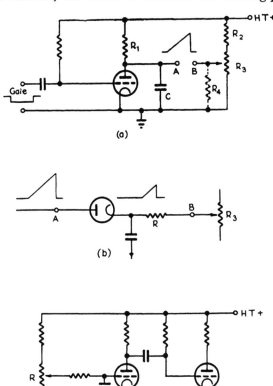

Figure 298.—Resistance capacity ranging circuits. (a) Typical bridge circuit. (b) Pick-off diode. (c) Controllable gate length flip-flop.

of a sawtooth which is required to give a certain fixed voltage at that instant.

Fig. 298 (a) represents a simplification of an extremely common type of bridge circuit used with Class A displays. A sawtooth is produced at A and compared with an accurately known DC voltage at B, derived from the precision potentiometer R_3. If A and B are each connected to a grid of a long tail sweep amplifier (section 6) and the potentiometer R_3 is adjusted so that an echo appears at a predetermined point on the cathode ray tube screen (determined for equal voltages at A and B), then the range may be read off directly on a dial attached to the moving arm of R_3. For good accuracy the dotted resistance R_4 is often inserted to compensate for the first order curvature in the sawtooth, compensation being exact when it is equal to $2(R_2 + R_3)$. Alternatively one of the methods of sawtooth linearisation described previously may be used in conjunction with a linear potentiometer. A set employing this type of ranging system but with a rather more complicated potentiometer technique has been constructed to give an accuracy of 0·05 per cent.[18] This is essentially a null method and it is often more convenient to produce a range marker at a specified range.[10] This is easily achieved by connecting a diode between points A and B as in Fig. 298 (b). When the voltage at A reaches the potential of B the diode begins to conduct and a potential as shown is developed across R and may be used to trigger a range-mark generator.

Another technique is represented by various forms of controllable gate generator, of which the simplest is a cathode coupled flip-flop in which the trailing edge of the gate triggers a marker generator. The grid of the initially non-conducting tube is returned to a variable voltage [Fig. 298 (c)], and it can be shown that for certain combinations of resistors the gate length is quite closely proportional to this voltage, so that it is possible to calibrate the potentiometer R directly in terms of range.[8] Accuracies better than 0·5 per cent are possible by this method, but it suffers from microphonic effects.

The phantastron (section 5) is also used in this type of system and can give accuracies approaching 0·1 per cent with fairly frequent calibration and could be relied on to 0·5 per cent over long periods.

[18] J. H. Piddington and L. U. Hibbard, "A precision time-base and amplifier developed for radar range measurement," *J. Instn. Elect. Engrs.* Vol. 93, Part IIIA, 1946, pp. 1602-10.

10. Some Circuit Arrangements

The operation of the basic circuits into which most display systems can be resolved having been explained, it will now be shown how these circuits are connected together in some typical systems.   In order to make the operation of the circuits more obvious, nothing but the bare essentials has been given by omitting range switching and monitoring facilities.   The importance of the latter should not be under-estimated, however, since the complexity of the circuits and the number of the components require easy and systematic fault finding procedures.   To assist in this, all tube currents are normally monitored and all important waveforms are brought out at test points.

Class A Display (Fig. 299)

This is a display giving Class A indication on a 5 inch cathode ray tube.   Maximum range is 180 miles with provision for expanding any selected portion of the sweep by a factor of eight.   It is usable with a pulse repetition frequency up to 400 cycles per second and has a range accuracy of about 1 per cent.   Range is measured by lining up the echo on a hair line across the tube face by rotating the range potentiometer.

Operation of the circuit is made clear by the block diagram. The tube V_1 is an anode-coupled flip-flop giving the required gate length of approximately 1,900 microseconds, the negative output of the tube being utilised to cut off the clamp tube V_2 and the positive output to brighten the cathode ray tube for the duration of the sweep.   The sawtooth produced at the anode of V_2 is fed to the long-tail amplifier V_3, V_4 and thence to the horizontal deflection plates of the cathode ray tube.   Sweep expansion is achieved by increasing the voltage across the charging resistor and range potentiometer by means of the ganged switches.   A point of interest here is that the circuit arrangement adopted, unlike that shown in Fig. 298 (a) keeps the comparison potential constant and alters the potential at which the sawtooth commences.   This has the merit that the long-tail amplifier tubes are always working under the

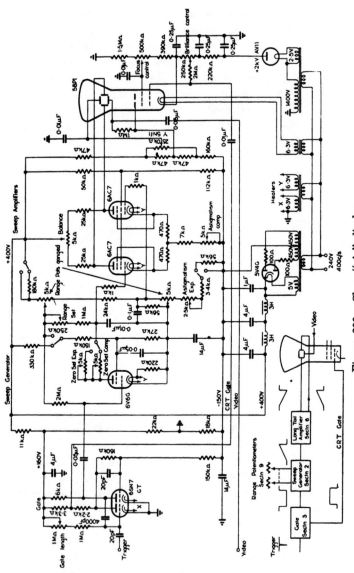

Figure 299.—Class " A " display.

same conditions when a range measurement is being made, and the mean potential of the deflecting plates is constant, leading to improved focus. Adjustable controls in the gate and amplifier circuits are all pre-set and need very little attention.

PPI with Rotating Deflection Coils (Fig. 300)

This represents a very flexible type of PPI circuit. The example shown in Fig. 300 has a sweep length of 120 miles on a 7 inch cathode ray tube with range marks at 20 mile intervals, the performance being independent of pulse repetition frequency up to about 600 cycles per second. The deflection coils are rotated in synchronism with the aerial by means of a selsyn transmission system.

The block diagram indicates that the techniques used are mostly those which have already been described in detail. Some points worth noting however are, firstly, the application of the cathode ray tube gate via an isolating stage in order to avoid any interference with the gate length by the video signals, secondly, the provision of coarse and fine bias controls on the cathode ray tube grid (which is desirable owing to the large variations in grid base found with these tubes) and thirdly, the method of mixing the video signals and range marks. This method is non-additive and prevents overdriving of the cathode ray tube screen when an echo coincides with a range marker.

PPI with Fixed Deflection Coils (Fig. 302)

It is not always possible or desirable to employ rotating deflection coils in a PPI system so various methods have been developed for producing a rotating sweep with fixed coils. The principle is simply to modulate the velocity of sweeps operating in a number of fixed directions by the direction cosines of the angles between the required sweep and these directions. The procedure is illustrated in Fig. 301 where the rotating sweep is built up out of two sweeps at right angles. It is seen that it is necessary to modulate these sweeps so that their velocity is proportional to the sine and cosine of the angle which the

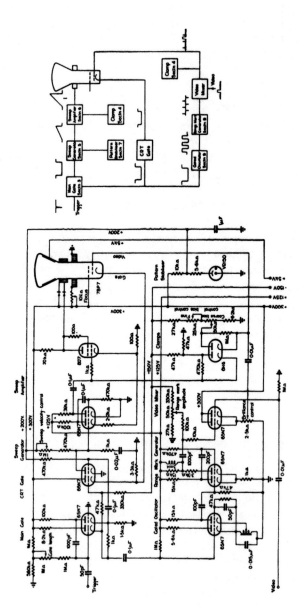

Figure 300.—PPI with rotating deflection coils.

required velocity sweep makes with one of the component sweep directions.

A number of techniques is available for performing this function one of the most common of which employs a selsyn control transformer and is described here.   Capacitance voltage dividers and potentiometers have also been used to derive the sinusoidal modulation.   The output of a gate generator in the circuit of Fig. 302 is applied to the cathode ray tube and to a sawtooth generator for a period of approximately 640 microseconds (60 miles).   The resulting sawtooth is amplified by an 807 power tube and fed through the rotor of a two phase

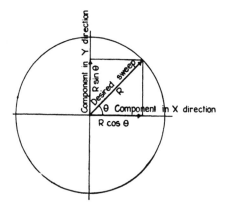

Figure 301.—Illustrating the production of a rotating sweep with two fixed deflection coils at right angles.

selsyn.   The rotor is geared directly to the aerial so that the stator outputs have sawtooth waveforms, the velocities of which are proportional to the required sine and cosine functions of the aerial position.   These waveforms are passed through the double-ended clamps to the 6V6 sweep amplifiers feeding the deflection coils.   The sweep direction is therefore synchronised to the aerial direction.

This type of system is inherently less accurate than one employing a rotating yoke, since range and bearing are not displayed directly in polar coordinates as measured, but are

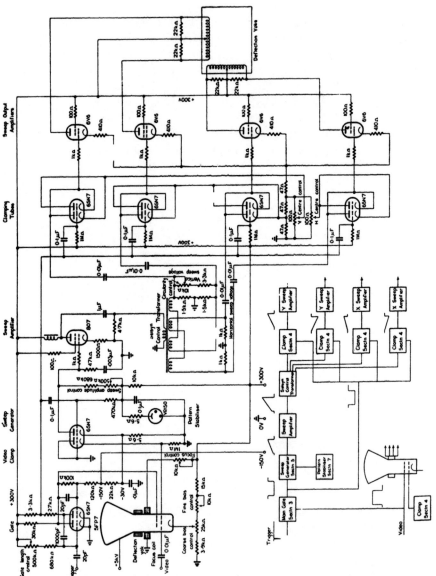

Figure 302.—PPI with fixed deflection coils.

first converted to Cartesian coordinates which are then recombined to give the polar form. The possibility of error is therefore greater.

Range Azimuth Display (Fig. 303)

Some of the circuit arrangements employed in high resolution sets are illustrated in the example of Fig. 303. It consists of an intensity modulated display in which range is measured along the Y axis and azimuth angle along the X axis so that the information obtained is similar to that obtained from a PPI, but the picture is distorted. The aerial is scanned back and forth through a small angle and a voltage proportional to its angular deviation from the central position is obtained from a potentiometer coupled to the aerial motion. This azimuth voltage is fed to the long-tail X axis amplifiers so that the horizontal displacement of the vertical range sweep is always proportional to azimuth angle. The range sweep is a little over 1,000 yards long and can be delayed by any amount up to 40,000 yards, the intention being to use this display in conjunction with another having a slow sweep covering the whole of this distance. It is useful with sets employing a very short pulse length of the order of 0·1 microseconds or less and therefore having a high resolving power.

The operation of the delayed sweep generator can be followed with the aid of the block diagram. A trigger pulse, delayed by a time proportional to the setting of the range potentiometer, is produced by the linear sawtooth generator and pick-off diode. This is amplified and used to trigger the delayed gate which switches on the cathode ray tube and initiates the sweep. A low impedance drive is employed here and it results in a sweep which is linear to better than 5 per cent. In order to utilise the full area of the tube face the sweep is made to commence considerably below the centre by means of an initial deflection coil carrying direct current and wound over the range coils. A range marker is produced in

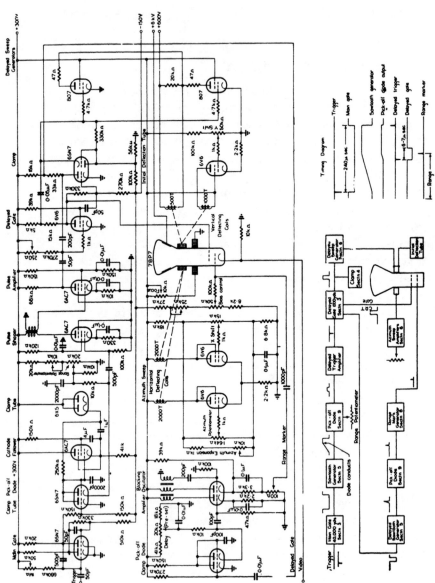

Figure 303.—A delayed sweep range-azimuth display.

the centre of the cathode ray tube by means of a blocking oscillator fired by a fixed delay circuit initiated by the variable delay trigger. When this marker coincides with an echo the range may be read directly off the range potentiometer. Accuracy with this circuit depends to a large extent on the accuracy of the range potentiometer, with a lower limit of about ± 25 yards if the error in the potentiometer is negligible and if the circuits have been carefully temperature compensated.

CHAPTER XV

AUTOMATIC RANGING CIRCUITS

1. Applications of Automatic Ranging Circuits

As described in Chapter X, a cathode ray tube is normally employed in radar sets for display purposes, but there are occasions when it is desirable to make the radar information available in the form of a meter indication or the angular displacement of a shaft. This occurs in certain airborne gun-directing sets which use automatic circuits to feed range directly to the gun computer, thus eliminating the radar operator. Again, in the field of navigational aids for civil aircraft meter indicators for the pilot are preferable to cathode ray indicators for a number of reasons. Among these are the economy of dash-board space achieved by the meter, the elimination of difficulties produced by large variations in ambient light intensity, and smooth presentation of information to the pilot who cannot afford to look at a cathode ray tube for signals which may be fading or obscured by interference.

Meter indicators may be used with either self-synchronised or externally synchronised radars. The first type covers distance-measuring radars in which a measurement is made of the time interval between the emission of pulses from a local transmitter and the reception of echoes from targets or of beacon responses. The second includes the hyperbolic navigational systems in which the measurement of the time delay between pulse transmissions received from fixed ground stations locates an aircraft or ship on a known line. Examples of the two applications are outlined in Chapter XVIII.

In both of these cases meter-indicating circuits operate in virtually the same fashion, so that the following description will be based on a self-synchronised system with a short note on the adaptation to an externally synchronised system.

Operation in the presence of complex echo patterns, such as are obtained from ground reflections, is of course beyond the capabilities of an automatic circuit and the application is therefore limited to those radars in which only discrete targets or beacon responses are picked up.

2. General Principles

Under ideal conditions the measurement of time delay between two trains of pulses, as shown in Fig. 304 (a), might be achieved very simply by feeding them as trigger pulses to a suitable multivibrator circuit, thereby producing a rectangular voltage wave [Fig. 304 (b)] of duration equal to the time delay. The

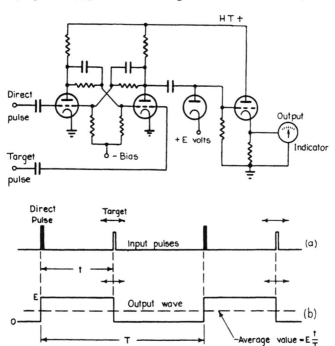

Figure 304.—Simple method of measuring range when only one target pulse and no interference is present.

average value of such a rectangular wave is proportional to its duration so that its application to a suitably calibrated moving-coil meter would give a direct indication of distance. However under practical conditions the picture of received pulses is more likely to be as shown in Fig. 305 where there is more than one target or beacon from which pulses arrive somewhat intermittently, and accompanied by varying degrees of interference. In this case the simple directly-actuated circuit described above will fail completely, and it is clear that a practical circuit must have properties which allow it to pick out the right pulses, and to give a smooth output reading with intermittent input information. The development of such a circuit follows logically from a study of the more familiar and flexible cathode ray tube display. The latter has some form of recurrent time base on which is displayed the video

a - direct pulse
b,c - targets
r - random interference

Figure 305.—Typical pattern of received pulses under practical conditions.

output from the receiver during a certain time interval after the transmitter pulse. In using this display the eye searches along the time base for a synchronised signal, and completely disregards all random pulses. On finding the required signal the search is stopped and its distance is measured off the appropriate scale. Should the signal fade for short intervals its position is remembered and it is picked up again when it reappears. The velocity of the target may also have to be remembered if its movement during fades is appreciable.

An auto ranging unit will therefore possess the following electronic counterparts of the above visual processes (Fig. 306) :

(a) A short gate pulse, generated after each transmitter pulse and movable over the time delay interval appropriate to the radar (the "eye" mechanism).

(b) A gated or coincidence amplifier to which this gate pulse is fed together with the video output of the receiver, and which only delivers an output when the gate pulse coincides in time with a video pulse. This is the mechanism of "seeing" a pulse. Actually it turns out that two gate pulses are needed, one just behind the other and feeding two gated amplifiers, to gain information as to the direction of movement of the target.

(c) A variable delay circuit for setting the position of the gates, controllable by means of a DC voltage, which is also applied to a meter calibrated in terms of range.

(d) A slow sweep voltage generator to be connected to the delay circuit so that the gates can be swept in and out in search of a target.

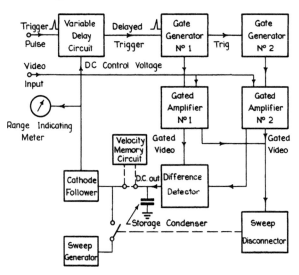

Figure 306.—Block schematic diagram showing the component parts of an automatic ranging unit.

(e) A sweep disconnector to stop the search when a target is found. This will include some form of pulse counter to discriminate against random pulses, and there will also be a suitable delay before the search is switched on again after a target disappears.

(f) A tracking mechanism consisting of a difference detector fed from the two gated amplifiers, and providing an increasing or decreasing DC output according as the target overlaps one gate more than the other. This, when applied to the delay circuit, will keep the gates moving along with the target.

The detector must have a large storage condenser which will maintain a substantially constant potential during signal fades, thus giving position memory. A cathode follower will be necessary to provide DC isolation between the condenser and the metering circuits.

An alternative tracking method which may be adopted in special circumstances is to provide the DC control voltage from a motor driven potentiometer, the motor speed and direction being controlled by a differential field system fed from the rectified gated amplifier outputs.

(*g*) If more than one target is present a trip button will be required to release the circuit from one target and allow it to look for the next one.

(*h*) An additional velocity memory circuit may have to be introduced if the motion of the target during periods of fading is liable to take it out of the gates.

There are many features of a " servo " mechanism in the above items, and the general theory of servos[1] is applicable to some aspects of the circuit design.

3. Circuit Design for a Self-Synchronised Radar

Variable Delay Circuits

The delay circuit is the heart of an auto ranging system and is discussed first. Any of the types described in Chapter XIV may be used although the phantastron is usually chosen because it combines good stability with reasonable economy in the number of tubes required. Two typical circuits for a 100 mile maximum delay phantastron are given in Fig. 307 using a 6SA7 pentagrid tube and a CV138 (miniature pentode) respectively. Minimum delay is commonly about 1 per cent of the maximum. Where the control voltage is directly metered, as in these circuits, the anode load of the phantastron may be made much smaller than the conventional value because a little current flowing through the anode diode will have no effect on the metering circuits. The cathode follower between

---

[1] L. A. Maccoll, *Fundamental Theory of Servomechanisms*, Van Nostrand, New York, 1945; H. M. James, N. B. Nichols, R. S. Phillips (Eds.), *Theory of Servomechanisms* (Massachusetts Institute of Technology, Radiation Laboratory Series, Vol. 25), McGraw-Hill, New York, 1947.

anode and grid may then be eliminated. Miniature tubes in general seem to be much less stable in delay circuits than their full sized counterparts, probably because of their more delicate construction. A 6SN7 twin triode used as a delay multivibrator will give reasonable stability, but a 6J6 miniature tube in a similar circuit may suffer from drifts of 2 to 3 per cent of the maximum delay.

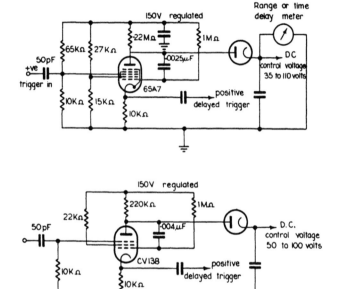

Figure 307.—Typical phantastron variable delay circuits.

Gate Generators

Any of the methods of generating short pulses described in previous chapters may be used to produce the delayed gates. In general the blocking oscillator method is the most useful as it gives a pulse at low impedance with a small high tension drain. A typical circuit for generating a pair of 4 microsecond gates in rapid succession is given in Fig. 308. Here the tail of the first pulse triggers the second oscillator. It is also sometimes possible to use the positive overswing on the anode winding of a single blocking oscillator as the second gate pulse. The required length of the gate pulses is a compromise between two factors. Firstly a high speed of search is naturally desired and this dictates a wide gate so that targets can

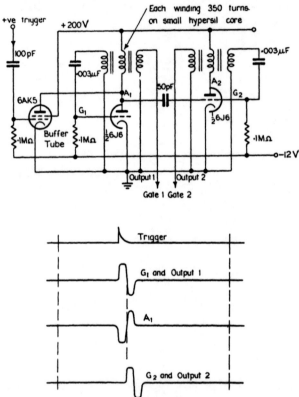

Figure 308.—Circuit for generating short gate pulses using a blocking oscillator.

be " seen " before the gate runs through them. Secondly a narrow gate is required to prevent interference getting through sufficiently often to actuate the search disconnector. The probability of this occurrence may be calculated for any given set-up, and should be made suitably low by choosing the correct gate length.

Gated Amplifiers

Any multiple grid tube will serve as a gated amplifier. Positive
video signals are fed to one grid and positive gate pulses to

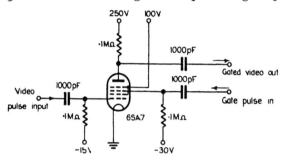

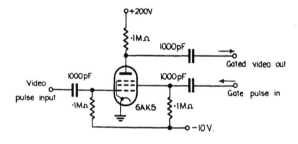

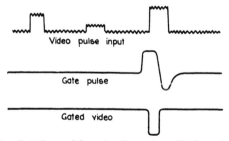

Figure 309.—Gated amplifier circuits using 6SA7 and 6AK5 tubes.

another grid, both having a cut-off bias applied. The result
is that anode current can flow only when both grids are " on "
simultaneously and the output at the anode consists of negative

pulses whose duration corresponds with the overlap between gate and video pulses. These are hereafter called "gated video pulses." Fig. 309 gives typical circuits and waveforms. Where a screen grid is used for gating in tubes in which the suppressor is internally connected the gate pulse generator must be capable of supplying the screen current required by the tube.

Difference Detector

The difference detector is required to produce a rising or falling DC output according as one or other of the gated amplifiers is producing longer gated video pulses. This potential being fed back to the delay circuit will then tend to hold the gates centrally over the target. Fig. 310 (a) shows

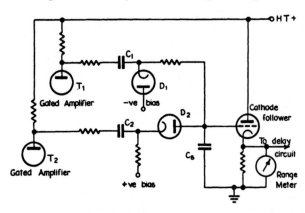

Figure 310 (a).—Difference detector circuit using a pair of diodes to produce a DC voltage suitable for maintaining the variable delay circuit in track with the target.

a circuit using two diodes feeding in opposite directions to a common storage condenser. Diode D_1 conducts for the duration of each gated video pulse from T_1, and the condenser C_1 acquires a proportional charge which subsequently is fed into the storage condenser C_s. Similarly D_2 conducts for the duration of each gated video pulse from T_2, taking a proportional charge out of C_s. The net effect is that the potential on C_s remains constant if the two gated amplifiers produce equal pulses, increases if the upper one produces longer pulses, or decreases if the lower one does so. In the absence of signals both diodes are non-conducting so that the

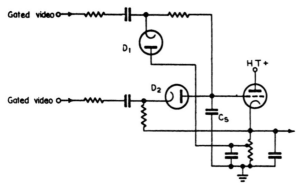

Figure 310 (b).—Circuit of Fig. 310 (a) modified to produce an approximately constant DC bias on the diodes.

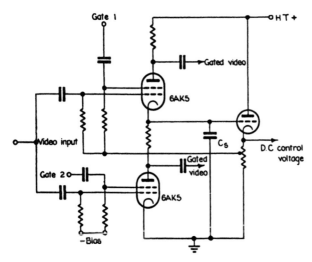

Figure 310 (c).—Difference detector using two pentodes.

condenser C_s maintains its potential constant to the extent determined by leakage, which must be kept small. In this circuit the bias on the diodes varies according to the potential on C_s so that there are varying delay voltages to be overcome before the diodes conduct. This effect can be minimised by connecting the diodes to the cathode follower load as in Fig. 310 (b).

When a relatively small range of potential is required from C_s it is possible to eliminate the diodes and use the arrangement of Fig. 310 (c). Here the gated amplifiers are connected one above the other, one feeding charge into and the other taking charge out of the storage condenser. With this connection, however, the gated video pulses vary in amplitude.

Search Sweep Generator

Searching may be accomplished by connecting the storage condenser C_s into a suitable relaxation oscillator circuit. Fig. 311 (a) illustrates a typical circuit in which the storage condenser is charged at a slow rate through a high resistance connected to the high tension supply and a thyratron is arranged to discharge the condenser when the gates have been swept out to the maximum time delay. The grid of the thyratron is returned to a point on the cathode load of the cathode follower to make its point of firing more definite than would be obtained with a fixed bias.

An interesting sweep circuit using a pentode tube is shown in Fig. 311 (b). This is termed a "Miller-transitron,"[2] and it combines both a transitron and a feedback time constant action. In operation the grid potential tends to rise by charging through the grid leak R, but with an amplified time constant ARC, where A is the stage amplification. The potential at the grid therefore rises linearly and the corresponding fall in anode potential is also linear. This fall continues until at low anode potentials the screen current increases and causes the screen potential to fall. The screen is coupled to the suppressor so that the potential of the latter also falls causing a reduction in anode current with a corresponding increase in screen

---

[2] B. H. Briggs, "The Miller integrator," *Electronic Engng.*, Vol. 20, 1948, p. 282.

current.   At a certain point this action becomes regenerative, the suppressor is cut off altogether and the anode returns to the high tension level.   The suppressor then recovers according to its time constant and when anode current can flow again

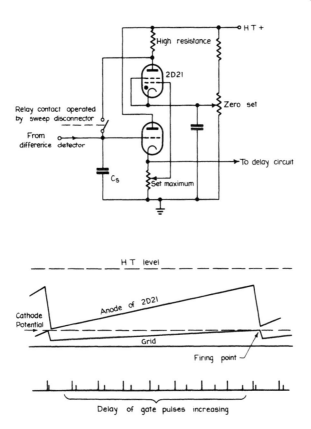

Figure 311 (a).—Circuit for producing a slow sawtooth sweep voltage. This is applied to the variable delay circuit to cause it to search for a target.

the anode and grid potentials drop rapidly together by an amount approximately equal to the grid base of the valve. The screen is then restored to normal potential and the cycle repeats from this point.   Values shown on this circuit will give a sweep time of about 10 seconds with a short recovery time. The grid leak may be open circuited at any point on the sweep and the anode potential will remain constant except

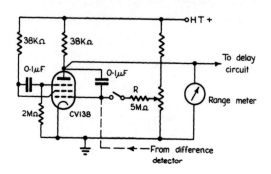

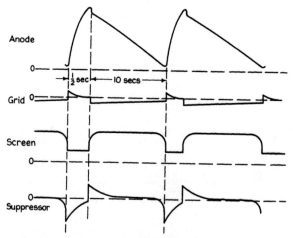

Figure 311 (*b*).—Alternative sweep generator using a pentode in a "Miller-transitron" circuit.

for small leakage effects. The condenser *C* then acts as a large storage condenser and the difference detector can be fed directly into the grid. Thus sweeping and tracking are accomplished with a single valve. There is however a slight disadvantage in that the sweep goes negative and therefore the search proceeds from maximum to minimum range.

"Hesitation" Search Circuit

A further refinement which may be incorporated into the design is what may be termed a "hesitation" search circuit. This circuit causes the search to be slowed down wherever the gates encounter a video pulse so that in effect they "stop to look at" the latter and move on if it is an isolated interference pulse.

To achieve this in the case of a search proceeding from minimum to maximum range, the output of the late-gated amplifier is disconnected from the difference detector during the searching phase, using an auxiliary pair of contacts on the sweep-disconnector relay. Circuit constants are so chosen that the difference current—now undirectional and opposed to the sweep current—is approximately equal to the latter. This causes the sweep to slow up whenever the early gate meets a video pulse. Exact adjustment will halt the sweep completely and over-adjustment will cause the gate to "bounce against" the video pulse.

The latter state of affairs is, of course, not desirable since excessive slowing up of the sweep will result when interference pulses are numerous.

When the "hesitation" circuit is employed a longer sequence of pulses will become available to actuate the sweep-disconnector circuit so that the latter can be designed for better discrimination against interference pulses, or conversely the nominal sweep rate may be considerably speeded up while maintaining the same discrimination against interference.

The circuit will therefore be seen to offer considerable advantages in applications where clutter is bad.

Sweep Disconnector

A sweep disconnector circuit is illustrated in Fig. 312 suitable for a radar working with a pulse repetition frequency of

100 cycles per second on beacons with a 4 microsecond pulse. Pulses from both of the gated amplifiers are fed to the diode T_1 whose charging time constant R_1C_1 is such that a gated video pulse builds up a positive potential of about four volts at the grid of T_2 for this particular example. The discharge time constant R_2C_1 is such that about two volts of this leaks away in a repetition period. As the gates are moved out in range they make chance encounters with random pulses which produce momentary positive pulses at the grid of T_2, but the effect of these pulses is not cumulative since they are fairly widely separated. However, when the gates encounter the beacon pulse a rapid succession of gated video pulses is produced and the potential at the grid of T_2 builds up faster than it leaks away. The cathode of T_2 follows this rise and causes T_3 to conduct, closing the relay in its anode circuit and disconnecting the sweep. If the signal fades the potential at the grid of T_2 drops but the cathode cannot follow it down because of the long time constant R_3C_2, and the relay remains closed. If the signal disappears altogether the relay opens after several seconds.

When "on-off" coding of the beacon is used the code may be reproduced by a neon lamp in the anode circuit of a valve T_4 whose grid is connected to the grid of T_2.

Trip Button

By momentarily shorting the cathode condenser of T_2 the relay K_1 can be made to open and allow the search to carry on to pick up targets at greater distances.

Complete Circuit

Fig. 313 shows the complete circuit of an experimental automatic ranging unit designed for a 100 mile interrogator-responsor set, and using miniature valves. A picture of the unit and its indicator is given in Plate XXI.

Analysis of Tracking Action

The simple automatic tracking circuit described in the foregoing has been developed by wholly intuitive methods and it is instructive at this stage to study the tracking action analytically. Considering, first, the case of a stationary target, a displacement of the gates from the equilibrium position will result in the difference detector producing a current proportional to the displacement since the effect depends on the *difference* in the lengths of the two gated video pulses. The storage condenser C integrates this current and the potential V which controls the variable delay circuit is given by

$$V = \frac{1}{C} \int i, \text{ where } i = \text{difference current}$$

or
$$\frac{dV}{dt} = \frac{i}{C}. \tag{1}$$

Now if R and R' are respectively the target range and the indicated range, we have

$$R' = K_1 V \tag{2}$$

and
$$i = K_2 (R - R'), \tag{3}$$

where $K_1 =$ control factor of the variable delay circuit in miles per volt, say,

$\quad K_2 =$ proportionality factor of the difference detector in amperes per mile of misalignment.

Hence from (1), (2) and (3)

$$\frac{1}{K_1} \frac{dR'}{dt} = \frac{K_2}{C} (R - R')$$

i.e.
$$\frac{C}{K_1 K_2} \frac{dR'}{dt} + R' = R.$$

Writing $T = C/K_1 K_2$, the solution of this equation is
$$R' = R (1 - e^{-t/T}).$$

Thus the quantity T may be considered as the time constant of the tracking circuit and is a measure of the time taken to respond to any sudden target displacements. An extension of the analysis to the case of a target moving with constant velocity U results in the steady-state solution

$$R = R' - TU.$$

Thus there is a range lag of magnitude TU when tracking at constant velocity.

BR

As a numerical example the case of a typical circuit with a 100 mile maximum range will be considered. Accurate calculation of the constant K_2 is somewhat involved since it depends on the characteristics of the gated amplifiers and series impedances in the difference detector circuit. However, each gated video amplifier will pass a current of the order of 10 milliamperes peak into the difference detector. These currents

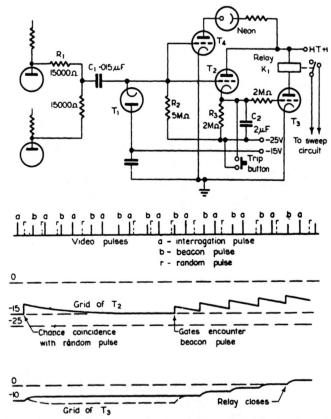

Figure 312.—Circuit for disconnecting the search sweep from the variable delay circuit when a target is found.

cancel each other under equilibrium conditions but, with a range misalignment of 1 microsecond, one gated video pulse

PLATE XXI.—An auto-ranging unit and indicator, using miniature tubes.

facing p. 500]

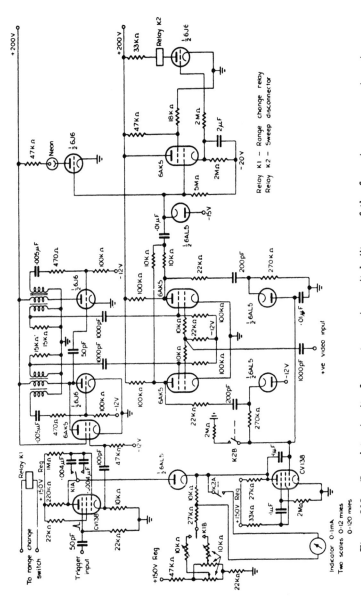

Figure 313.—Complete circuit of an auto-ranging unit built up of the foregoing component parts.

will be shortened by 1 microsecond and the other lengthened
by 1 microsecond. Thus a difference current of 10 mA peak will
flow for 2 microseconds. At a pulse repetition frequency of
100 per second the resulting average current works out at
2 microamperes.

Hence $K_2 = 2 \times 10^{-6}$ amperes per microsecond misalignment or approximately 2×10^{-5} amperes per mile misalignment. A typical value of K_1 will be 1 mile per volt, and with

$$C = 1 \text{ microfarad}$$

$$T = C/K_1K_2 = \cdot 05 \text{ second.}$$

Velocity Memory and Measurement of Rate

In the simple tracking circuits already described, control is
established through a single stage of integration. By introducing a second stage of integration between the difference

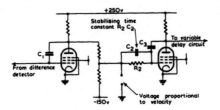

Figure 314.—Velocity memory circuit for maintaining the tracking motion
of the variable delay circuit during periods of fading.

detector and the variable delay circuit it becomes possible to
impart the property of velocity memory. However, a successful circuit cannot be developed by intuitive methods alone,
since the tracking equation involves second order terms, and
instability may result if the components are not properly pro-
portioned. A detailed discussion of double integration networks is beyond the scope of this chapter and the reader
wishing to investigate the subject further may consult
references (1) and (2) of the bibliography. (See end of
chapter.)

Fig. 314 shows a typical velocity memory circuit. The
tracking action of this circuit is such that any range misalign-

ment causes a difference current to flow into C_1, thus changing the output voltage V_1 of the first integrator. The latter varies the output voltage V_2 of the second integrator in such a direction as to correct the range misalignment. The rate of change of V_2 and hence the tracking velocity is proportional to V_1, so that with a target moving at constant velocity the range misalignment is cancelled when V_1 reaches the appropriate steady value. Under these conditions no difference current flows and if the signal fades tracking continues at the same velocity, i.e. the circuit possesses velocity memory. Another convenient feature is that the voltage V_1 is a measure of velocity.

In the case of the simple tracking circuit with position memory it is possible to measure velocity by differentiating the range control voltage. A basic differentiating circuit is presented in Fig. 315. This circuit gives an output voltage $R_1C_1 \, dV_c/dt$, where V_c = control voltage, and a measure of velocity is therefore obtained. Large values of R_1C_1 increase the voltage output but, at the same time, increase the sluggishness of the circuit. Hence the value of R_1C_1 must be chosen

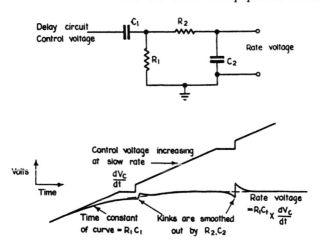

Figure 315.—Circuit providing a direct indication of target velocity.

to suit particular requirements. Since the curve of control voltage versus time may have discontinuities if signals fade, a suitable series time constant R_2C_2 may be inserted to smooth the output voltage.

Sensitivity to Weak Signals

By careful adjustments an automatic ranging circuit can be made to lock on and track with a signal to noise power ratio somewhat less than unity, but for reliable operation the signal to noise ratio should be greater than 2 : 1.

4. Circuit Design for an Externally Synchronised Radar

In an externally synchronised system such as the multiple track system of Chapter XVIII, p. 552, the time delay is

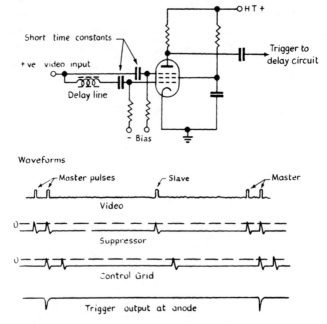

Figure 316 (*a*).—Discriminator responsive to double pulse transmission from master station.

measured between two trains of pulses usually known as the "master" and "slave" transmissions, both of which appear at a common receiver output. It is desired to trigger the delay circuit by means of the master pulses and lock the gates on to the slave pulses, at the same time ensuring that interference pulses do not cause an undue rate of triggering. Selection of the master pulses may be effected by giving them a

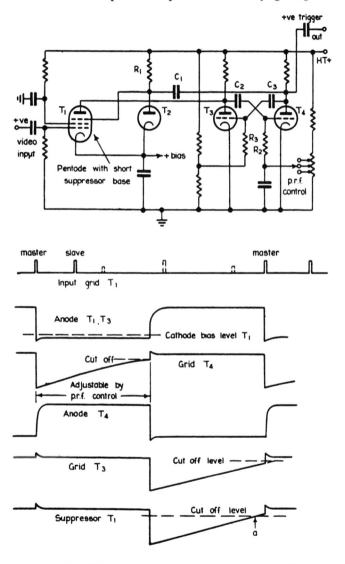

Figure 316 (b).—Selector responsive to pulse repetition frequency of master station.

special characteristic or an exact repetition frequency, and by using a discriminating circuit on the receiver output which is sensitive to this pulse characteristic or repetition frequency within a narrow tolerance.

An arrangement of the first type is illustrated in Fig. 316 (*a*) in which the master station transmission consists of double pulses. Positive pulses are fed direct to the suppressor of a pentode tube and also to the control grid through a network which introduces a delay equal to the pulse spacing. Both grids have cut-off bias applied, so that a pulse is produced in the plate circuit only when the second of the two pulses arrives at the suppressor simultaneously with the delayed arrival of the first pulse at the control grid. Applying the pulses to the grids through a short time constant differentiator ensures that only the leading edges of the pulses are effective at the grids. This makes the discrimination sharper and prevents a single long pulse from actuating the circuit. The pulse output from the anode may then be used to trigger the delay circuit.

The alternative circuit using a selector sensitive to repetition frequency is shown in Fig. 316 (*b*). In this circuit T_3 and T_4 form a multivibrator using stable components so that its natural frequency will hold to within a few cycles per second over long periods. This is set at a slightly lower frequency than that of the ground station to which it is desired to synchronise, by means of a DC bias control on T_4. Positive video pulses are fed to the grid of T_1 whose anode is connected to the anode of T_3 and whose suppressor is capacity coupled to the anode of T_4. Assume for the moment that the multivibrator has been synchronised by a " master " pulse, so that the anode of T_3 has swung negative and that of T_4 positive (see waveforms). The anode of T_1 is now below its cathode potential while the suppressor is held at cathode potential by the diode T_2. Under these conditions pulses on the grid of T_1 have no effect in the anode circuit. When the multivibrator completes a half-cycle determined by the time constant R_2C_2 and the DC bias control, potentials reverse and the anodes of T_1 and T_3 are now positive, while the anode of T_4 and suppressor of T_1 swing negative. Pulses on the grid of T_1

still have no effect in the anode circuit because the suppressor
is now cut off. However the suppressor recovers according
to the time constant R_1C_1, which is adjusted so that the cut
off level is crossed (point " a ") just before the next master
pulse arrives. The multivibrator itself is at this stage very
near changeover and any pulse on the grid of T_1 produces a
negative pulse at the anode of T_3, tripping the multivibrator.
With careful adjustment of the time constant R_1C_1 the circuit
can be made quite selective so that, for example, pulse repetition frequencies more than 10 cycles per second in 2,000 off
the correct frequency will not synchronise the multivibrator,
although an adjustment for $\pm$ 20 cycles selectivity in 2,000,
i.e. 1 per cent, is regarded as a good working figure. With the
arrangement described above it is of course possible for the
circuit to lock on to the " slave " pulses. This can be avoided
by a scheme such as missing out alternate slave pulses.

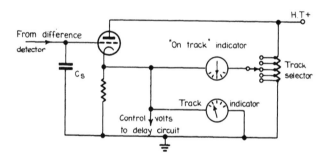

Figure 317.—Indicating system to facilitate navigation on any selected
track of a multiple track system.

With both of these methods it is possible for the delay
circuit to be triggered by a fortuitous arrangement of interference pulses but, provided the rate of triggering from this
source does not exceed a few per cent of the rate of triggering
by the desired pulses, no harm will be done since the delayed
gates which are normally locked to the slave pulses will on
these occasions " see " very few pulses—again by chance
coincidence with interference pulses. The net effect of these
pulses on the tracking will cancel out since there will on the

average be as many coincidences with one gate as with the other.

The indicator in this system may be marked in terms of time delay or suitably numbered tracks and to facilitate navigation on any desired track a track-selecting switch may be provided together with a centre-zero instrument for indicating deviation to port or starboard of the selected track. This is very simply accomplished by a bridge network as shown in Fig. 317.

BIBLIOGRAPHY

F. C. Williams, F. J. U. Ritson, T. Kilburn, "Automatic strobes and recurrence-frequency selectors," *J. Instn. Elect. Engrs.*, Vol. 93, Part III A, 1946, pp. 1275-300.

B. Chance, R. I. Hulziser, E. F. MacNichol and F. C. Williams (Eds.), *Electronic Time Measurements* (Massachusetts Institute of Technology, Radiation Laboratory Series, Vol. 20), McGraw-Hill, New York, 1949.

B. Chance, "Time modulation" and "Time demodulation," *Proc. I.R.E.*, Vol. 35, 1947, pp. 1039-49.

V. D. Burgmann, "Distance measuring equipment for aircraft navigation," *Proc. Instn. Elect. Engrs.*, Vol. 96, Part III, 1949, pp. 395-402.

CHAPTER XVI

RADAR SYSTEMS

BEFORE a radar set can be designed a statement is required of the functions which the system has to perform. This statement is commonly known as the "operational requirement." During the war the armed forces indicated their operational requirements for radar to perform such functions as air warning, gun laying on land, sea or in the air, air to air interception, ground controlled interception, air to surface vessel search, sea search, air and sea navigation, and a host of other important roles. When the requirement has been specified it is possible to consider the design of a particular set in terms of the known technical limitations of radar. At present, for example, the upper limit of transmitter power is as much as 1,000 kilowatts at wavelengths down to 10 centimetres and the corresponding receiver sensitivity is of the order of 10^{-13} watts per megacycle bandwidth. Below 10 centimetres, on the other hand, systems are limited to transmitter powers of the order of some hundreds of kilowatts, and correspondingly smaller receiver sensitivity. Although the present state of knowledge makes possible the construction of a set of any desired frequency and power within the above range, components are often available only at a number of discrete frequencies.

The characteristics of a radar system may be defined as those features of a set which have to be varied to meet the operational requirement. They can be divided into "functional" and "operating" characteristics. The functional characteristics are such features as coverage, presentation and general construction, and are directly related to the function the set is required to perform, while the operating characteristics are deduced from them in the process of design. Published examples of a number of typical radar systems will be found in the literature.[1]

---

[1] For example, Ch. XV of *Radar System Engineering*, ed. L. N. Ridenour (M.I.T. Radiation Laboratory Series, Vol. 1), McGraw-Hill, New York, 1947.

1. Functional Characteristics

Coverage

The coverage of a radar set is the term applied to the volume of space in which a given target can be located by the set. The coverage diagram is geometrically similar to the polar diagram of the aerial expressed in field strength and varies for different targets. A typical coverage diagram is shown in Chapter XVII, Fig. 322.

Because the echo from a target is subject to fading, the boundaries of a coverage diagram are not precisely defined unless a figure is quoted giving the percentage of occasions on which a target is just detectable at the range indicated by the diagram. Depending on the particular operational use of the equipment, it is usual to use a figure of 50 per cent as this criterion when knowledge of the exact pick-up range of a target is not vital and 100 per cent when it is.

Floodlighting and Scanning

Coverage of the specified volume is produced in one of two ways, either by floodlighting or by scanning. Floodlighting was common with the older low frequency radar sets of 100 megacycles per second and less when the aerial was not highly directive. Bearing and elevation measurements were then made by the ordinary direction finding methods, using a goniometer or some form of amplitude or phase comparison.

The broadcast technique gave good service until the advent of microwaves made the use of narrow beams possible. In order to search the whole volume set down by the required coverage diagram, the microwave beam then had to be moved about in a methodical manner. This process is called scanning and it may take many forms.

A common type of scan uses what is known as a " fan " beam, that is a pattern which corresponds to the coverage diagram in the vertical plane and is narrow in the horizontal plane. The aerial is rotated about a vertical axis to sweep the required volume. Such scans are used on the ground for aircraft search, in the air for surface vessel search and terrain recognition, and at sea for surface search. Another example of a fan beam system is one in which the aerial pattern is broad in the horizontal plane and narrow in the vertical plane. When oscillated about a horizontal axis it is used for height finding of aircraft.

Other types of scan use aerial patterns which are narrow in both planes and are referred to as " pencil " beam systems. These generally require more complex scanning motions but provide more precise information about the target location. A helical scan is an example of this type and is produced by slow turning about one axis and rapid rotation about the other. This has been used in air to air interception sets and in later forms of combined aircraft warning and height finding sets.

The conical scan is another useful form which is described adequately by the name and is used in automatic following sets. The aerial system describes a rapid conical motion, and error signals produced by displacement of the target from the axis of rotation are used to keep the aerial continuously pointing at the target.

The scanning period is the interval of time taken by the aerial to return to its original position after completely scanning the required volume of space. This time is therefore a measure of the frequency with which information is obtained on a particular target and is often referred to as the plotting rate. The length of the scanning period has a marked effect on the design of the set. The shorter the period the more difficult the mechanical construction of the aerial becomes ; on the other hand, the user has certain requirements in respect of accuracy, clarity of presentation, and movement of the target between cycles which must be met by sufficiently rapid scanning.

Presentation

Presentation is the term used for the way in which the radar information is given to the user. This is usually done by means of a cathode ray tube, but for certain purposes a meter, a light or an aural indication may be preferable.

The Class A display has already been described (Chapter II, Fig. 4) and is found in floodlight systems where bearing measurements are not taken directly from the display. The PPI display (Chapter II, Plate I) and variations of it are commonly used with " fan " beam scans. One important variant is the Class B display in which range and azimuth are given in rectangular coordinates. It is often used in airborne sets with a forward view of 180 degrees or less. The time base is generated in the normal way and sweeps vertically up the face of the cathode ray tube with intensity modulation of the echoes. The whole time base is then swept sideways with a displacement proportional to the azimuth angle of the scanning aerial. In this way a distorted but recognisable radar picture is formed from which accurate data can be obtained.

Many other forms of display have been used but they are all straightforward developments from the two basic types, Class A and PPI. Some examples of these are : spiral or circular display to give a long time base for accurate range measurement, and azimuth versus elevation display to give steering direction for use in fighter interception.

Accuracy and Resolution

The accuracy of a radar set is given either by stating the allowable error in range, bearing and elevation or any other coordinates such as ground range, bearing and height, appropriate to the operational use. With some sets the relative accuracy tends to become more important than the absolute. For instance in shell splash spotting in naval gunnery, once the shells are falling near the target the corrections to the shell splash position are the real purpose of the set. The direction of a pursuing fighter relative to a target bomber in a ground controlled interception system is another example.

Relative accuracy is mainly determined by the resolution of the system, and by resolution is meant the ability of the radar set to distinguish between two targets in range and in angular separation. The range resolution is of the same order as the pulse length ; the angular resolution is of the same magnitude as the aerial beam width.

Construction

As well as having the right operating characteristics, equipment needs to meet requirements for size, shape, weight, power consumption and ruggedness to a degree suited to the user. Airborne sets require to be suitable for distribution about the aircraft in light units designed for centralised controls and operable at all flying heights ; ground sets require to be rugged and weather-proofed ; shipboard sets must be rugged to withstand gun blast and suitable for installation in offices scattered throughout the ship and often a considerable distance from the aerial system.

2. Operating Characteristics·

As in any engineering design, the choice of the operating characteristics of a system to meet a given operational requirement represents a compromise between conflicting factors. The principal operating characteristics of a radar set are frequency, aerial size and pattern, scanning motion, transmitter power, receiver sensitivity, pulse length and pulse repetition frequency.

Perhaps the most important consideration is the coverage required by the user and this cannot in general be met exactly by design but a close approximation is often possible by a careful selection of parameters, particularly aerial pattern and frequency.

In the case of floodlighting there is no scanning and the subsequent design is straightforward. In the more usual case when scanning is used, a tentative selection of the method must be made before proceeding to decide the aerial pattern, frequency etc. In a scanning system a number of factors must be given careful consideration :

(a) A complete cycle of scan must be finished in a time determined by the user's requirements in respect of target movement, accuracy, and appearance of the display.

(b) The beam widths in bearing and elevation must be correctly chosen for the required angular accuracy and resolution.

(c) The passage of the beam through a target must not be so rapid that insufficient echoes are received for efficient display. At least five and preferably more than ten echoes per target is a desirable design figure.

(d) The scanning motion and maximum speed of scan must be chosen so that optimum visibility[2] of the echo is possible on the display. In other words due attention must be paid to scanning losses.

(e) The scanning motion must be mechanically feasible.

After selecting the method of scanning, a pulse length is chosen at least adequate for the required range resolution, together with a repetition frequency which gives a mean transmitter power consistent with requirements for power consumption, size and weight. The operating frequency, aerial design and transmitter power can then be selected to give the aerial pattern required by the scanning motion already chosen (Chapter VII). In making this calculation it is clearly economical to assume the use of a receiver of maximum possible sensitivity for the selected frequency.

It is also most economical within limits to use as large an aerial aperture as possible rather than obtain the performance by using a powerful transmitter. In the case of ground and shipboard radars close attention must be paid to aerial height particularly for systems which are intended to detect targets close to the surface of the earth, when it is important to place the aerial as high as possible. The choice of frequency is further modified by consideration of scattering from particles such as raindrops and attenuation in the atmosphere (Chapter II, section 7), particularly above about 3000 megacycles per second in frequency. Effects above this frequency can be serious enough to restrict their operational use to applications where reliable all weather ranges of less than ten miles are required or where performance is only sought in clear weathe·

[2] Ruby Payne-Scott, "The visibility of small signals on radar P.P.I. displays," *Proc. I.R.E.*, Vol. 36, 1948, pp. 180-96.

CHAPTER XVII

MILITARY RADAR SYSTEMS

The physical and electrical characteristics of the many radar systems which were developed to meet a wide range of military needs were determined, in large part, by the conditions under which they were to be used. There are, for example, generally fewer restrictions on the size, weight and power consumption of land-based installations, and hence this group includes the largest and most powerful radars and those of highest performance and greatest complexity. Considerations of size and weight are, on the other hand, of paramount importance in the design of radars for use in aircraft. It is possible here to describe briefly only a representative sample of the systems which were extensively used during World War II. It is convenient to treat them according to whether they were originally intended as ground-based, naval or airborne installations.

1. Ground-Based

The earliest conception of radar in Great Britain was to provide warning from the ground of the approach of hostile aircraft so that the defences could be alerted in adequate time. To this were soon added the provision of range, bearing and elevation data accurate enough for the control of anti-aircraft gunnery (and searchlights), warning against the approach of surface targets, and the control of fire from coast defence batteries.

Air-Warning Radar

The function of air-warning radar is to provide information on the presence and movements of all aircraft in the vicinity of a defended area in the form of range and bearing data, and preferably also of height, on each aircraft or group of aircraft.

As will be seen later, the fitting of secondary radar enables friendly and hostile aircraft to be readily distinguished. The prime requirement in an air-warning station is a large detection range combined with adequate cover over the area to be served. Range and bearing accuracy are of secondary importance provided this is adequate to allow the speed and course of targets to be determined. A measure of height to within a thousand feet is usually satisfactory. Since air-warning stations serve merely as the eyes of a defended area, their effectiveness is dependent upon the prompt dissemination of the information gathered. This is handled through centrally located filter and operations rooms.

CH (Chain Home). The earliest air-warning radar to see combat service (the British CH) made use of a fixed transmitting

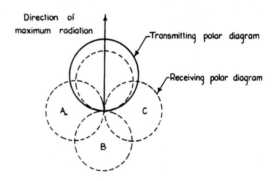

Figure 318.—Transmitting and receiving polar diagrams for a CH station. Receiving lobes *A*, *B* or *C* may be deleted at will for sense finding.

aerial to "floodlight" an area roughly quarter spherical in shape, the greater part of the energy being confined within 20 degrees of the horizontal. For reception a series of three fixed aerials, fitted with parasitic reflectors which could be switched in or out as desired, was used to produce a variety of polar diagrams. By comparing the signal strengths received with the polar diagram in various positions the quadrant in which the target lay could be uniquely determined (see Fig. 318) and its bearing measured. The main receiving array consisted of a pair, or stack of pairs, of crossed dipoles, these being connected by balanced feeders to the opposing coils of a goniometer. Range was read on a scale on a 12-inch oscillograph and azimuth by rotating the goniometer to the position of minimum signal. The maximum reliable range on a single aircraft was about 120 miles and on a concentration of heavy bombers about 180 miles. Bearing could be measured to about 1½ degrees under most conditions. Ground reflections naturally gave rise to large gaps in the vertical coverage of each station, but these were filled by using an alternative set of aerials at a different height. The resultant coverage in the vertical plane with and without "gap filling" is shown in Fig. 319.

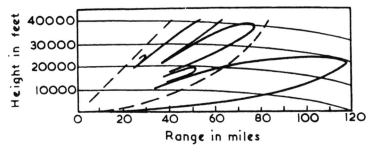

Figure 319.—Vertical coverage diagram for CH station on large aircraft.
——— with main aerial; - - - - - with gap-filling aerial.

Height finding was accomplished by comparing the echo strengths derived from two receiving aerials at different heights (see Fig. 320). Since ground reflections were involved in both transmission and reception, very careful siting of CH stations was essential, but under good conditions height could be measured to roughly ± 1,000 feet throughout most of the region covered.

The Home Chain operated at frequencies within the 20-50 megacycles-per-second band, because at the time of its original development in 1935 this represented the workable limit of available receiver and transmitter techniques. The aerial systems were therefore large and costly (see Plate XXII). Early transmitters were fitted with large water-cooled continuously evacuated valves and had a peak pulse output of about 150 kilowatts. Later, as valve techniques improved, sealed-off valves with peak powers of up to 1 megawatt came into use, though it is interesting to note that a power output of this order was obtained by a CH station as early as 1940.

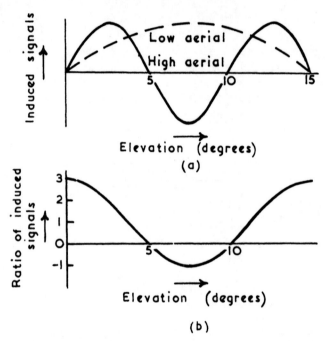

Figure 320.—Elevation finding with aerials at different heights (CH). (a) Signal strengths with high and low aerials. (b) The ratio of these. Elevation angle is determined uniquely up to 7·5 degrees.

For normal operation a pulse length of about 12 microseconds and receiver bandwidth of 150 kilocycles per second were usual, but longer pulse lengths (with corresponding narrower bandwidth) were available to provide greater searching range, and shorter pulses and wider bandwidth when better resolution on targets was required. The pulse repetition frequency was necessarily low (25 or 12½ pulses per second) to obviate clutter from ionospheric echoes on subsequent sweeps of the time base.

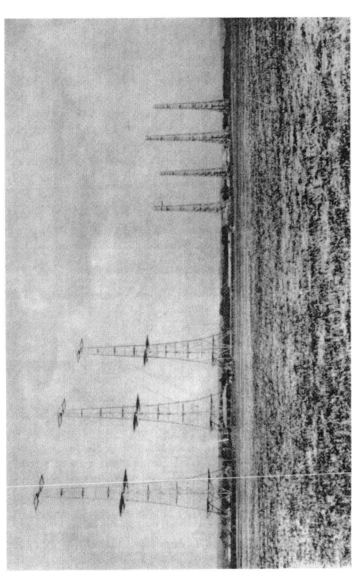

PLATE XXII.—A typical CH station with transmitting aerials on left and receiving aerials on right.

(a) A recent Australian helically scanning microwave early
warning radar.

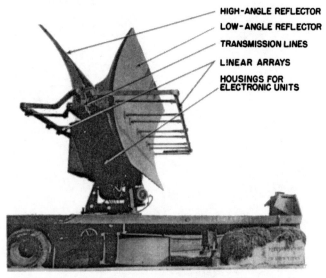

HIGH-ANGLE REFLECTOR
LOW-ANGLE REFLECTOR
TRANSMISSION LINES
LINEAR ARRAYS
HOUSINGS FOR
ELECTRONIC UNITS

(b) A truck-mounted version of American MEW radar (Microwave Early
Warning) or AN/CPS-1.

PLATE XXIII.

The major disadvantages of the CH system were its relatively short range on low-flying aircraft and its inability to handle large numbers of aircraft. It was used continuously and successfully, however, throughout World War II and formed the basis of the air defence of Great Britain.

MEW (Microwave Early Warning). The advent of the cavity magnetron and the feasibility of high-power outputs at very much shorter wavelengths made possible the production of narrow "searchlight" beams, as distinct from the "floodlight" illumination of the CH system. Troubles due to ground reflections were therefore eliminated except at very low angles of shoot. Fig. 321 shows a typical lobe pattern

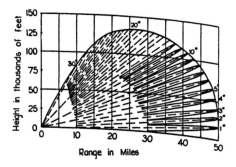

Figure 321.—Lobe pattern in the vertical plane of a typical microwave early warning radar.

for a microwave warning radar. The difficulty of searching a large volume of space with a narrow beam was attacked by using a vertically fanned beam rotating in azimuth; by scanning a narrow beam slowly in elevation and rapidly in azimuth or vice versa (as, for example, in the helically scanned Australian LW/AWH Mark II, illustrated in Plate XXIII (a)); or by using several aerials mounted on a common turntable. One of the most efficient of the microwave air warning radars is the American MEW (Microwave Early Warning) or AN/CPS-1, the truck-mounted version of which is shown in Plate XXIII (b). It was used with considerable success in the later stages of World War II.

MEW operates in the wavelength range 10·3 to 10·8 centimetres and has a peak power output of about 1 megawatt at a pulse repetition frequency of 320 per second. A pulse length of 1 microsecond is used. In order to provide maximum detection range on low-flying aircraft and to make possible continuous tracking of an aircraft once it has been located, two separate high-gain aerials mounted back to back on the same turntable and fed from separate transmitters are used. The lower beam is derived from a semi-cylindrical paraboloid

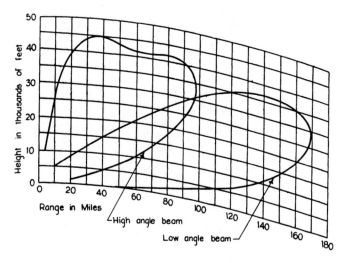

Figure 322.—Vertical coverage diagram for MEW (Microwave Early Warning) radar against medium bomber.

of dimensions 25 feet by 8 feet and a gain of 10,000, while the second antenna, for high-angle cover, has an aperture 25 feet by 5 feet (gain 3,500) and is shaped to produce a cosecant squared beam. The antenna system may be rotated in azimuth at a rate of 1, 2 or 4 revolutions per minute and provides all-round cover which is virtually free of gaps (see Fig. 322), combined with good warning against aircraft at all normal heights. No indication of the height of aircraft located is, however, given. The detail which can be seen on a high-power microwave set is illustrated in Plate XXIV.

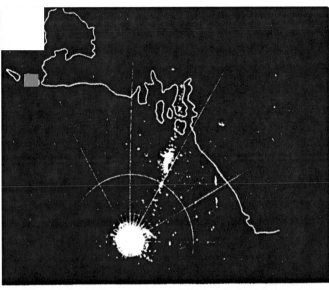

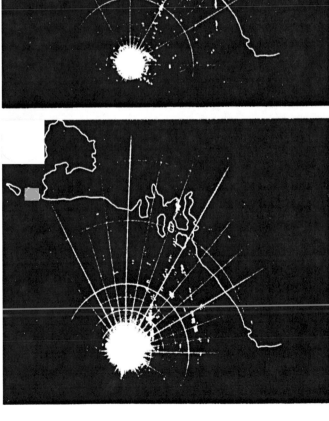

PLATE XXIV.—Microwave early warning radar PPI pictures showing a large number of aircraft over the North Sea and the change in the pattern after a 15 minute interval.

[*By courtesy of the U.S. Army Air Force.*]

A light-weight air-transportable air warning radar designed for use in offensive operations—the Australian LW/AW Mark IA.

PLATE XXV.

MEW used a multiple display system in order that the great amount of information collected by the radar could be effectively used. Up to five separately manned range-azimuth displays were available for detailed position-reporting, while for control purposes off-centre PPI's were used. No difficulties were encountered in controlling 20 or more simultaneous flights. Although the all-up weight of a station was in the vicinity of 50 tons, the absence of costly aerial structures and stringent siting requirements made it possible to use MEW in offensive operations. Nevertheless, careful siting for such a high-powered radar as MEW is required to obtain freedom from ground clutter, a shallow saucer-shaped depression where the ground horizon is relatively close being preferred. It was one of the first sets to use Moving Target Indication (MTI), a device in which the receiver output from each complete scanning is subtracted, after an appropriate delay, from that from the succeeding sweep and difference signals only appear on the display. Permanent echoes are cancelled out and virtually only moving targets are seen.

The Australian LW/AW Mark I. An example of a lightweight set specifically designed for use in an offensive campaign such as that of the Allies in the south-west Pacific area, was the Australian LW/AW Mark IA (see Plate XXV). Essential requirements were that the complete station should be readily transportable by barge or small ship, by normal transport aircraft, or, if necessary, by hand portage, and be capable of erection and operation within a few hours of arrival at its site.

The LW/AW Mark I had a peak output of 10 kilowatts at 200 megacycles per second, unusually long pulse length (22·5 microseconds), and a correspondingly narrow bandwidth ($\pm$ 70 kilocycles per second) at the receiver. Although this was only a low-power system, the combination of long pulse length and narrow receiver bandwidth gave an excellent performance. The type of lobe pattern obtained is shown in Fig. 323. Reliable warning ranges of 80 to 100 miles and more were consistently obtained. Particular attention was given to the requirements for operation under tropical conditions, and upwards of 150 of these radars were used by the Australian and United States Forces with considerable success in the south-west Pacific campaign.

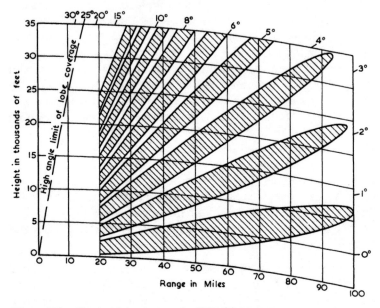

Figure 323.—Vertical lobe pattern for LW/AW Mark IA air warning radar (200 megacycles per second) at a height of 100 feet, overlooking sea.

Height Finding

Early methods of height finding at metre wavelengths were dependent upon use of the interference pattern produced as a result of reflection from a ground plane in the near vicinity of the aerials. Even though great care was taken in-siting, individual calibration of each station with the aid of aircraft flying at known altitudes was necessary; and gaps in the vertical coverage were inevitable. The development of microwave techniques made possible a much more satisfactory method of height finding employing narrow "searchlight" beams, either pencil or fan-shaped.

A microwave height finder of this type, such as the "CMH" radar shown in Plate XXVI (a), is almost completely free from siting effects and gives reliable heights down to angles of elevation of ¾ degree or less. The beam is made very narrow in the vertical plane, but fanned out in the horizontal

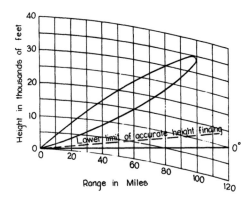

Figure 324.—Microwave height finder.   Vertical coverage diagram.

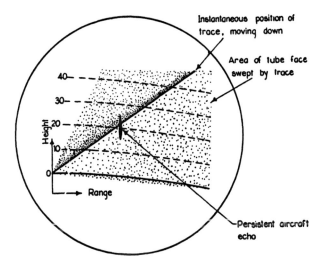

Figure 325.—Range-height display used with microwave height-finding radar.

plane; Fig. 324 shows a typical vertical coverage diagram.   It is scanned automatically up and down through the target, the target height being read off directly from a range-height type of display, shown diagrammatically in Fig. 325. A height-finding radar of this kind is generally used in conjunction with some form of search radar, the latter providing azimuth information on which target heights are required.

Ground Control of Interception (GCI)

A change in tactics by the enemy in 1940—from day to night bombing raids—led to the use of ground radar for directing fighters close enough to the enemy to enable engagement to take place. The requirements of such a ground-controlled system were virtually continuous information on the position of friendly and hostile aircraft both in plan and in height, their relative heights to within $\pm$ 500 feet, and a direct communication link between ground controller and fighter pilot. The majority of ground-controlled interception during the recent war was carried out by metre-wave stations, an example of one such radar being shown in Plate XXVI (*b*). The original British GCI system embodied height-finding provision on the same principle as used in CH but was fitted with an aerial system which produced a relatively narrow beam in the horizontal plane and rotated continuously. The display embodied one of the most revolutionary devices in the evolution of radar, the Plan Position Indicator (PPI). This provided the ground controller with a plan-picture of all aircraft in his area, so that he was able to direct fighters under his control until they were in the right position and close enough to engage the enemy visually or with their own AI radar.

The coming of centimetre radar with its narrow beams capable of greater scanning rates gave higher precision and better resolution between targets, but necessitated the complete separation of azimuth and height-finding functions. A number of such systems were under development toward the close of the war but, except for the combination of the MEW and CMH systems described above, few saw operational service. Their method of use was fundamentally the same as that so successfully employed during the early stages of the Battle of Britain.

(b) Ground-controlled interception radar.

PLATE XXVI.

(a) Microwave height finder.

(a) A 150-centimetre searchlight adapted for radar control (SLC).

(b) General view of a microwave fire control radar (American SCR-584). The aerial is lowered into the trailer during transportation.

By courtesy of the U.S. Army Signal Corps.

PLATE XXVII.

Control of Searchlights

Considerations of urgency in the early days of World War II made necessary the addition of radar facilities to equipment which was already available and in service. Plate XXVII (*a*) shows how it was adapted for the control of searchlights. This SLC (Searchlight Control) set was one of the first to make use of a split-beam technique (see Fig. 326). Five Yagi aerials were mounted around the searchlight barrel, the uppermost being used for transmitting and the remaining four, in pairs, to swing the received polar diagram alternately left-right and up-down. The received signal strengths with the beam swung to left and right were compared on a centre-zero or cathode-ray indicator; and similarly for the up-down positions. The

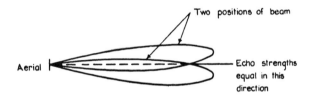

Figure 326.—Use of beam switching to provide accurate direction finding.

direction of escape of the target in two directions at right angles was thus indicated, and manual corrections to the pointing of the searchlight in azimuth and elevation were made independently by two operators. A third operator followed the target in range on an A-scope with a movable strobe, the azimuth and elevation indicators being operated only by echo signals appearing in the narrow range-interval covered by the strobe. SLC operated at a frequency of about 200 megacycles per second with a peak power of 5 to 10 kilowatts and a pulse length of 2 to 3 microseconds. Under favourable conditions a medium bomber could be located at a range of 25,000 yards and followed in from about 15,000 yards.

The use of radar for the control of searchlights was of minor operational importance in the later stages of the war. Microwave fire control systems are well suited for this purpose and were, in fact, used to some extent.

Fire Control

The determination of range by radar methods eliminates the baseline necessary for optical range finding, provides an answer under all conditions of visibility, and one which is generally more accurate and whose accuracy, moreover, is independent of distance. Angles of azimuth and elevation can also be measured to a satisfactory precision at the same site, and hence radar was widely and successfully used for the direction of gunfire against hostile targets at sea and in the air, particularly following the development of techniques for use at microwave frequencies. An example of a modern radar of this type is the American SCR-584 (see Plate XXVII (b)).

The SCR-584 operates in the 10 centimetre band, has a peak pulse power of 300 kilowatts and a pulse length of 0·8 microseconds, with a pulse repetition frequency of about 1,700 per second. The dual function of first searching for a target and then, once a target has been located, of providing a steady flow of range, azimuth and elevation data on that target is carried out with a single parabolic aerial 6 feet in diameter. For search purposes the antenna is scanned helically through 360 degrees and over a range of elevation angles up to 10 degrees. Warning is provided on a single aircraft up to a range of 70,000 yards The same antenna is used to provide a conical beam once a target has been located. This is obtained by rapidly spinning the paraboloid (see Fig. 327). The error signals in azimuth and elevation derived from this conical scan operate servomechanisms in such a way as to correct the pointing error and thus ensure fully automatic following of the target and a steady output of its present position data. Tracking in range is done manually by an operator who is provided with two displays. The first shows "coarse" range to a maximum of 32,000 yards and the second a magnified 2,000 yard interval of this. When a cursor on the coarse display is placed over a target echo, the "fine" display shows a magnified trace of this echo, together with an interval of 1,000 yards on either

side of it. Continuous target range is fed out when a cursor
on the "fine" tube is maintained exactly over the leading edge
of the echo. Azimuth and elevation circuits are operated only
by echo signals which fall within a narrow range immediately
following that of the cursor. Transmitting selsyns and potentiometers provide information on azimuth, elevation, slant

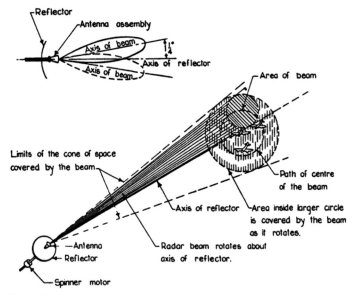

Figure 327.—Conical scanning system used in SCR-584 gunnery radar.
A sinusoidal error signal is produced when the target lies off the axis
of the reflector.

range and height in electrical form suitable for direct transmission to predictors. Under good conditions the range
accuracy with SCR-584 was ± 25 yards and that of the
angular measurements 0·06 degrees.

2. Naval

The use of radar at sea revolutionised the science of naval
warfare, principally because of the facilities it provided for
continuous and accurate measurement of distance and bearing

at all useful ranges, in complete darkness or under any conditions of visibility. It thus became possible to locate and engage the enemy with telling effect, for example, as in the Battle of Cape Otranto, under conditions which previously would have precluded any such offensive action. Naval radars serve the same basic functions as their land-based counterparts but with additional applications not present on land, such as for navigation, particularly in coastal or confined waters; for station-keeping when sailing in formation or in convoy; and for collision warning. A modern capital ship carries upwards of thirty radar sets, each fulfilling a separate function.

While naval radars are also electronically similar to their corresponding installations on land their general design is necessarily determined by the special conditions which exist on a ship. Space, for example, is generally at a premium, and there are limits therefore to the size and shape of equipment for installation below deck, to the size and weight of aerial structures and to the height above deck at which they may be mounted, while the general layout must be such as to permit adequate servicing in the restricted area available. The equipment itself must be designed to withstand the severe accelerations met with during gunfire or rough weather, and to operate reliably over long periods with the minimum of attention under extreme conditions of temperature and humidity, varying from those of tropical to sub-zero climates. Special consideration must be given to the protection of all parts, particularly those which are fully exposed such as aerials and feeders, from the corrosive effects of salt spray. The problems of accurate position finding at sea are made more difficult because of the natural motions of a ship—pitch, roll, yaw and turn—and for the most precise applications it is essential to stabilise the aerial mountings against these accelerations, or alternatively to mount the aerials themselves on gun directors which are already stabilised.

A very high degree of co-ordination in the handling of data provided by the various radar sets with which a fighting ship is fitted is essential. An attack may come from any quarter and in several different forms simultaneously, the objective, unlike an attack on land, being the radar site itself. The warning system is naturally the heart of a ship's defence; from the information it provides the navigation of the ship is determined, targets are allocated between the available fire power and their approximate range and bearing supplied so that they may be taken over by the more specialised radars which usually provide only limited searching facilities. It is usual, therefore, to relay radar information fairly generally in PPI form throughout a ship, specifically to those centres concerned with navigation, control of fighters, defence against aerial attack and sea defence.

Air Warning and Fighter Control

Naval air-warning and fighter control systems are similar to their equivalent installations on shore. It is usual, however, to stabilise the display itself so that the direction of North always appears at the 12 o'clock position on the oscillograph face irrespective of how the ship may be lying. The direction of the ship's head is indicated by brightening the scan momentarily at that bearing, or by other means.

Plate XXVIII (a) shows a typical metre-wave air-warning radar aerial mounted at a masthead. The performance of naval air-warning radars is equivalent to that of land-based installations at microwave frequencies, but at metre wavelengths reduced coverage on low-flying aircraft is obtained because of limitations placed on the height above sea level at which aerials may be mounted. For this reason sea-search sets are usually relied upon to provide warning also against low-flying aircraft.

Sea Search

The function of sea-search radars is primarily to provide information on the presence and disposition of surface vessels in their vicinity. A continuously rotating aerial combined with a PPI type of display is therefore employed, and at microwave frequencies detection of a ship at horizon range—about 20 miles for a typical installation on a cruiser—is possible. At metre wavelengths the performance is considerably reduced because limitations in available height for the aerial make it impracticable to depress sufficiently the first lobe of the aerial pattern. Sea reflections, on the other hand, are more serious at the shorter wavelengths, and it is necessary to use swept-gain or similar techniques at the receiver to produce a usable performance at closer ranges. High accuracy in range and azimuth determinations is not required, since engagement of the targets located is subsequently taken over by more precise radars designed for that purpose. To guard against loss of echoes due to the natural motions of the ship it is necessary at microwave frequencies either to stabilise the aerial mounts or to provide a beam which is fanned out in the vertical plane. Plate XXVIII (b) shows a typical microwave sea-search aerial and mount. The characteristics of the centimetre wave sets render them suitable also for the detection of low-flying aircraft, and of surrounding coast lines and land masses. They therefore fulfil a useful subsidiary role in the navigation of the ship.

Fire Control

Naval gunnery installations are broadly classified according to the purpose for which they are intended: for long-range low-angle firing against surface targets or low-flying aircraft, for high-angle firing against relatively high-flying aircraft, or for close-range defence.

Radars for the control of a ship's main (low-angle) armament are similar in function and performance to those used in conjunction with coast-defence installations except that precise stabilisation of the aerial is essential in order to counter the natural motions of the ship. This is almost universally achieved by mounting the aerial on the gun director itself (see Plate XXIX (a)). And because of the narrower beams, higher angular accuracy and reduced size and weight feasible at microwave frequencies, modern fire-control radars operate at centimetre wavelengths. They are, however, unsuited for general searching functions and depend upon the supply of "putting on" information from other radars within the ship.

(a) Metre-wave air-warning radar aerial. The interrogator aerial is mounted at the masthead (p. 529).

(b) Microwave sea-search radar aerial and mount (p. 530).

PLATE XXVIII.

(*a*) Main armament director with fire control radar aerials mounted above.

(*b*) Heavy anti-aircraft director with fire control radar built in.

PLATE XXIX.

Systems for use in conjunction with high-angle gunnery are also usually mounted on the director itself (see Plate XXIX (b)). The aerial is provided with a conical or equivalent form of scanning so that elevation as well as range and bearing data are available, these generally being fed direct to an appropriate predictor. In some instances provision is made for firing the guns automatically when a predetermined range corresponding to that of the fuse setting has been reached by the target aircraft.

Close-range weapons are intended for action against dive bombers, suicide planes, torpedo-carriers, controlled bombs, etc., usually at ranges of considerably less than 5,000 yards. Radar systems for use with these weapons are of two types. In the first of these, see Plate XXX (a), range only is measured and an indication given (or the gun fired automatically) when a predetermined range has been reached, the gun being directed manually at the target aircraft throughout. The second is a later development in which the radar forms an essential part of the gun mounting and provides blind-firing facilities. The example shown in Plate XXX (b) is of a 3-centimetre system using a 2 feet diameter paraboloid which is rotated rapidly, the mirror axis being skewed from the axis of rotation to provide a conical scan. Preliminary target data is supplied from the ship's air-warning system, and the conical scan of about 5 degrees is adequate to locate the target intended. The system "locks-on" to the target at a range of about 7,000 yards, and the gun is automatically directed at and follows the target aircraft until the predetermined firing range is reached.

3. Airborne

The earliest operational requirement for radar in aircraft was to permit night fighters to "home" on to enemy bombers, but even before this problem had been satisfactorily solved a further need had arisen and been met, namely, the detection of surface targets from the air. It was effectively used as an aid to navigation and to blind bombing, and for relatively subsidiary purposes such as warning of the approach of fighter aircraft from the rear, and against collision. Equipment was also developed for the automatic training and firing of gunnery from aircraft, and later projects included a complete airborne radar warning station, its purpose being to provide the extended detection ranges on aircraft or ships which are possible at great heights. Neither of these, however, saw combat service during World War II.

Although the operational requirements for airborne radar were thus generally similar to those on land or at sea, the strict limitations permissible in size, weight and power consumption of equipment for use in aircraft places airborne radar in a special class. Equipment of high performance which was yet extremely compact and light in weight resulted. Major advances followed the introduction of microwave techniques and the feasibility of carrying high-gain directive aerial systems, which could be faired into the aircraft structure or mounted in "blisters" in such a way as to provide excellent radar performance without detracting appreciably from the flight characteristics of the aircraft. Aircraft radar must be capable of operation at the low pressures encountered at high altitudes, and it is not uncommon to pressurise those units in which troubles from flash-over may occur. Miniaturised components are used extensively to reduce size and weight.

Aircraft Interception (AI)

AI is the name given to the equipment developed specifically to counter the enemy's switch from day- to night-bombing raids during the second phase of the Battle of Britain. It is intended for use by defensive night fighters to enable them to locate and home upon hostile bombers, particularly in the final stages of interception. Simplicity and accuracy of angular presentation and small minimum range are therefore of prime importance, while the range of detection and cone of search should clearly be as great as possible. The early AI systems, operating at frequencies around 200 megacycles per second, made a substantial contribution to success against these night bombing attacks, but their maximum range was inadequate. With the aerial systems used at these longer wavelengths—a

(a) Pom-pom director using Yagi aerials (600 megacycles per second).

facing p. 532]

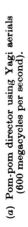

(b) A self-contained 40-millimetre gun-mounting with 3-centimetre radar equipment.

[British Crown Copyright.

PLATE XXX.

single dipole and reflector for transmitting, and two similar pairs for receiving, providing split beams for homing in azimuth and elevation (see Fig. 328)—it was impossible to eliminate clutter from ground returns received via the side lobes, and the maximum range was limited to that of the height at which the fighter was operating.

Satisfactory microwave systems using narrow beams and substantially free from this limitation were developed in Great Britain and the U.S.A. The American SCR-720 is an example

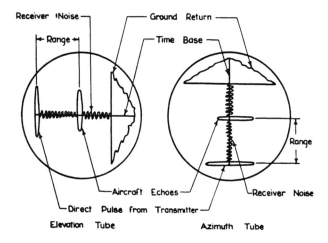

Figure 328.—Display system used with long-wave Aircraft Interception radar (AI Mark IV).

of a successful 10 centimetre AI equipment. A cone of semi-angle 25 degrees in front of the aircraft is scanned helically by slowly tilting the antenna dish (a 29-inch diameter para-boloid) as it rotates rapidly about a vertical axis, the beam produced (about 10 degrees wide) being narrow enough to allow a target aircraft to be picked up and followed at dis-tances up to several times the aircraft's height. A pick-up range of about 10 miles could be obtained with a power output of 200 kilowatts, while the minimum range (with microsecond pulses) was about 150 yards. Two intensity-modulated dis-plays are provided, one showing range against azimuth and the other elevation against azimuth. These are normally under the control of a radar operator, and the only information presented to the pilot is a repeat of the latter display for echoes falling in a restricted range interval (selected on the range-azimuth tube), together with a meter-indication of range.

The installed weight of the SCR-720 equipment, including cabling, was over 1,000 pounds; it was therefore not suitable for use in single-engined aircraft, particularly carrier-based fighters. For this purpose 3-centimetre systems using a smaller dish (e.g. the American AN/APS-6) and with an all-up weight in the vicinity of 150 pounds were designed. Although not actually used in combat during World War II an automatic tracking radar was also developed in which the error,signals in azimuth and elevation were used to cause the antenna to lock-on to and automatically follow a selected target, range, azimuth and elevation data being fed out continuously.

Detection of Surface Vessels (ASV)

The first airborne radar to be used operationally for the detection of surface vessels was the British ASV Mark II, a long-wave set operating on a frequency of 176 megacycles per second. During the initial stages of searching for targets, transmitter and receiver were connected to Yagi aerial systems which produced broadside beams alternately to port and starboard, the direction of deflection of the echoes on a Class "A" display being switched in synchronism so that the position of a target relative to the aircraft's line of flight could be seen. For the final homing two further aerials pointing forward and having overlapping beams were brought into use (see Fig. 329). This system suffered from the defects inherent in the use of the longer wavelength, but nevertheless was widely used and was in part responsible for successes obtained in the early stages of the Battle of the Atlantic. At a height of 5,000 feet a detection range of 50 miles could be obtained on a 10,000-ton ship and of 5 miles on a surfaced submarine. ASV Mark II was naturally ineffective against partially submerged submarines.

The development of microwave ASV followed closely upon that of microwave AI, the radar equipment being very similar except for aerial and display details. It was possible to obtain coverage over the complete field of view by scanning the antenna through 360 degrees while searching, and to switch to a more rapidly executed sector-scan when homing on to a selected target. An example of a 10-centimetre ASV system which was widely used in heavier aircraft was the American SCR-717B. A number of successful 3-centimetre systems of high performance were evolved, of which the American AN/APS-3(ASD-1), Plate XXXI, is representative. This is a smaller, lighter and more compact set which is suitable for

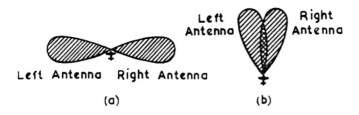

Figure 329.—Aerial patterns in a horizontal plane of ASV Mk. II.
(a) during search; (b) for homing.

fitting to single-seater aircraft. The antenna system is an 18-inch paraboloid, which is generally mounted in a nacelle in the leading edge of a wing. The reflector executes a wide sector-scan of 160 degrees, centred dead ahead, about 35 times per minute, and tilts automatically through an angle of 2 degrees during the sweep. It may be set at will to operate at a fixed angle of elevation or depression of up to 8 degrees. In the interests of compactness and simplicity in operation the general search functions of ASD were therefore sacrificed somewhat in favour of good homing facilities. The full scan of 160 degrees was used in the initial stages of searching, and the central portion of this (± 30 degrees) could be expanded when homing. Targets were indicated on a range-azimuth type of display, provided with 4, 10, 40, 80 and 120-mile ranges, the latter for use only on beacons. With ASD a submarine periscope or Schnorkel could be located at about 8 miles.

Navigation (H_2S)

At microwave frequencies ground reflections vary in strength characteristically with the nature of the terrain. Echoes from water surfaces are relatively weak, those from open country are stronger, while the numerous flat walls, roof tops, etc., in towns and cities reflect a high proportion of the radio energy incident upon them. The contrast is sufficiently striking to enable ground features to be readily identified (see Plate XXXII). H_2S is the code name given to a series of radars which were designed specifically for observing ground echoes. With the PPI type of display used the pattern of the country over which an H_2S equipped aircraft is flying can be recognised in such detail that the system serves as an excellent navigational aid and, in particular, makes possible the precision bombing of built-up areas at night or through overcast.

Electronically H_2S is very similar to other microwave airborne radars, but differs again in antenna design and in the display. The first H_2S sets operated at a wavelength of 9 centimetres, but 3-centimetre and 1-centimetre versions (sometimes referred to as H_2X and H_2K respectively) which gave greatly improved definition were extensively used. The reflector produces a beam which is narrow in the horizontal plane and shaped in the vertical plane according to a cosecant squared law so that the echo strength varies with the terrain and the height of the aircraft, and is independent of range. It is fed with a waveguide radiator (illustrated in Plate XXXIII) and rotated about a vertical axis at 60 revolutions per minute. Time-base distortion is deliberately introduced in the PPI display so that the picture painted is of ground range (instead of slant range) against bearing. The display is stabilised to show true north always at the 12 o'clock position so that it may readily be compared with a ground map of the area. The aerial system is itself stabilised against pitch and roll of the aircraft in order that the picture may be unimpaired during steep turns or other evasive manœuvres which may be neces-

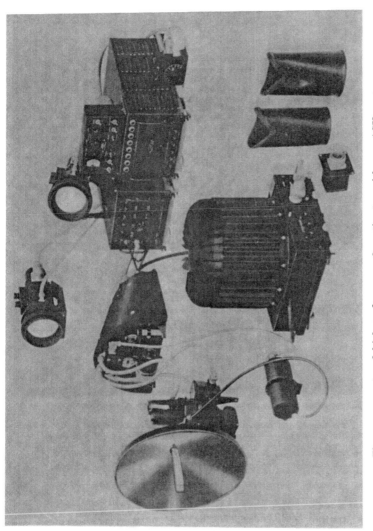

The components of high-performance 3-centimetre airborne ASV system (American AN/APS–3, or ASD–1). [By courtesy of the U.S. Air Force.]

PLATE XXXI.

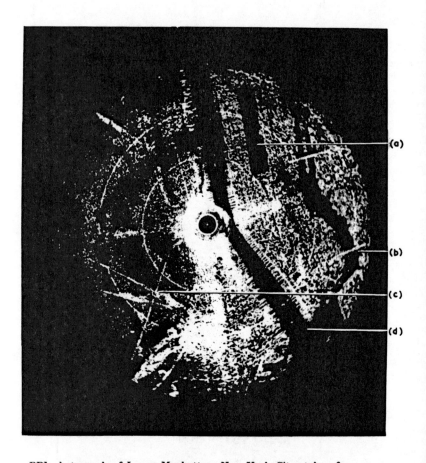

PPl photograph of Lower Manhattan, New York City, taken from an altitude of 4,000 feet with an H₄S type of equipment (U.S. H₂K).
Note: (a) Central Park, (b) bridges over East River, (c) railways, (d) Hudson River and wharves.

[*By courtesy of the Radiation Laboratory, Massachusetts Institute of Technology.*

PLATE XXXII.

sary over the target area. With a peak pulse output of 50 kilowatts H_2X (3-centimetre) can "see" a moderate-sized town at a range of the order of 30 miles.

4. Secondary Radar

In the primary radar systems so far discussed a target is detected solely from the energy which is scattered by its surfaces and thus plays no co-operative part in its location. A secondary radar system is one in which a particular object to be located is provided with a combination of receiver and transmitter known as a "responder", which, when triggered or "interrogated" by a primary radar, emits a characteristic reply signal of its own. The responder not only generates much more energy than would be returned by simple reflection, but can be made to provide a coded and hence distinctive signal from which it may readily be identified. The chief applications of secondary radar systems are as aids to navigation and as a means of distinguishing friendly aircraft or ships.

Beacons

Because a responder is capable of radiating a signal which is very much stronger than the natural echo of the site upon which it is placed, it provides a convenient radio "land mark," analogous to a lighthouse, which may be seen and identified at extreme ranges. When so used it is generally referred to as a beacon. Beacons serve as excellent navigational aids. They are used for homing to base, e.g. to airports or to aircraft carriers, to obtain a position fix (by range measurement from two or more), and for locating relatively inconspicuous positions, e.g. for the purpose of dropping supplies. Several blind bombing systems of high accuracy were developed for the bombing of small targets such as bridges, railway centres, etc., in which the bomb release point is dependent upon the precise determination of range from two fixed points. In the British "H" system distance measurements from two beacons at known points on the ground are made from the bombing aircraft (Fig. 330) in "Oboe," a variant of this, the aircraft carries a responder beacon and range measurements are made at ground stations (Fig. 331).

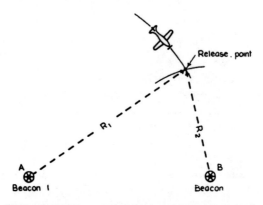

Figure 330.—*H* system. For blind bombing, the aircraft flies on a circular course which passes through the bomb release point. The range to beacon 1 is maintained equal to R_1 and bombs are released when the range to beacon 2 equals R_2.

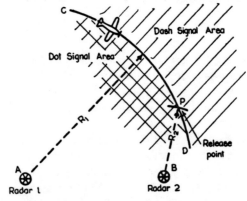

Figure 331.—Oboe. The radar station at *A* directs the aircraft so that the range R_1 remains constant. The station at *B* gives the bomb aimer a signal when the release point *P* is reached.

Beacons are used extensively as aids to navigation in association with search radars, both long wave and microwave. The response is usually initiated by the primary radar pulses and appears as an echo at the appropriate range on the radar

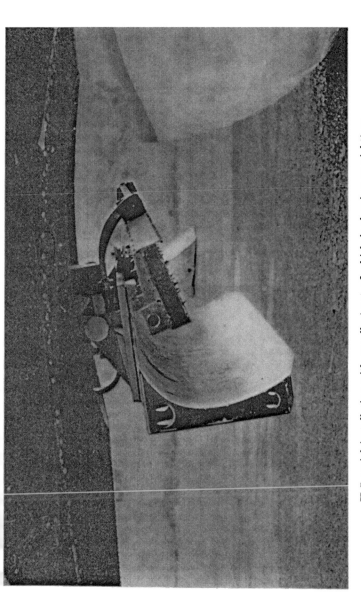

H₂S aerial installation with nacelle (part of which is showing on right) removed. The reflector profile is designed to produce a cosecant law. The feed is a waveguide with slot radiators.

PLATE XXXIII.

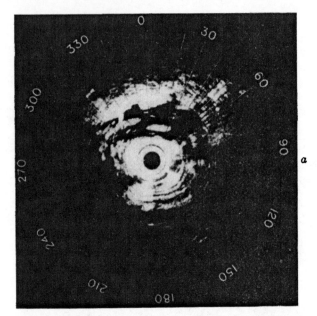

(a)

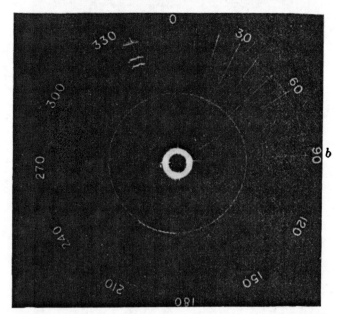

(b)

Radar and beacon signals of a 3-centimetre radar set:
(a) shows only the radar echoes, (b) shows the beacon reply; both were received at nearly the same position.

PLATE XXXIV.

screen, either amplitude-modulated or followed by an additional echo pattern (according to the type of display used) to identify the beacon being interrogated. It is usually transmitted on an adjacent frequency channel, and in such cases the output from the special beacon receiver may be displayed superimposed upon the normal radar echoes, or alone (see Plate XXXIV). In the latter case clutter from ground reflections in the vicinity of the primary radar is eliminated and a clear picture of the beacon response obtained.

Responder beacon techniques have found many peace-time applications in the field of civil aviation and large-scale mapping and surveying. These are described in more detail in Chapter XVIII.

IFF

The widespread use of radar for the control of interception by fighters and for the automatic direction of gunfire raised the problem of distinguishing friendly from hostile targets, and one of the most important military applications of secondary radar was in the form of IFF—a means of Identification, Friend or Foe. For this purpose friendly aircraft and ships were fitted with responder beacons. Search and fire control primary radars were provided with means for challenging or interrogating these beacons, and the characteristically coded echo from ships or aircraft fitted with IFF offered a ready means for distinguishing them from hostile targets visible on the radar screen at the same time. The earliest IFF systems, used, for example, with the British CH air-warning radars, were interrogated by the primary radar pulses themselves, and it was necessary to sweep the IFF receiver over the range of frequencies covered by the ground stations. This method was unsatisfactory, however, for use with stations employing narrow-beam slow-speed rotating aerial systems and a separate lower-power transmitter with a relatively omni-directional aerial was usually fitted as an adjunct to the primary radar for the specific purpose of IFF interrogation.

The widespread use of IFF in an active theatre of war

involves the highest degree of inter-Service and inter-Allied co-ordination and raises technical and political problems of great complexity. Although these problems were not completely solved during World War II, one uniform IFF system was in operation in nearly all theatres. Substantially the same techniques are likely to find peace-time application to the control of air traffic in the vicinity of busy air terminals, where the problems are similar but of lesser complexity.

BIBLIOGRAPHY

Proceedings at the Radiolocation Convention, March-May 1946, *J. Instn. Elect. Engrs.*, Vol. 93, Part IIIA, 1946, pp. 1-1486.

L. N. Ridenour (Ed.), *Radar System Engineering* (Massachusetts Institute of Technology, Radiation Laboratory Series, Vol. 1), McGraw-Hill, New York, 1947.

D. G. Fink, *Radar Engineering*, McGraw-Hill, New York, 1947.

Bell Telephone Laboratories, *Radar Systems and Components*, D. Van Nostrand, New York, 1949.

H. E. Penrose, *Principles and Practice of Radar*, Newnes, London, 1949.

CHAPTER XVIII

CIVIL APPLICATIONS OF RADAR

1. Aerial Navigation

The potentialities of radar as an aid to navigation became apparent quite early in World War II, and since the end of the war the chief practical applications of the fundamental pulse techniques of radar have been in the field of navigation.

Aerial navigation may conveniently be subdivided into three main branches:

(i) Long-range navigation.
(ii) Short-range navigation.
(iii) Navigation during the approach and landing phases.

Long-range navigation is the term generally applied to cases in which the aircraft is flying at ranges greater than about 100 miles from the navigational point or points of reference on the ground. Short-range navigation applies to the area within a radius of 100 miles of the reference point other than that occupied by aircraft manœuvring to approach and land. The last-mentioned area is sometimes called the airport control area.

The division of aerial navigation into the three categories mentioned is useful in helping to classify the numerous systems which exist and to understand the problems involved. The divisions are not, however, clear-cut, and some systems fall into more than one division. The features of short-range navigation which particularly distinguish it from its long-range counterpart are that it applies to an area which is frequently characterised by traffic lanes converging on an airport, with consequent increase in traffic density and changing aircraft altitudes. Also it so happens that the extent of short-range navigation corresponds approximately with the coverage of radio aids operating in the VHF and higher frequency bands, for which propagation occurs within the line of sight.

Pulse techniques have several fundamental advantages for navigation as compared with CW methods. In particular, they make possible great precision without ambiguity, the use of high peak transmitter output with only small mean power dissipation, and high discrimination against unwanted reflected signals and against interference. To realise these advantages it is, however, necessary to use considerably greater bandwidths than in CW systems; the corresponding implication is that more information is transmitted in the pulse systems.

Long-Range Navigation

In this field there are two pulse systems which were developed and used extensively in wartime and have had considerable application for civil purposes since then. They are the British system known as Gee and the American development, Loran. Of these, the former is more properly described as a medium-range system, having a maximum range less than that of Loran but greater than that of the short-range systems to be described later.

Gee. Gee belongs to the group which has become known as hyperbolic navigation systems. Such systems consist basically of a spaced pair of omni-directional transmitting stations located at known fixed points, the two transmissions being synchronised. In the case of pulse transmissions, a receiver at any point within the coverage area of both stations receives pairs of pulses, the time interval between the constituent pulses of a pair being dependent on the position of the receiver relative to the two ground stations. The locus of points of constant time interval is a hyperbola (strictly a hyperbolic surface), and different assigned values of time interval give rise to a family of hyperbolae, whence such navigational systems derive their name. Particular hyperbolae of the family are usually selected and defined as "lines" or "tracks" and allotted identifying characters. These lines can then be superimposed on maps to provide the necessary correlation between the geography of the region concerned and the navigational system. Fig. 332 represents a map of portion of the English Channel on which Gee lines have been superimposed.

The Gee receiver and the timing circuits required to measure the time interval between pulses of a pair are carried in the aircraft. The measurement is made in terms of the "time-units" in which the Gee lines on the map are defined. From a pair of Gee ground stations an aircraft can determine a "position line," at some point on which the aircraft is situated. If a second Gee pair is available another position line may be obtained, the intersection of this with the first resulting in a fix of the aircraft position.

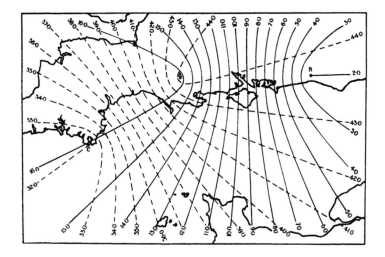

Figure 332.—Map of southern England showing a superimposed Gee lattice.

Usually one master station is associated with three slave stations to form a Gee chain of three pairs, the slave stations commonly being located at approximately equal intervals around a rough circle centred on the master station. A common value of separation between master and slave stations is 75 miles. The ground transmitters employ pulse powers of about 300 kilowatts.

The airborne equipment comprises a receiver and a timing unit with cathode-ray-tube display. The display consists of

two horizontal traces; on the upper trace appear pulses from
the master and one slave station, while on the lower trace the
master pulse is repeated but with that of another slave station.
Calibration markers can be superimposed on the traces. Simultaneous observations may therefore be made on two Gee pairs,
yielding two position lines and hence an instantaneous fix, a
facility which increases the accuracy and the rapidity with
which observations of position can be made.

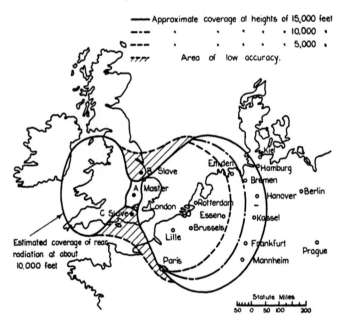

Figure 333.—Coverage of a typical Gee system in northern Europe.

Gee operates in the frequency band 20 to 80 megacycles per
second, where the propagation extends somewhat beyond the
line of sight. The radius of coverage of a Gee chain is of the
order of 150 miles at low altitudes, increasing to about 450 miles
at a height of 30,000 feet. The accuracy of a fix depends in
part on such geometrical considerations as range from the
stations, the particular position lines used (lines nearest to
the perpendicular bisector of the master-slave baseline give

the greatest accuracy for a given precision in measurement of time difference) and the angle of cut of the position lines. The error of fix varies from about 200 yards close to the stations to about 5 miles at the extreme range of 450 miles. Fig. 333 illustrates the coverage of a typical Gee chain.

Loran. Loran is an important long-range navigational aid; it is based on the hyperbolic principle already described. To achieve the desired range, Loran operates at frequencies of the order used for long-distance communication, i.e. much lower than those employed for other pulse navigational aids. As in Gee, measurement is made in the aircraft of the time interval between pulses from pairs of synchronised ground stations, and lines of position thus determined; the technical details of the system differ, however, from those of Gee.

Loran was designed to provide a coverage by the ground stations of 500 miles or more and careful selection of the transmission frequency was necessary to ensure satisfactory propagation. Both ground-wave and sky-wave propagation are used, the former by day and both modes by night. Ground-wave attenuation decreases as the frequency is reduced. Using sky-wave propagation it has been found possible to obtain reliable results when reflection from the E layer of the ionosphere is involved; pulses reflected from the F layer, however, at times suffer distortion and unpredictable delays and are not used. Another consideration affecting transmission frequency is that, if adequate accuracy of measurement of time interval between pulses is to be achieved in the aircraft, the leading edge of the pulses must be relatively sharp and well defined, necessitating the use of a wide bandwidth which is difficult to realise if the frequency is too low. As a compromise between all the relevant factors, of which the foregoing are the more important, the frequencies used lie in the band 1·7 to 2·0 megacycles per second.

The daytime range of a Loran station is approximately 500 miles over sea and 200 miles over land, the latter offering greater attenuation to the ground wave. At night, sky-wave propagation extends the range to about 1,400 miles. These figures are approximate only, being dependent on ionospheric conditions, degree of electrical interference, and the nature of the terrain.

The transmitters used have powers normally of about 100 kilowatts. Continuous reliable synchronisation, using the ground wave, between master and slave stations can be achieved with a separation, known as the baseline, of up to 500 miles over water. A long baseline is desirable from the aspect of accuracy but a low value is more suitable from the point of view of coverage; a usual compromise is about 300 miles.

Synchronisation of master and slave stations has occasionally been carried out using sky-wave propagation between widely separated stations, e.g. 1,000 miles apart. The accuracy of synchronisation is somewhat impaired, but this is offset by such advantages as extended coverage of the system and improvement in the "angle of cut" of Loran lines and hence in fixing accuracy.

The operator in the aircraft uses a cathode-ray display with two horizontal time bases vertically displaced. The master pulse appears on the upper time base and the slave pulse on the lower. For purposes of measurement, the traces are brought together and shifted so that the pulses are superposed. The time interval between pulses is subsequently read off with the aid of calibration markers synchronised with the time base. Considerable care is required in identifying and superposing pulses. In particular, it is necessary to distinguish the ground wave or E-layer pulses which are to be used. Also, there are present at times unwanted synchronised pulses, e.g. those propagated by the F layer. Changing distortion of the pulse shapes may also occur.

Discrimination between different pairs of Loran stations is effected both by the use of a number of radio-frequency channels and by variation of the pulse repetition frequency from one pair to another. As a rough figure, the accuracy of a position line obtained from a Loran pair using sky-wave propagation may be taken as 1 mile in the most favourable direction, i.e. along the perpendicular bisector of the line joining master and slave. The corresponding accuracy using ground waves will normally be several times better than this. On other bearings the accuracy falls as the cosine of the bearing with respect to the perpendicular bisector of the master-slave line.

Some development has been carried out on a Loran system operating on a frequency of 180 kilocycles per second. Such a frequency would have advantages in respect of reliable long-range propagation by ground waves. The system has not, however, so far come into general use.

Short-Range Navigation

There is no universal agreement on the kind of radio navigation system most suitable for short-range navigation, nor on the basic geometrical characteristics of such a system. There are geographical, economic, military and historical reasons for this lack of agreement. It has, however, been agreed by the International Civil Aviation Organisation that for international use a system based on polar coordinates (r, θ) will be adopted for the immediate future; that is to say, the basic navigational data will be in the form of range and bearing from known points, such points usually being terminals or reporting points along air routes. The internal navigation system in America, Australia and elsewhere tends to be of this form. The principal alternative to the r-θ type of system is one in which the co-ordinates form an approximately rectangular pattern, giving coverage over an area rather than along particular routes.

The pulse systems which are in use as short-range navigational aids fall into two general categories, namely, hyperbolic systems and those based on the echoing process. Hyperbolic systems have already been discussed in connection with Gee and Loran; the special forms which they take as short-range aids will be described in this section. The echoing process, on which the other group of aids is based, is the fundamental process of radar by which information about the distance of an object is obtained with a single measurement. This ability to measure distance directly has provided the navigator with an important new facility, as heretofore range from a given point was found only by first obtaining a fix with position lines.

Basic Technique of Distance Measuring by Radar. The aircraft carries a pulse transmitter which may be either part of a separate radar equipment or one used for distance measuring only. The pulses radiated from the aircraft, called interrogating pulses, reach the ground where they are either directly reflected by ground objects or picked up by the receiver of a "responder beacon." In the latter case, the beacon is "triggered" and transmits pulses (called responses) in correspondence with those received by it. The beacon pulses or direct echoes return to the aircraft, where timing arrangements are provided in the equipment which allow the time interval between each outgoing interrogation and the corresponding reply to be measured (see Fig. 334). Since the velocity of

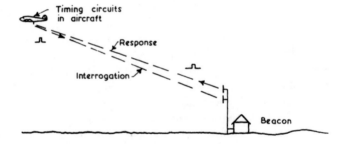

Figure 334.—Illustration of basic principle of distance measurement by interrogation of a radar beacon.

propagation is known, the range to the beacon or reflecting object can be found directly.

The method of distance measurement by simple echoes is applicable only where echoes are returned from some known geographical point and the display is such as to allow those echoes to be identified. For example, if a PPI display (Chapter II) is available in the aircraft, echoes from prominent coastal features such as headlands and harbours can readily be identified and ranges to those points quickly measured. It frequently happens, however, that points of navigational interest (e.g. airports) are not characterised by identifiable radar echoes; it is then necessary to place beacons at those points and to use the responder technique. The beacon response may be identified in a number of ways, e.g. by a particular frequency of transmission, by pulse grouping, or by a combination of such methods. Important additional advantages in using a responder are the increased range over which distance can be measured, since the beacon transmitter may be arranged to have considerable power, and also the possibility of using automatic ranging circuits (Chapter XV) in the aircraft, the response pulses being suitable for actuating such circuits.

Civil Applications of Distance Measuring. Since the war, considerable development has been directed towards applying distance measuring techniques to the problem of civil aviation. This development has been carried on in Britain, the United States, Canada and Australia. The aim has been to evolve a compact, light-weight equipment, automatic in operation, which will provide the pilot with direct information on a meter as to his range from any one of a number of known ground stations. The lines along which development has proceeded are illustrated by Australian and American equipment which have been designed for airline use.

Australian Distance Measuring Equipment.[1,2] Equipment designed for use on internal airlines in Australia operates in the 200 megacycles-per-second frequency band. The airborne unit comprises interrogating transmitter, receiver, and automatic ranging circuits contained within a single unit. The pilot is provided with a distance meter having alternative scales of 0 to 10 and 0 to 100 nautical miles and with a 12-channel beacon selector switch. Plate XXXV (a) shows the main unit and pilot's indicator. Once the pilot has selected the channel appropriate to the beacon whose range is required, the operation is entirely automatic. An electronic strobe searches for and "locks" on to the beacon pulse in the manner described in Chapter XV. The pointer of the pilot's needle then flies up and indicates the instantaneous range of the

---

[1] V. D. Burgmann, "Distance measuring equipment for aircraft navigation," *Proc. Instn. Elect. Engrs.*, Vol. 96, Part III, 1949, pp. 395-402.

[2] D. G. Lindsay, J. P. Blom and J. D. Gilchrist, "Distance measuring equipment for civil aircraft," *Proc. I.R.E. Aust.*, Vol. 11, 1950, pp. 307-15; Vol. 12, 1951, pp. 9-20.

beacon, and thenceforward continues to do so. As a check on the pilot's selection of the correct beacon, a lamp flashes a characteristic code sequence several times a minute. Should it be desired to find the range to another beacon, the channel selector switch is reset, when the automatic ranging process recommences and the new range is indicated.

The interrogating transmitter has a peak power of about 400 watts and a pulse repetition frequency of 100 pulses per second. The facility of 12 channels is provided by making the interrogation consist of pairs of accurately spaced pulses, the value of the spacing being dependent on the setting of the channel selector switch, and by using at each beacon a pulse-spacing discriminator which will pass only pulse-pairs having a particular spacing. By this method any one of a number of beacons up to twelve which may be within range can be selected at will. The sensitivity of the airborne receiver is about $4 \cdot 5 \times 10^{-12}$ watts.

Transmission and reception are effected on a single quarter-wave aerial by means of a coaxial-line duplexer. The total weight of the airborne installation is about 35 pounds.

The ground beacon uses a 5-kilowatt transmitter and a sensitive receiver. It is capable of providing responses to 100 aircraft simultaneously and is arranged for continuous un-attended operation. At each installation the beacon is dupli-cated, with automatic change-over in the event of a failure. Again, a common aerial is used for transmission and reception. A network of some 80 beacon installations is envisaged to cover the principal Australian air routes.

The accuracy of the system is of the order of 2 per cent or 1 mile, whichever is greater, on the 0 to 100 mile scale, and 2 per cent or $\frac{1}{8}$ mile, whichever is greater, on the 0 to 10 mile scale.

American Distance Measuring Equipment.[3] The equipment designed for airline use in America operates in the 1,000 mega-cycles-per-second frequency band. Choice of this frequency was dictated largely by the need for as many as 100 discrete

[3] C. J. Hirsch, "Pulse-multiplex system for distance measuring equipment," *Proc. I.R.E.*, Vol. 37, 1949, pp. 1236-42.

(a) Australian distance measuring equipment as fitted in aircraft (Channel-selector switch not shown).

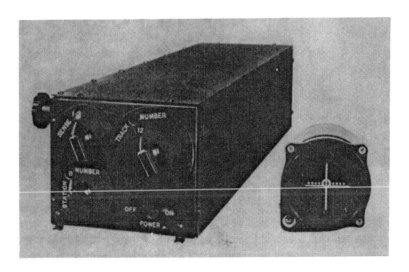

(b) Simplified lightweight version of airborne equipment for Multiple Track Range system.

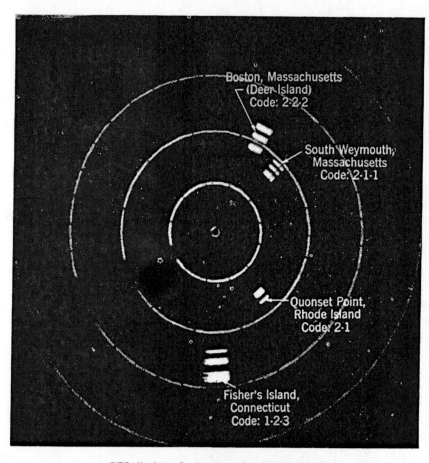

Boston, Massachusetts
(Deer Island)
Code: 2-2-2

South Weymouth,
Massachusetts
Code: 2-1-1

Quonset Point,
Rhode Island
Code: 2-1

Fisher's Island,
Connecticut
Code: 1-2-3

PPI display of microwave beacon responses.

[By permission from "Radar Aids to Navigation", ed.
by J.S. Hall, M.I.T. Rad. Lab. Series. Copyright,
1947. McGraw-Hill Book Co. Inc.

PLATE XXXVI.

operating channels and the fact that insufficient spectrum space was available at lower frequencies to accommodate this number. As in the Australian airborne equipment, interrogating transmitter, receiver and automatic ranging circuits are contained within a single unit. The pilot is provided with a meter directly indicating beacon range in miles, and with a beacon selector switch; the desired channel having been selected, the ranging action is fully automatic. A second meter indicates rate of change of distance. The total weight of the airborne installation is about 45 pounds. A very extensive network of distance measuring beacons is planned for the United States, a large proportion of which are to be operated in conjunction with CW omni-directional radio ranges to provide a polar coordinate navigational system.

Distance Measurement using Microwave Radar. An interesting technique for distance measurement using airborne microwave radar and a PPI display (Chapter XVII) was developed during the war and has been tested for civil use. Microwave beacons, arranged to respond on a frequency slightly different from that used for interrogation, are interrogated by the aircraft, and the PPI display, instead of indicating the normal radar picture, can be switched to show the beacon responses as spots of light at the appropriate range and bearing. Thus, both polar coordinates are determined by a single system. Individual beacons are distinguished by the use of multiple responses. Plate XXXVI shows the appearance of a typical display.

Hyperbolic Systems for Short-Range Navigation. The principle of hyperbolic navigation, described earlier in connection with Gee and Loran, has important applications to short-range navigation. The Gee system itself is in limited use in northern Europe as an aid to civil flying. Several proposals have also been made for hyperbolic systems for short-range use in which the baseline length is greatly reduced from the value commonly used in Gee. Of these, that which has been subjected to the greatest amount of development and testing is the Australian Multiple Track Range system.

The Australian Multiple Track Range.[4] This equipment, developed since 1945, provides, in a form suitable for use in civil aviation, the radial component of a polar coordinate navigational system.

The Multiple Track Range operates in the 200 megacycles-per-second frequency band. It employs one pair only of synchronised pulse transmitters, the master and slave stations. The baseline separating the pair has a value usually between 3 and 9 miles, which is much less than the baselines of Gee and Loran systems. In any hyperbolic system, at distances of more than one baseline length from the centre of the system, the hyperbolic lines or tracks become nearly straight lines radiating from the centre of the system and almost coincident with the hyperbolic asymptotes, which latter can for most navigational purposes be regarded as the actual tracks. By reason of the relatively short baseline used in the Multiple Track Range, straight radial tracks are obtained over the greater part of the area covered by the stations. By selection of suitable values of pulse-time interval, up to 40 radial tracks (i.e. 20 hyperbolae) can be provided in a system having a baseline of 3 miles; the tracks are identified by numbering them 1, 2, ... As in Gee and Loran, for a given precision of timing, position-line accuracy is greatest along the perpendicular bisector of the baseline and least along the extension of the baseline. Fig. 335 shows the track systems provided by three Multiple Track Range installations along an air route.

The airborne equipment comprises a receiver and automatic ranging circuits (see Chapter XV), designed to measure the master-slave time interval. The pilot is provided with two meters, a track indicator and a left-right meter, and a control box. A particular Multiple Track Range system (i.e. a pair of stations) having been selected by means of a channel switch on the control box, the ranging circuit automatically measures the time interval between master and slave pulses and displays the result on the track indicator, the scale of which is calibrated in track numbers. By reference to a map on which the track system has been superimposed the pilot can deter-

[4] M. Beard, "Multiple Track Range," *Proc. Instn. Elect. Engrs.*, Vol. 96, Part III, 1949, pp. 245-51.

mine his bearing from the centre of the system. If it is desired
to fly along a particular track, a track selector switch on the
control box is set appropriately and the left-right indicator
then shows the departure of the aircraft from the selected
track. Individual Multiple Track Range systems are distinguished by the use of a double-pulse technique, the master
transmission in all cases consisting of pairs of spaced pulses

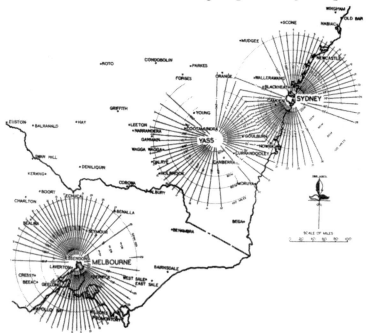

Figure 335.—Track systems provided by Multiple Track Range installations on an air route.

and the value of the spacing being characteristic of a particular
system. The channel selector in the airborne equipment consists of a double-pulse discriminator which allows for the
selection of any particular track system up to a total of 12 in
number. A complete airborne installation weighs about
35 pounds.

A simplified version of the airborne equipment has been
developed which does not have the facility of direct indication

of track number, but which is, however, greatly reduced in weight from the standard version, its installed weight being 12 pounds. It is illustrated in Plate XXXV (b). To determine track number the operator rotates a selector knob which has the effect of varying the time delay of an electronic strobe. When the strobe delay is equal to that between master and slave pulses, a light comes on and the selector knob then indicates the appropriate track number. At the same time a centre-zero meter becomes operative, and the indicated track is being followed when the needle is kept centred. To fly any desired track the selector knob is set accordingly and the aircraft flown until the track is reached, when the light comes on; the centre-zero indicator then shows any departure from the selected track. This simplified version of the airborne equipment is very suitable for providing general track guidance in light aircraft. It is also applicable as, and has the necessary accuracy for, an aid to final approach to a runway (i.e. as a "localiser") if the ground stations are so disposed that the centre (straight) track of the system is aligned with the runway.

The accuracy of bearing provided by the Multiple Track Range system is about 1 degree for most tracks, but deteriorates to 4 or 5 degrees at the extreme tracks (i.e. those nearest to the extension of the baseline).

The ground transmitters employ powers of 1 to 3 kilowatts. The slave station automatically synchronises its transmission with that of the master and both stations are designed for continuous unattended operation.

Radar Aids to Approach and Landing

Some of the short-range aids already described are applicable to navigation in the airfield approach area by virtue of their precision and form of display. The requirement in this area during the approach is that the position of the aircraft be known to the pilot at all times with an accuracy corresponding to a time interval (assuming a typical aircraft approach speed) which is short compared with the minimum time separation at which it is desired to land aircraft. If this requirement is met, the aircraft can perform "holding" procedures and make its way to the entry or "gate" of the final letdown path while maintaining a safe separation from other aircraft in the area. During the final stages of the landing a considerably higher order of precision than that of aids already described is required if the landing is to be made by instruments and not visually.

Such aids as Distance Measuring Equipment, Gee and the Multiple Track Range have been successfully used to navigate aircraft with precision in the approach area and it appears almost certain that equipment of such kinds will be required if delayed landings are to be avoided at airports with high traffic density. Considerable assistance can be provided to aircraft during the final stages of approach and landing by means of Ground Controlled Approach or GCA techniques. These are described in more detail below.

Airways Control. Ground radars can be designed with characteristics which make them applicable to navigation. Large aerials may be used giving high resolution and gain, and very high values of transmitter power may be achieved, so that the radar can have an extensive range. The position of an aircraft can therefore be determined by such a radar with fair precision within a wide area. Since this information is derived on the ground it is not directly available to the pilot of the aircraft, but can be communicated to him by a ground controller. Such radars therefore do not normally constitute a primary navigational aid but serve the important purpose of surveillance over the movement of aircraft navigating by other means, and can also provide navigational assistance in the event of failure of primary aids.

The typical range of a large ground radar is of the order of 80 to 100 miles (assuming line-of-sight propagation) on all-metal aircraft of the type normally used for civil transport, and position may be determined without difficulty to 1 mile in range and 1 degree in azimuth, provided that the aircraft echo is not obscured by "clutter" due to echoes from ground objects or from rain.

If full advantage is to be taken of the capabilities of a powerful ground radar it must be sited in such a way as to avoid screening by hills or buildings, at least in those regions where radar coverage is important. Such a site may be at some distance from the point at which the radar information is required, as for example an airport control tower. It is there-fore necessary to provide means for transmitting the radar picture overland for a distance of some miles; this is normally done by radio or cable.

The method of displaying to the ground controller the in-formation provided by such a radar calls for special con-sideration. The display should be visible in full daylight with-out the use of a visor, a condition not normally fulfilled by the cathode-ray tube. Again, the radar picture will normally contain much information which is irrelevant to the controller and which will confuse or distract hím. Also, the radar picture provides information only on the instantaneous positions of aircraft, whereas it is of importance for the controller to know the tracks they have been making good. For these reasons, techniques have been developed by which one or more operators observe the normal radar picture in a darkened enclosure and plot the tracks of aircraft within the area covered. These plots are automatically transferred from the cathode-ray tube by an electro-mechanical linkage to a large screen in the controller's room where the tracks, suitably identified, are clearly visible in daylight.

Methods have been envisaged which allow the display of the ground radar to be transmitted to aircraft, when it becomes directly applicable for navigation; a ground-to-air television link is one such method.

Airport Control Radar. Irrespective of the kinds of primary navigational aid used by aircraft in the vicinity of an airport, it is now accepted that a surveillance radar should be used at the airport to provide the traffic controller with a general picture of the traffic situation and to assist him in guarding against collision dangers. Such a radar requires 360 degree coverage up to a height of about 5,000 feet and a range of 20 miles approximately. Ground "clutter" is very undesirable in such applications and the present trend is to use the tech-nique of moving-target indication.

In order to obtain suitable coverage by the airport radar it may be necessary to locate it at some distance (possibly up to a mile) from the control tower. The problem of transmitting the radar information to the airport controller thus arises here, as in the case of the Airways Control radar previously discussed, and a radio or cable link is used. As in that case also the relevant radar information—if it is to be in a form useful to the controller—must be selected from the cathode-ray-tube screen and presented as a display visible in daylight.

Ground Controlled Approach. An important use of ground radar as an aid to final approach and landing is in the technique known as Ground Controlled Approach (GCA), which was developed initially for military use. Two radars are used, located at the up-wind end of the runway in use. One is a search radar, having a range of about 20 miles and coverage of 360 degrees, whose purpose is to check aircraft movements within the approach area and guide those aircraft desiring to land to a position suitable for commencing the final let-down; this is done by ground-to-air radio communication. The aircraft, having been brought to such a point, is "handed over" to the controller of the second, or precision, radar. This radar has two narrow-beam aerials (beamwidths of 0·9 and 0·6 degree) which scan respectively limited horizontal and vertical sectors about the final let-down path. The position of the aircraft with respect to this path as it makes its descent can thus be accurately determined by the radar controller, and any necessary corrections passed rapidly by radio to the pilot, who is thus "talked down." Using GCA, aircraft may with safety be brought down to 100 feet above the airstrip, and in military use blind touchdowns have been made. The rate at which the controller can instruct descending aircraft limits the handling capacity of GCA but landing rates in excess of one aircraft every 3 minutes have been achieved.

In GCA for civil use, the controllers are located in the airport control tower and the information from the radar at the airstrip is sent to the tower by radio or landline.

2. Marine Navigation

Two types of navigational aids using pulse techniques have found important marine applications during and since World War II. The first is the long-range navigational system Loran which has been described above; the second is the microwave shipboard radar which is being used to an increasing extent to provide collision warning at sea in bad weather, for off-shore navigation, and for navigation in harbours and estuaries. In addition to these primary navigational aids, ground-based radar is, as in the case of civil aviation, beginning to fill the important role of surveillance of shipping movements in confined waters and to provide emergency navigational assistance to vessels which, because of failure of primary aids or for other reasons, are unable to navigate.

Shipboard Microwave Radar

Microwave radar is finding increasing application to marine navigation in coastal waters, harbours, lakes, rivers and all similar situations where a recognisable pattern of radar echoes can be obtained. The PPI type of presentation is almost universally used. Depending on the degree of resolution of the radar and the minimum range at which it is capable of detecting and displaying an echo, it may be applied to collision warning, general navigation, or pilotage in narrow waters, the last-mentioned application requiring the greatest resolution.

Present-day marine radars are designed to provide facilities for general navigation and for pilotage in narrow waters, the facility of collision warning being obtained concurrently. They usually operate in the 3-centimetre band, although some American radars operate at 10 centimetres. The latter wavelength is less attenuated when propagation is through rain but does not permit of the degree of resolution possible with 3-centimetre equipment. The radar aerial is mounted as high in the ship as possible, e.g. above the bridge or at or near the masthead, and is sometimes gyroscopically stabilised against the effect of pitch and roll. Plate XXXVII shows a shipboard antenna installation. The transmitter and receiver may

Aerial installation of a shipboard navigational radar.

Plate XXXVII.

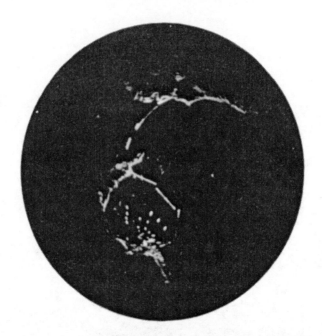

(a) Weymouth harbour as seen on the radar screen of an approaching ship. The coastline, breakwaters, and vessels in the harbour are clearly delineated.

[By courtesy of Kelvin and Hughes Ltd.

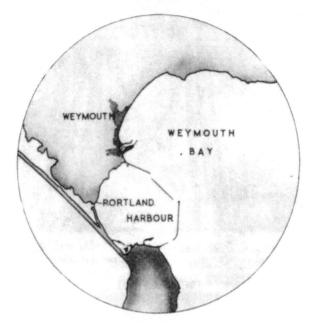

(b) An actual map of the area.

[By courtesy of Kelvin and Hughes Ltd.

PLATE XXXVIII.

be combined with the aerial unit or connected to the latter by waveguide. The PPI display and control unit are installed on the bridge.

A typical specification for a shipboard radar, assuming the aerial to be mounted 40 feet above the water line, calls for a detection range of at least 20 miles on ground features 200 feet or more in height, and of 2 miles on navigation buoys; while the performance at minimum range should allow the echo from a small buoy at a distance of 50 yards to be distinguished. The discrimination in range between small objects should be 100 yards, and in bearing 3 degrees. At least four range scales (1·5, 5, 10 and 30 miles) are called for, the range accuracy being ± 5 per cent of the maximum range on any scale. The display screen must not be less than 5 inches in diameter and the design of the whole equipment such that it should be capable of operation and interpretation by ships' officers on the bridge.

Applications of Shipboard Radar

A modern microwave radar provides the navigator with a two-dimensional plan on which other vessels, coastal features, harbour outlines (including piers, wharves, bridges, etc.) and any reflecting objects of appreciable size within radar range can be distinguished. The range of detection of buoys and other small objects can be increased several times by fitting them with corner reflectors, while radar beacons may be installed at points such as lightships or headlands for which a prominent and reliable indication on the radar scope is necessary, or positive identification of the point—obtained by coding the beacon—is required.

The possession of such a picture makes possible the navigation of vessels with safety at sea or in coastal waters under all conditions of visibility and without reduction of speed. It is also of the greatest assistance to the free movement of shipping under similar conditions in estuaries, harbours and channels. However, the extent to which radar can be used to navigate through narrow channels, to anchor in restricted waters or, in extreme cases, to berth under conditions of very poor visibility depends on such factors as the experience of the master or pilot in radar navigation, his knowledge of the route being traversed, his ability to interpret the corresponding radar pattern, and his confidence in and the performance of the radar. Of these (assuming a radar of adequate performance) familiarity with the characteristic appearance on the radar screen of topographical features in the area being traversed is probably the most important. Plate XXXVIII shows a typical display obtained on a vessel approaching a harbour, together with a map of the area. The superposition on the radar screen of a transparent sheet to the same scale, showing the shore outlines, channel markings and other features of interest to the navigator and thus allowing the ship's position to be determined quickly and precisely, is a refinement of the display which is envisaged.

Marine Use of Port Radar

The use of shore-based radar of high performance provides, under all weather conditions, means of observing the movement of vessels into and out of the port and of checking the position of channel markers such as buoys, so that the state of the port and its approaches can at all times be quickly ascertained.

Such a radar can also be used to advise vessels using the port of the position and movement of other vessels in the channel, and so avoid delays which would otherwise result from uncertainty as to the state of the channel, and to provide navigational assistance to pilots desiring to bring vessels into or out of port in heavy fog conditions. For these applications, an adequate two-way radio communication link between ship and radar controller is necessary.

A specially designed shore radar for port control, recently installed at Liverpool, England, is representative of present-day technique in this field. This equipment operates in the 3-centimetre wavelength band. The aerial is at a height of 80 feet and rotates at 10 revolutions per minute; its horizontal beamwidth is 0·7 degree, providing good azimuthal resolution. The transmitter power is 30 kilowatts and pulse length 0·25 microseconds which permits good resolution in range. The display is an elaborate one, consisting of four large-scale (2 inches to 1 mile) overlapping displays covering the port approach, a total of some 16 miles in length, as well as a general display of the whole port area. VHF communication to vessels is used. The radar is operated at 4-hourly intervals in clear weather, and continuously under fog conditions. It has proved to be of considerable value in reducing shipping delays and increasing the safety of navigation generally in the port area.

3. Radar Aids to Surveying

The process of surveying inaccessible and undeveloped country has in the past been a lengthy and arduous procedure due to the difficulty of operating ground parties in such terrain. The advent of aerial photography has reduced the labour and increased the rate at which an area is covered, but for reasonable accuracy it is still necessary to establish bench marks and levels at many points within the surveyed area.

Radar can contribute to the problem of surveying undeveloped terrain because radio waves are transmitted easily and reliably over distances greater than the visual, and because of the ability to measure distance accurately from one or other end of the baseline. It is not necessary to traverse the baseline and the intervening ground need not be flat or even accessible. The accuracy of measurement in military practice was of the order of ± 20 yards in 200 miles and recent developments[5] show that this figure can be improved to ± 5 yards. The limitation is now primarily due to variations in atmospheric propagation as was remarked on page 7.

Full advantage of radar methods is not obtained if measurements are confined to the ground and the most profitable method of making use of them is in the air. Various sug-

[5] C. I. Aslakson, "Tactical use of Shoran and accuracy obtained," *Photogrammetric Engineering*, Vol. 12, 1946, pp. 379-81; C. A. Hart, "Some aspects of the influence on geodesy of accurate range measurement by radio methods with special reference to radar techniques," *Bulletin Géodésique*, nouvelle serie, No. 10, 1948, pp. 307-52; and J. Warner, "The application of radar to surveying," *Empire Survey Review*, Vol. 10, 1950, pp. 338-48.

gestions have been put forward for radio methods of precision distance measurement on the ground using phase comparison methods but they suffer from defects due to the ground-wave propagation velocity being a function of varying earth constants. Therefore, the applications of greatest interest are the conduct of large-scale triangulation by means of distance measurements in aircraft, and the ground control of survey aircraft.

The technique employed for long-base triangulation is analogous to the H technique described in Chapter XVII,

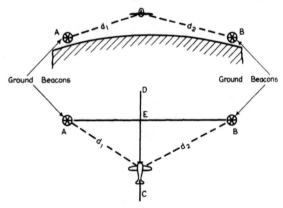

Figure 336.—The distance between A and B is measured by flying along a path such as CD and making a continuous record of d_1 and d_2. The minimum value of d_1+d_2 must correspond to the point E and the distance AB can then be calculated.

section 4. Beacons are placed on the ground from 300 to 500 miles apart. An aircraft flies across the line between them and measures simultaneously the distance to each of them. At the crossing point these two distances, together with the aircraft altitude, allow a computation to be made of the ground distance between the beacons as shown in Fig. 336. At a flying height of 30,000 feet a beacon separation of 500 miles is possible and the error of measurement should not exceed $\pm$ 10 yards. With this as baseline, a triangulation with sides each some 500 miles long may be built up, giving an exceptionally open grid which can be used to link up points over a large continent or even between land masses via an island chain.

A problem that would arise in the use of radar in the measurement of long lines is breaking down the triangulation nets into smaller units. If this were done by normal ground methods, it is doubtful if much advantage would be gained by the use of radar in the first instance. Since, however, it is possible to fix accurately the position of the aircraft at any point within range of the two radar beacons, it is possible to fix the position of a point on the ground vertically below. With the new gyro-stabilised camera mounts recently developed, this should allow points on any air photograph to be located to about ± 5 yards provided the mean of a number of fixations is used.

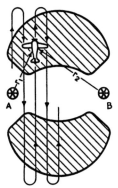

Figure 337.—For the ground control of survey aircraft, beacons are placed at the points A and B and the aircraft flies a rectangular grid as shown in the figure. The accuracy of a fix within the shaded areas may be as high as ± 20 yards.

The ground control of survey aircraft is accomplished by the H method in practically unmodified form. One of the many procedures which may be adopted[6] is to place the ground beacons A and B (Fig. 337) at two known points about 200 miles apart in the terrain to be surveyed. The aircraft then flies a rectangular grid, and each time a survey photograph is taken a record is obtained of distances r_1 and r_2 to each of the beacons. The accuracy of a single fix within the shaded areas can be as high as ± 20 yards. Due to unfavourable geometry the

[6] C. A. Hart, "Surveying from air photography fixed by remote radar control," *Empire Survey Review*, Vol. 9, 1947, pp. 71-83.

accuracy elsewhere is lower, but for the aircraft track shown the position of photographs taken in the unshaded area can be determined by interpolation. In this way a single pair of ground stations can be used for the survey of a total area of some 100,000 square miles.

In addition to its use for fixing the position of the aerial photographs, the same radar may, of course, be used as a navigational aid to provide track guidance for the pilot of the survey aircraft. The normal radar equipment that would be used, namely, Gee-H or Shoran, provides track guidance only in the form of a series of concentric circles round either ground beacon. However, instruments have been developed[7] for use with Shoran to provide straight-line track guidance in any direction within the operational area of the radar equipment. Other navigational aids, such as Decca which provides a hyperbolic system of tracks by means of a phase comparison of two CW transmissions, have also been used as navigational aids for aerial surveying.

The use of radar as a navigational aid and position-fixing device during aerial photography results in a considerable saving in the time required to cover a given area and in the amount of photographic material used.

BIBLIOGRAPHY

J. S. Hall (Ed.), *Radar Aids to Navigation* (Massachusetts Institute of Technology, Radiation Laboratory Series, Vol. 2), McGraw-Hill, New York, 1947.

A. Roberts (Ed.), *Radar Beacons* (Massachusetts Institute of Technology, Radiation Laboratory Series, Vol. 3), McGraw-Hill, New York, 1947.

J. A. Pierce, A. A. McKenzie and R. H. Woodward (Eds.), *Loran* (Massachusetts Institute of Technology, Radiation Laboratory Series, Vol. 4), McGraw-Hill, New York, 1948.

Proceedings at the Radiolocation Convention, March-May 1946, *J. Instn. Elect. Engrs.*, Vol. 93, Part IIIA, pp. 331-48; 449-512.

R. A. Smith, *Radio Aids to Navigation*, Cambridge University Press, 1947.

R. F. Hansford, "The development of shipborne navigational radar," *J. Instn. Navig.*, Vol. 1, 1948, pp. 118-47.

---

[7] R. C. Richardson and J. Warner, "Straight flight of aircraft equipped with radar-operated pilot's indicator," *Photogrammetric Engineering*, Vol. 16, 1950, pp. 544-49.

CHAPTER XIX

APPLICATIONS OF RADAR TO PHYSICAL SCIENCE

WE have so far been concerned with a detailed description of the various techniques on which radar is based and with a survey of the manner in which they have been applied initially for military purposes and subsequently in civil spheres. In many respects more interesting are the applications of the same fundamental techniques to various branches of the physical sciences themselves. These have come into prominence in more recent years and are providing not only alternative experimental methods, for example, for the measurement of such fundamental physical quantities as the velocity of electromagnetic waves, but, in addition, access to information which was previously inaccessible or unable to be uniquely determined by other means, for example, concerning some characteristics of the solar corona.

The scientific applications of radar fall naturally into two classes: those in which radar *systems* are used in whole or in part as new tools for research, and those in which the *techniques* of radar are adapted in various ways to the solution of essentially physical problems. The advances in modern physical science discussed below are representative of those which have only become possible through the advent of radar. There are numerous other fields in which instrumentation and experimental method are benefiting to an increasing extent as knowledge of the techniques become more widely disseminated.

1. Applications of Radar Systems

Ionospheric Research

The earliest use of pulse methods to demonstrate the existence of an ionospheric reflecting layer and to measure its height was in the classical Breit and Tuve experiment of 1925. In the years that followed, considerable research was devoted, parti-

cularly by Appleton, to the development of improved techniques, and the pulse method became well established as the
standard method for investigating the ionosphere. It was un-
questionably the knowledge and experience gained in the use
of these techniques for ionospheric research over a decade or
so which made possible the rapid development of radar in the
few years just prior to 1939.

Ionospheric research has now, in turn, benefited from wartime developments in the radar field. The most significant
advances have come from application of the responder beacon
technique, that is, from the use of a master transmitter to
trigger one or more suitably placed similar transmitters. If
these be situated within ground range of the master a virtually
instantaneous picture of the ionosphere over an extended area
can be obtained. This method has been used[1] to study vertical
and horizontal movements of clouds of ionisation within the
layers and has led to the identification of cellular waves within
the ionosphere[2] which move with a horizontal velocity of the
order of 300 to 600 kilometres per hour. Sky-wave synchronisation can be used to provide to and fro transmissions over
long paths and hence a ready means for investigating the
properties and behaviour of the ionosphere for radio waves
incident obliquely upon it.

Moon Echoes

Radar echoes from the Moon have been received by workers
in the United States,[3] Hungary[4] and Australia.[5] Since the
Moon's mean distance is 3.8×10^5 kilometres, an echo begins
to arrive about $2\frac{1}{2}$ seconds after the start of the transmitted

[1] G. H. Munro, "Travelling disturbances in the ionosphere," *Proc. Roy.
Soc.* A, Vol. 202, 1950, pp. 208-23.
[2] D. F. Martyn, "Cellular atmospheric waves in the ionosphere and tropo-
sphere," *Proc. Roy. Soc.* A, Vol. 201, 1950, pp. 216-33.
[3] J. Mofenson, "Radar echoes from the Moon," *Electronics*, Vol. 19, 1946,
pp. 92-7; and J. H. de Witt and E. K. Stodola, "Detection of radio signals
reflected from the Moon," *Proc. I.R.E.*, Vol. 37, 1949, pp. 229-42.
[4] Z. Bay, "Reflection of radio waves from the Moon," *Hungarica Acta
Physica*, Vol. 1, 1946, pp. 1-22.
[5] F. J. Kerr, C. A. Shain and C. S. Higgins, "Moon echoes and penetration
of the ionosphere," *Nature*, Vol. 163, 1949, pp. 310-3; and F. J. Kerr and
C. A. Shain, "Moon echoes and transmission through the ionosphere," *Proc.
I.R.E.*, Vol. 39, 1951, pp. 230-42.

pulse. Relatively long pulse lengths are necessary to carry the required amount of energy (10^4 joules as compared with normal radar practice of about 1 joule) and very narrow-band receivers are required to obtain the utmost sensitivity. The ideal intermediate-frequency bandwidth of one or two cycles per second is difficult to attain, and, moreover, transmitters cannot be made sufficiently stable for all the energy to be confined within such a narrow band. Hence a somewhat less effective alternative is employed, namely, an i.f. bandwidth of about 50 to 100 cycles per second, followed after detection by a narrow-band filter in the "video" circuit, to reduce the noise against which the echo must be detected. In tuning the receiver, allowance must be made for the Doppler shift arising from the Moon's line-of-sight velocity, which is due in part to the observer's motion on the rotating earth, and in part to the eccentricity of the Moon's orbit. Under the circumstances of the American and Australian experiments, this Doppler shift was of the order of 300 and 50 cycles per second respectively.

Moon echoes were first obtained in 1946 by the United States Army Signal Corps. The frequency used was 111·5 megacycles per second, with a transmitter power of 3 to 15 kilowatts, pulse length 0·05 second, and repetition period 4 seconds. The aerial was a 64-element broadside array, mounted on a rotatable tower, so that observations could be made near moonrise and moonset. Echoes were obtained on a number of occasions, sometimes up to 20 decibels above noise, but the intensities were found to be highly variable. The maximum intensity was of the order to be expected for a transparent atmosphere and interplanetary medium.

Shortly after the American work began, Bay, in Hungary, also obtained Moon echoes. Through lack of sensitivity in his receiving system, he was forced to use an integrating method, in which the presence of an echo from the Moon could just be detected by adding the impulses received over a period of 30 minutes. Thus, no information was obtained with regard to short-period variations in intensity of the echoes.

In the Australian experiments low frequencies were used (about 20 megacycles per second), with the primary object of investigating transmission through the ionosphere. This provided a new method of obtaining information on the upper portion of the F region, since waves capable of reaching beyond the level of maximum ionisation cannot otherwise be returned to Earth.

Special transmissions were made from the high-frequency broadcasting station, Radio Australia, generally on 21·54 megacycles per second, 70 kilowatts output, and echoes were received on conventional communications-type equipment. The pulse length was either $\frac{1}{4}$ second or 2·2 seconds, the latter giving the possibility of studying short-period amplitude variations of the echo. The repetition period was 6 seconds.

Echoes were received on 23 out of 30 attempts. Intensities, however, were lower than the theoretically expected values, and the minimum altitudes at which echoes were first detected from the rising Moon were unexpectedly high. The daily values of echo intensities and minimum altitudes for detection were found to correlate well with the critical frequency of the F_2 region, though they did not agree with values derived from existing theory. This result indicates that the anomalies arose in the F_2 region, but that the currently accepted model of the region is inadequate.

In addition to the day-to-day intensity variations, fading of two types, with periods of the order of minutes and seconds, was observed. The slower fading was probably due to ionospheric irregularities. The faster type, which is illustrated in Plate XXXIX (a), was shown to be associated with the Moon's libration. This result can be taken as evidence that the Moon reflects as a rough body at 20 megacycles per second, and has a bearing on proposals that have been made for communicating between two points on the Earth via the Moon. A reflected signal pulse shorter than 10 milliseconds could not be obtained.

Radar methods provide the possibility of an accurate, direct measurement of the Moon's distance. Short pulses and wide receiver bandwidths must be used for accurate ranging, but the corresponding echo intensity would then be much less than that obtained in any of the long-pulse experiments carried out

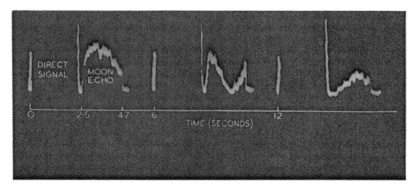

(a) Radar echoes from the Moon—a sample photographic record of three successive echoes obtained with 2·2 second pulses, and illustrating the rapid fading which is observed.

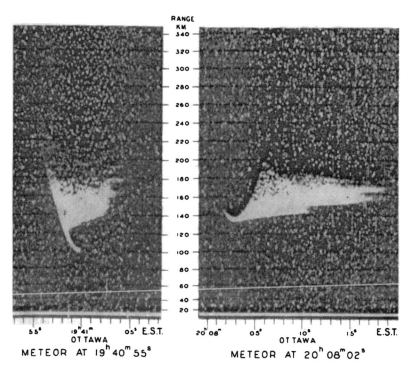

(b) Radar echoes from meteors at a wavelength of 9·2 metres. The echo from the meteor itself is visible first approaching and then receding. It is followed by an echo from an ionised region which persists for many seconds after the passage of the meteor.

PLATE XXXIX.

so far, since, with short pulses, only a small part of the Moon's surface is contributing to the echo at a given instant. More powerful equipment would therefore be needed for accurate distance measurement.

Meteor Echoes

The study of meteors by visual means is handicapped by their sporadic occurrence and by difficulties imposed by cloud and daylight. It had long been suspected that meteors were responsible for a substantial contribution to the ionisation of the E region but it was not until military-type radar equipment was applied to the study of echoes from meteoric ionisation[6] that simple effects could be identified and an individual correlation obtained between transient echoes and visible meteor showers. The use of radar techniques makes possible continuous observations independent of visual seeing conditions, and with a sensitivity greater than that obtainable by eye. (See Plate XXXIX (b).)

The number of transient echoes observed is found to be critically dependent on the radio frequency used. As the frequency is reduced, there is a large increase in the number of observed echoes at about 50 megacycles per second. Above this frequency there is a high correlation between transient echoes and visible meteors, but below, the echoes from visible meteors are lost among large numbers of echoes of similar appearance. These show seasonal and diurnal variations typical of meteors and are suspected to be due to vast numbers of tiny particles of meteoric dust, too small for optical detection.

In the 60 to 70 megacycles-per-second region the observations have been carried out, largely in England, using modified Army radars, with a peak power of about 150 kilowatts, and Yagi aerials which could be pointed anywhere in the sky. Most of the short-period echoes observed last for

* J. S. Hey and G. S. Stewart, "Radar observations of meteors," *Proc. Phys. Soc.*, Vol. 59, 1947, pp. 858-83; and A. C. B. Lovell *et al.*, "Meteors, comets and meteoric ionization;" *Reports on Progress in Physics*, Vol. 11, 1948, pp. 389-444.

only a fraction of a second, but a few persist for many seconds. The normal rate is one echo every few minutes, increasing markedly during meteor showers. The greatest rate so far reported was 168 echoes per minute at the peak of the Giacobinid shower of October 1946. These rates are several times the corresponding rates for visual meteors, indicating the existence of some which produce insufficient light for visual detection, but sufficient ionisation for radio detection. Ninety per cent of the transient echoes occur at heights between 87 and 108 kilometres, in good agreement with the known heights of occurrence of visible meteors.

Meteors are usually particles a fraction of a millimetre in diameter, whose reflecting power is far too small to give a detectable echo from the particle itself. The normal radar echo is due to reflection from the cylindrical column of ionised gas produced by the meteor as it travels through the air at a high velocity, and hence is much greater in a direction perpendicular to the cylinder of ionisation. This property has been used to determine the direction of arrival of streams of meteors, and, further, a diffraction effect which occurs during the formation of a meteor trail has been utilised to obtain the velocity of individual meteors. By these and other means, our knowledge of the meteor streams which intersect the orbit of the Earth has been greatly extended over the past few years. The most interesting result, physically, is that no meteor, either belonging to one of the streams or a sporadic one, has been found to have a velocity corresponding to a hyperbolic orbit about the Sun. Meteors, it seems, are members of the solar system. Radar, in the study of meteors, has made a notable contribution to astronomy. On the other hand, further investigation of the behaviour of the ionisation produced by meteors is likely to lead to substantial additions to our knowledge of the ionosphere.

Meteorology

The effects produced by water droplets on the scattering and attenuation of radar waves have already been discussed (Chapter II, section 7). During World War II microwave

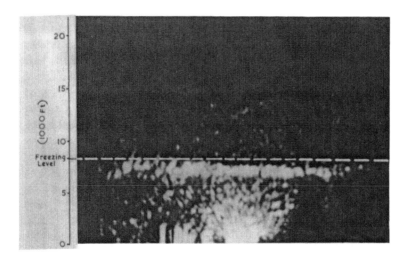

(a) The "bright band" structure characteristic of rain produced by the Bergeron process.

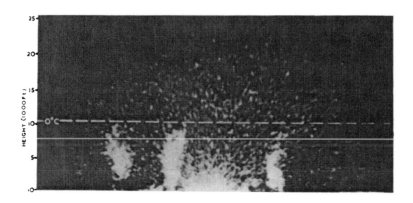

(b) The column structure typical of non-freezing rain produced by coalescence of cloud droplets.

facing p. 570]   PLATE XL.—Radar echoes from rain.

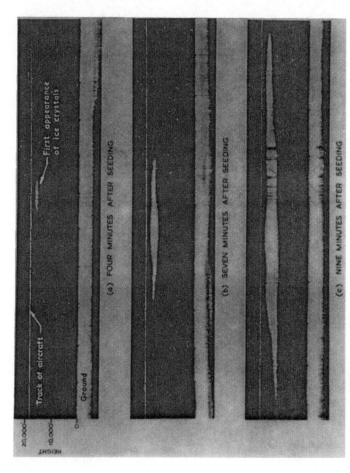

Radar pictures taken from an aircraft showing the development of rain areas within a cloud during an artificial rainmaking experiment.

PLATE XLI.

radars were sometimes used to detect the presence of rain storms (see Plate II) and to plot their movements. Modern radars carried by ships or aircraft are relied upon to provide this information as an essential aid to navigation, while on land the same information can serve as a useful subsidiary aid to short-term meteorological forecasting. High-power search radars are sometimes installed specifically for this purpose, or for following the course of special reflecting balloons so that wind velocities at high altitudes may be determined.

Because of the facilities they provide for "seeing" raindrops or ice and snow particles within clouds, radar methods are coming into use in certain phases of meteorological research, particularly for studying the physics of rain which occurs naturally or may be induced by artificial means. A centimetre wave radar is used either in aircraft or on the ground, a particularly effective form being one in which the aerial beam is rotated continuously about a horizontal axis.[7] This gives a side-elevation picture of rain echoes in a vertical plane through the radar set.

When natural rain is observed on such a system, one of two general types of pattern, illustrated in Plate XL (a) and (b) respectively, is usually obtained. The first shows a horizontal band structure with a marked increase of radar echo intensity occurring just below freezing level in the atmosphere. This pattern is obtained when rain forms by the Bergeron process. The faint echoes above the band (often referred to as a "bright" band or a "melting" band) are received from ice crystals and snowflakes, and those below the band from raindrops. The enhanced echo intensity just below freezing level is due to the increased reflectivity of the ice crystals or snowflakes at the instant of melting, which, in turn, is due to their being covered by a thin film of water at this stage.

The second type of echo pattern, illustrated in Plate XL (b), is of the column type and is obtained from convective showers in clouds which do not necessarily extend up to freezing level in the atmosphere. Aircraft flights through certain of these clouds established that the temperatures were everywhere

[7] E. G. Bowen, "Radar observations of rain and their relation to mechanisms of rain formation," *J. Atmos. Terr. Phys.*, Vol. 1, 1951, pp. 125-40.

warmer than 0 degrees Centigrade. It is clear that ice crystals play no part in the formation of the rain in these cases and it has been shown that it is due to coalescence of cloud droplets to form large raindrops.[7] Similar echoes are sometimes obtained from convective clouds which extend many thousands of feet above freezing level and investigation suggests that in many of these cases also the rain forms by coalescence, both the cloud droplets and the larger raindrops remaining in the supercooled state.

Radar observations also play an indispensable part in controlled experiments in which dry ice, silver iodide, water droplets, or other substances are introduced into cloud areas in an attempt to induce rainfall by artificial means. They are used initially to provide definite evidence that the cloud to be treated does not already contain raindrops, and subsequently to allow the development of any precipitation to be followed in detail. Data on raindrop size and rainfall intensity at any stage are obtained from measurement of the intensity of the rain echoes observed. Plate XLI shows a series of pictures of the radar screen during a typical cloud-seeding experiment.

2. Applications of the Techniques of Radar

Radio Astronomy

The pioneer discovery that radio waves of extra-terrestrial origin could be received at the Earth's surface was made by Jansky in 1932. They appeared to be coming from the direction of the Milky Way. Subsequent studies, notably by Reber, confirmed that their distribution over the sky conformed reasonably well with what was known of the distribution of matter in the Galaxy. Radio waves from the Sun were not identified until 1942 when they were detected as a form of interference by radar observers in England,[8] and

[8] J. S. Hey, "Solar radiations in the 4-6 metre radio wavelength band," *Nature*, Vol. 157, 1946, pp. 47-8; and E. V. Appleton, "Departure of long-wave solar radiation from black-body intensity," *Nature*, Vol. 156, 1945, pp. 534-5.

observed independently by Southworth[9] in America. Previous attempts to detect radio-frequency radiation from the Sun had been unsuccessful but Southworth made use of new microwave radar equipment which had been developed for military purposes and observed steady radiation corresponding in intensity to that from a black body at a temperature of 18,000 degrees Kelvin. The Sun's optical temperature is about 6,000 degrees Kelvin.

These discoveries were of fundamental importance and marked the opening of a new era in astrophysical investigations. Since the end of World War II radio methods have been extensively used in this new field of "Radio Astronomy"[10] to provide information which is supplementary and often additional to that obtained by the more conventional astronomical techniques using light waves. Two distinct methods are in use: the "echo" method of radar and the "radio-frequency radiometer" technique. The former has been applied only to the nearest astronomical objects, meteors and the Moon, and has already been discussed. In the second, a radio aerial and receiver are used to measure the radio-frequency radiation spontaneously emitted by the various heavenly bodies.

The received radiation is found to have the random waveform characteristic of fluctuation noise, and hence that from the Sun is frequently referred to as "solar noise." Extraterrestrial radio-frequency radiation of other than solar origin is referred to as "cosmic noise," "galactic noise" being used for that which arises within our own Galaxy. Its intensity is specified either in terms of a flux density or an equivalent temperature, i.e. the temperature of a black body subtending the same solid angle as the actual source which would yield the observed radiation intensity.

Observations have been taken on wavelengths from 1 centimetre to 15 metres. Over this range the aerials and receivers

[9] G. Southworth, "Microwave radiation from the Sun," J. Franklin Inst., Vol. 239, 1945, pp. 285-97; and erratum issued with Vol. 241, March 1946.
[10] For a general survey, see J. S. Hey, "Radio astronomy," Mon. Not. R. Astr. Soc. Vol. 109, 1949, pp. 179-214; and M. Ryle, "Radio astronomy," Reports on Progress in Physics, Vol. 13, 1950, pp. 184-246.

employed follow normal radar design but the radar display unit is replaced by a recording milliammeter operated from the output of the second detector. A distinctive feature of this equipment is that exceedingly minute powers, sometimes only 0·1 per cent of that generated in the receiver itself, are able to be measured. Elaborate arrangements are therefore necessary to minimise spontaneous changes in output which could mask the desired ones. The incident radiation is measured in terms of the small change in output current which it causes, calibration being effected by connecting the receiver alternately to the aerial and to a known noise source, such as a resistor at known temperature. The smallest power detectable corresponds to radiation from a body at a temperature of only a few degrees above absolute zero.

The angular resolution attainable in radio astronomy, however, is markedly inferior to that of optical astronomy because feasible ratios of wavelength to aperture are vastly greater. A single directional aerial may be used to obtain relatively rough position information by maximising on the source (see Fig. 338). The accuracy obtainable is, at best, of the order of a degree. For more accurate determinations of direction of arrival, interference methods have been developed. One technique which has been used with success in Australia is in the form of a Lloyd's mirror interferometer. A single aerial pointing horizontally is mounted on a cliff site overlooking the sea and interference is obtained between the direct and sea-reflected rays when a source of radio noise rises above the sea. With this method, which is illustrated in Fig. 339, only one aerial is required and the problem of identifying a source is somewhat simplified, but very careful correction for atmospheric refraction effects is necessary. Under favourable conditions a "point" source may be located within less than 10 minutes of arc. Another method, which is less susceptible to errors due to refraction, is a radio version of Michelson's stellar interferometer, in which two or more horizontally spaced aerials, sometimes in two planes at right angles, are combined. A position accuracy of several minutes of arc is attainable. This form is illustrated in Fig. 340. In both

methods the position of a source is calculated from the interference produced as the source moves through the lobe pattern
of the combined aerial system. These methods are applicable
therefore only to enduring sources of radio-frequency radiation
which do not change appreciably during the time necessary

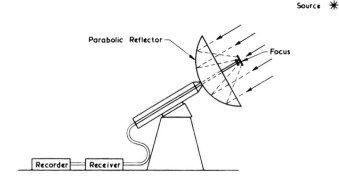

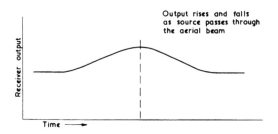

Figure 338.—Diagrammatic representation of a radio-frequency radiometer, in the form of a "radio telescope," using a single aerial provided
with right ascension and declination axes. Accuracy of position finding
in both coordinates is determined by the beam width of the aerial.

to complete an observation. In a new form of spaced-aerial
interferometer which has recently been developed, particularly
for the location of transitory disturbances,[11] virtually in-

---

[11] A. G. Little and R. Payne-Scott, "The position and movement on the
solar disk of sources of radiation at a frequency of 97 Mc/s. I. Equipment,"
Aust. J. Sci. Res. A, Vol. 4, 1951, pp. 489–507; and R. Payne-Scott and
A. G. Little, "The position and movement on the solar disk of sources of
radiation at a frequency of 97 Mc/s. III. Outbursts," *Aust. J. Sci. Res.* A,
Vol. 5, 1952, pp. 32–46.

stantaneous position-finding is achieved by using an electrical
device to sweep the lobe pattern of the aerial system across
the source at a fast rate.   An angular accuracy of about
2 minutes of arc may be obtained.

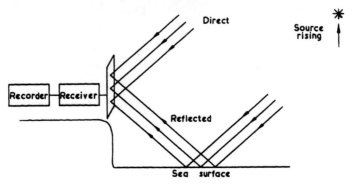

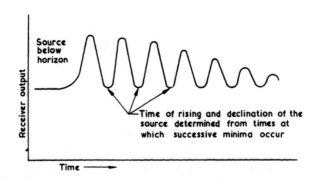

Figure 339.—A radio version of a Lloyd's mirror interferometer using
a single aerial and the sea as a reflector. The declination of a source is
inferred from time of rising (or setting) as determined from the interference pattern; right ascension from the time of culmination, i.e. midway
between times of rising and setting.

The polarisation of radio-frequency radiation is determined
by comparing the amplitude and phase of signals received on
two identical aerials which are plane-polarised in directions at
right angles.   By this means, plane and randomly polarised

radiation can be differentiated from that which is circularly polarised, and in the latter case the degree and sense of circular polarisation measured.

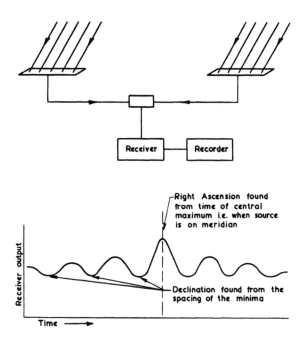

Figure 340.—Diagrammatic representation of a two-aerial vertical interferometer in an east-west plane. Declination is determined from the spacing of the minima of the interference pattern and right ascension from the time at which the central maximum occurs.

(a) Radio Waves from the Sun—Solar Noise

The outstanding feature of radio-frequency radiation from the Sun is that its intensity is highly variable.[12] These variations are greatest on wavelengths greater than about one metre

---

[12] J. L. Pawsey, "Solar radio-frequency radiation," *Proc. Instn. Elect. Engrs.*, Vol. 97, Part III, 1950, pp. 290-310.

where increases of ten to a thousand fold often persist for hours or days and short-period bursts of radiation of much greater intensity, but lasting only for seconds or minutes, are very common.   On shorter wavelengths, rapid variations are infrequent but there is still a day-to-day fluctuation of two or three to one in intensity.

Despite these complex variations it is possible to recognise a number of distinct components of the radiation.   Firstly, there is a general background or base level below which the intensity does not fall.   This is attributed to thermal radiation

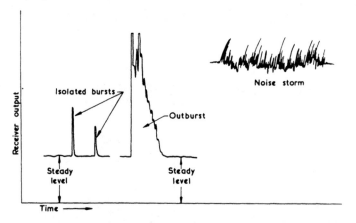

Figure 341.—The characteristic appearance of the various components of solar radio-frequency radiation at an observing frequency of 200 megacycles per second.

from the undisturbed Sun.   The other components originate in disturbances.   On the shorter wavelengths, the "slowly varying component" has been shown to be closely correlated with sunspot area.   At longer wavelengths, periods of marked disturbance lasting for days, "noise storms," are associated with particular large sunspots, and "outbursts," the most intense disturbances observed, with the great solar flares. Other bursts which have certain features in common have been grouped under the term "isolated bursts."   Fig. 341 illustrates diagrammatically the characteristic features of records of these components at a typical observing frequency of 200 megacycles per second.

Thermal Radiation. Solar radio-frequency radiation has been observed throughout the wavelength range from 1 centimetre to 4 metres and above. Pawsey and Yabsley have shown that over this range there is a background component which is relatively steady, shows random waveform characteristics, random polarisation and more or less uniform distribution over the Sun's disc. This is identified with radiation arising from thermal agitation of electrons and ions.

The Sun is known from optical observations to be surrounded by an atmosphere which is highly ionised, at high temperatures and decreases in density with decreasing height. Magneto-ionic theory has been applied, independently by Martyn[13] and Ginsburg,[14] to the thermal emission and propagation of radio-frequency radiation in such a medium. Martyn deduced that radiation in the metre wavelength region should originate chiefly in the outermost regions known as the corona, and the shorter radio-frequency waves at much lower levels, in the chromosphere; and that throughout most of the range the emissivity should approach the black-body value. This is illustrated diagrammatically in Fig. 342.

Observed values of the intensity of thermal radiation from the Sun, expressed as apparent disc temperatures, at various wavelengths are shown in Fig. 343. The striking feature of these results is that the temperatures, which at a wavelength of 1 centimetre approximate to the optical value of 6,000 degrees Kelvin, progressively increase above this value at longer wavelengths. This progression is, of course, associated with the movement of the region of origin from the low chromosphere to the high corona and provides very direct evidence that the kinetic temperature of the corona is about 10^6 degrees Kelvin. This remarkable result had previously been deduced from optical evidence. Most of the observations have been made around a period of sunspot maximum but there is evidence that temperatures (or electron densities) in parts of the corona

---

[13] D. F. Martyn, "Temperature radiation from the quiet Sun in the radio spectrum," *Nature*, Vol. 158, 1946, pp. 632-3.
[14] V. L. Ginsburg, "On solar radiation in the radio spectrum," *C.R. Acad. Sci. U.R.S.S.*, Vol. 52, 1946, pp. 487-90.

may vary over a long period, possibly with the sunspot cycle.[15]

The feature of the use of radio observations for the study of the solar atmosphere is that the bulk of the radio-frequency radiation comes from the Sun's atmosphere itself. Optical

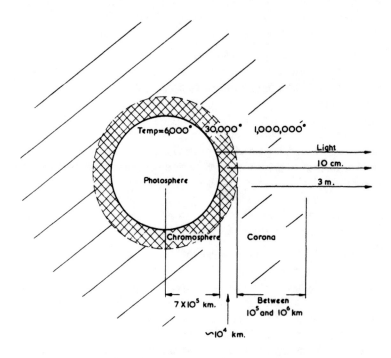

Figure 342.—Approximate temperatures and levels of origin of radio-frequency waves in the solar atmosphere.

observations of the same regions are difficult to make because of the much more intense radiation from the photosphere; in fact, until recently, it was possible to study the corona only at times of total solar eclipse.

[15] W. N. Christiansen and J. V. Hindman, "A long-period change in radio-frequency radiation from the 'quiet' Sun at decimetre wavelengths," *Nature*, Vol. 167, 1951, pp. 635–7.

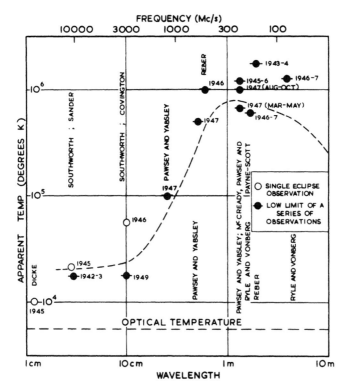

Figure 343.—Available estimates of apparent temperature due to the thermal component of solar noise. The dotted line is Martyn's prediction. Filled circles indicate results based on series of measurements; open circles, single observations at times of solar eclipse.

"**Noise Storms.**" An outstanding feature of solar radio-frequency radiation at the longer wavelengths (order of metres) is that the intensity may increase by a very large factor for periods of hours or days at a time, and fluctuate violently during this period. This "noise storm," as it has been termed by analogy with a magnetic storm, is correlated with sunspot activity (see Fig. 344). The radiation received during a noise storm is sometimes referred to as "enhanced radiation." It is circularly polarised and has been shown by accurate direction-finding methods to come from a small area on the Sun in the vicinity of a large sunspot.[12] *Giant* sunspots have always been associated with intense radiation but, on the other hand, it is not yet possible to decide in advance which of the smaller spots will be associated with a noise storm.

From the improbably high temperature of up to 10^{10} degrees Kelvin which would be needed to account for it and from the

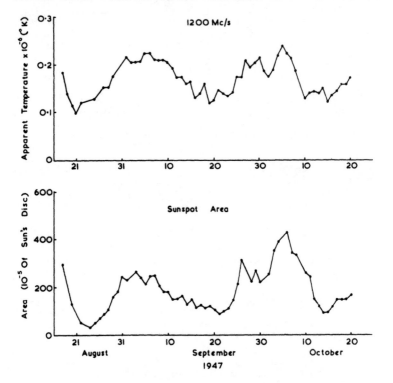

Figure 344.—Correlation between the daily means of intensity of solar noise at a frequency of 1200 megacycles per second and sunspot area.

way in which it fluctuates, it is clear that this radiation is not of thermal origin. It is only observed at the longer wavelengths and therefore originates high in the solar atmosphere, probably as a result of enormous electrical disturbances occurring in the vicinity of the active regions of which sunspots are a visible manifestation.

Bursts. A burst is the name given to one form of short-term disturbance which lasts only for a very brief period—from fractions of a second up to perhaps 20 seconds—and appears almost exclusively at the longer (metre) wavelengths. When these short-period fluctuations are observed simultaneously over a range of wavelengths, at least two distinct types of bursts can be identified.[16] The first of these occurs only over a very narrow frequency band and shows the same characteristic polarisation as, and is generally associated with, noise-storm periods. "Storm" bursts appear to be due to momentary disturbances high in the corona in the vicinity of active areas on the Sun, almost certainly of electrical origin. It has in fact been suggested that a solar noise storm may merely be the resultant of a very large number of such bursts, the frequency of occurrence at times being so great that they give rise to a moderately steady resultant.

The second distinctive type occurs sporadically and shows no preference for noise storms. These "isolated" bursts are equally transitory, are not circularly polarised and often appear almost simultaneously over a range of wavelengths. Some of them conform to a distinct group in which appearance at high frequencies precedes that at lower frequencies by a time interval of the order of seconds.[16] The rate of "drift" of frequency is very rapid (20 megacycles per second per second). It cannot be explained as due to selective group retardation of waves emanating from a fixed source in the solar atmosphere. The frequency drift could be associated with the rapid motion of a source travelling outwards through the solar atmosphere with a velocity of the order of 30,000 kilometres per second, but if so, this velocity is very considerably greater than that of any known motion in the corona. An interesting feature of isolated bursts is that they decay exponentially with a time constant appropriate to the theoretical value for a transient disturbance at the expected level in the corona.

[16] J. P. Wild and L. L. McCready, "Observations of the spectrum of high-intensity solar radiation at metre wavelengths," *Aust. J. Sci. Res.* A, Vol. 3, 1950, pp. 387-98; and J. P. Wild, *ibid.*, pp. 541-57.

Outbursts. Outbursts show features similar to isolated bursts but they are generally much more intense and more complex and last for periods reckoned in minutes rather than seconds.[17] They do not show circular polarisation. They are associated with solar flares.

The outstanding characteristic is that the source frequently moves rapidly over the disc of the Sun, usually in a manner which suggests that the actual source is moving outwards with a radial velocity of the order of 1000 kilometres per second.[11] This movement has been observed using the rapid interference technique previously described, but it was not discovered in this way. It was discovered from the observation that the time of commencement of an outburst had a tendency to show a progressive delay as the observing frequency was decreased.[17] This was interpreted as being due to the movement of a disturbance upwards in the solar atmosphere with a velocity of this order. The interesting context to this hypothesis is that, at the time of certain solar flares, streams of particles are ejected from the Sun. These reach the Earth a day or so later (which implies a velocity of about 1000 kilometres per second) and give rise to aurorae and magnetic storms. It seems that in an outburst we are seeing the genesis of a magnetic storm.

(b) Galactic Noise

Galactic radio-frequency radiation was discovered completely unexpectedly in 1932 by Jansky in the course of tracking down sources of interference to short-wave (15 metre) communication circuits. It was not until 1940, however, that a systematic survey of its distribution was completed (by Reber). Since then a number of other accurate surveys of the intensity distribution of galactic noise over the celestial sphere have been carried out.[18] These have been made at a number of

[17] J. P. Wild, "Observations of the spectrum of high-intensity solar radiation at metre wavelengths. II. Outbursts," *Aust. J. Sci. Res.* A, Vol. 3, 1950, pp. 399-408.

[18] G. Reber, "Cosmic static," *Astrophys. J.*, Vol. 100, 1944, pp. 279-87; G. Reber, "Cosmic static," *Proc. I.R.E.*, Vol. 36, 1948, pp. 1215-8; J. S. Hey, S. J. Parsons and J. W. Phillips, "An investigation of galactic radiation in the radio spectrum," *Proc. Roy. Soc.* A, Vol. 192, 1948, pp. 425-45; and J. G. Bolton and K. C. Westfold, "Galactic radiation at radio frequencies. I. 100 Mc/s. survey," *Aust. J. Sci. Res.* A, Vol. 3, 1950, pp. 19-33.

wavelengths and it is clear that interstellar space may reasonably be assumed to be transparent for the higher frequencies commonly used in radio astronomy. This is in marked contrast to its behaviour at optical wavelengths where vast clouds of obscuring matter cut off our view of the central parts of the Galaxy. On the assumption that the source of galactic noise is distributed in a representative manner, a new method is thus provided for determining the principal structural features of our Galaxy. In this way, a new determination has been made of the position of its centre and of the Sun with

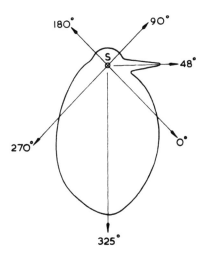

Figure 345.—A plot of cosmic noise intensity (excluding that from radio stars) in the galactic plane.

respect to the galactic plane. Results obtained at several radio frequencies are in reasonable agreement, placing the centre at about galactic longitude 330°, and galactic latitude −2°.

Fig. 345 shows the distribution around the galactic plane of the observed intensity at a frequency of 100 megacycles per second.[19] There is a possibility that the secondary maximum at galactic longitude 48° is due to a spiral arm of our Galaxy.

The origin of galactic noise is still a mystery. It is a curious feature that over the longer wavelengths it is more intense

[19] J. G. Bolton and K. C. Westfold, "Galactic radiation at radio frequencies. III. Galactic structure," *Aust. J. Sci. Res.* A, Vol. 3, 1950, pp. 251-64.

than solar noise, an inversion of the familiar large ratio of sunlight to starlight. Reber first suggested that galactic noise might be due to thermal emission from hot interstellar gas, but this theory is quantitatively unsound. A second suggestion was that it originated in processes in stars similar to those which give rise to intense solar noise. New evidence which may prove relevent is now available in the existence of a new class of astronomical object, the "radio stars."

Radio Stars. The early observations of galactic noise were made with equipment with poor resolving power which could not differentiate between origin in a continuous background or a number of localised regions, or even "point" sources such as the stars. In 1946, however, Hey, Parsons and Phillips[20] discovered that the radiation from a region in the constellation of Cygnus, unlike that from other directions, showed intensity fluctuations. This was shown by Bolton[21] to be associated with the existence of a point source of radiation in that constellation and he also discovered other similar sources. Some eighty of these discrete sources, often referred to as "radio stars," are now known, fifty in the northern and thirty in the southern hemisphere, but in only three cases has it been possible tentatively to identify any of them with visible objects. Of these, the Crab nebula, which is believed to be the remnants of a supernova of A.D. 1000, is the most interesting. Since this object is unique in the heavens, however, it does not help to identify radio stars with any particular stellar type. The intense radio stars are certainly not the bright visual ones and there is reason to suppose that they represent a previously unknown type of astronomical object having a high radio but low optical emission. The fluctuations which led to their discovery have now been shown to be due principally, or perhaps totally, to effects produced in the terrestrial atmosphere, analogous to those which give rise to the twinkling of visible stars. The mechanism of their radiation is not yet understood

[20] J. S. Hey, S. J. Parsons and J. W. Phillips, "Fluctuations in cosmic radiation at radio frequencies," *Nature*, Vol. 158, 1946, p. 234.
[21] J. G. Bolton and G. J. Stanley, "Observations of the variable source of cosmic radio-frequency radiation in the constellation of Cygnus," *Aust. J. Sci. Res.* A, Vol. 1, 1948, pp. 58-69.

nor can we be sure whether galactic noise is due entirely to radiation from the radio stars. It is possible that a further continuous background exists.

Linear Accelerators

The use of resonant circuits for the production of high voltages and their application to the acceleration of elementary particles at radio frequencies is not new. Sloan and Laurence in 1931 constructed a linear accelerator using a radio frequency of 10 megacycles per second, which accelerated mercury ions to an energy of 1·2 million electron volts; and later Sloan produced 800 kilovolt electrons with the aid of a single circuit resonant at 6 megacycles per second. Because of difficulties, some of them inherent in the use of relatively low radio frequencies, subsequent development tended towards the use of cyclical devices of which the cyclotron and betatron are typical.

Wartime radar developments have made available techniques for the generation of very high peak powers, in the form of pulses, at radio frequencies much higher than was previously feasible, and the application of these techniques to the development of linear accelerators has led to a revival of interest in this method of obtaining high-speed particles.[22] Three main lines have been followed: the use of resonant cavities and of either travelling waves or standing waves in a length of loaded waveguide.

Bowen, Pulley and Gooden[23] appear to have been the first to apply pulse techniques to a single resonant cavity for the acceleration of electrons. Their cavity, shown in Fig. 346, was resonant at a frequency of 1,200 megacycles per second and when fed from a 500 kilowatt magnetron produced electrons of energy 0·6 million volts. This work was further

[22] D. W. Fry and W. M. Walkinshaw, "Linear accelerators," *Reports on Progress in Physics*, Vol. 12, 1948-9, pp. 102-32.
[23] E. G. Bowen, O. O. Pulley and J. G. Gooden, "Application of pulse technique to the acceleration of elementary particles," *Nature*, Vol. 157, 1946, p. 840.

developed by Mills[24] who obtained a mean electron beam
current of 70 microamperes at a peak voltage of 1·1 megavolts.
An interesting application of this accelerator was as a source
of high-voltage X-rays, the arrangement used being shown
schematically in Fig. 347. The mean diameter of the beam
at the target was 0·5 millimetres (compared with a value of
about 5 millimetres which is common in commercial high-
voltage X-ray equipment) and advantage was taken of the
pulsed nature of the source to provide stroboscopic operation.

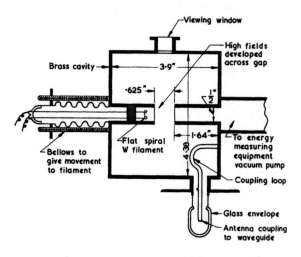

Figure 346.—Single-cavity electron accelerator for use at 1200 megacycles
per second.

It is possible also to use several cavities in cascade, and
successful multiple-cavity accelerators of this type have been
developed.

In travelling-wave accelerators a length of waveguide, usually
circular, matched to the source of radio-frequency power is
fitted with a series of metal irises or corrugations at appropriate
intervals, of the order of 3 to 4 per wavelength for maximum
efficiency. These are so arranged that the phase velocity in
the waveguide is equal to that of the accelerated particles.

    [24] B. Y. Mills, "A million-volt resonant cavity X-ray tube," *Proc. Instn.*
Elect. Engrs., Vol. 97, Part III, 1950, pp. 425-37.

Holes in the diaphragms provide coupling between the sections and allow also for the passage of the electron beam along the guide. As the wave propagates down the waveguide, power is dissipated in the walls and that reaching the end is absorbed in a matched dummy load. Fig. 348 illustrates schematically a linear accelerator of this type for producing 4 million volt electrons. A single source of radio-frequency power is employed. This is preferable to the use of a number of lower-powered oscillators spaced at intervals along the waveguide because of the difficulty in maintaining the necessary frequency stability and phase relationship between them.

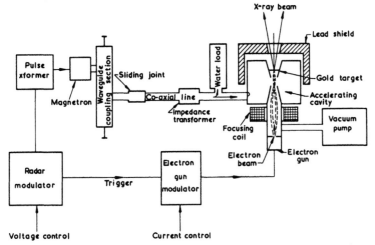

Figure 347.—Schematic arrangement of a million-volt resonant-cavity X-ray tube.

A standing-wave accelerator is generally similar in that a length of waveguide loaded with discs is used, but the end of the guide is closed and a standing-wave pattern instead of a travelling-wave pattern is therefore set up. In this type it has proved more feasible to employ multiple excitation. One such accelerator developed by Slater, for example, is made up of five resonant sections, each of 24 cavities, each section being fed at appropriate intervals along its length from four separate 1 megawatt 10 centimetre magnetrons, suitably locked in frequency and phase. Electrons of energy 28 million volts

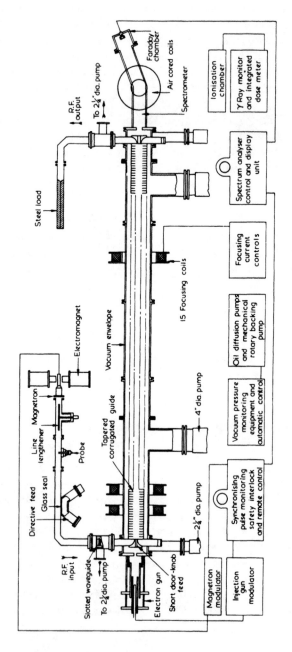

Figure 348.—Schematic arrangement of a travelling-wave linear accelerator, producing electrons of energy 4 million volts.

were produced in this accelerator. Both travelling-wave and standing-wave accelerators are necessarily more bulky than the single or multiple resonant cavity types, the overall length of the radio-frequency section of the former being reckoned in metres.

The same basic principles may be applied but with reduced efficiency to the acceleration of heavier particles, the longer transit times of the slower particles necessarily involving operation at lower radio frequencies. An example of the application of modern waveguide techniques to the production of high-speed protons is the Berkeley accelerator developed by Alvarez et al.[22] The resonator is 38 inches in diameter and 40 feet in length. It operates at a frequency of 200 megacycles per second with a peak power input of 1·5 megavolts and has an energy output of 32 million electron volts.

Microwave Spectroscopy

Transitions between one energy level and another within atoms and molecules give rise to the emission or absorption of radiation which shows a characteristic frequency spectrum because of the limited number of possible energy levels. As a rough generalisation, changes of state within the nucleus are associated with γ-ray spectra, those in the orbital electrons with X-ray and optical spectra, and those in the molecule with spectra in the infra-red and radio-frequency regions. A pioneer experiment was made in 1934 (by Williams) but radio spectroscopy has only come under active experimental investigation in recent years as a result of the application of wartime developments in radar techniques.[25] Most of the studies of radio spectra have been made in the microwave region where the spectral lines are more abundant and the methods generally more feasible than at lower frequencies. For this reason the field of investigation has become known as "microwave spectroscopy."

The study of microwave absorption spectra provides information on the structure of heavy molecules which is un-

[25] W. Gordy, "Microwave spectroscopy," Rev. Mod. Phys., Vol. 20, 1948, pp. 668-717; and L. Marton (Ed.), Advances in Electronics, Vol. 2, Academic Press, New York, 1950, pp. 299-362.

obtainable from spectra in other frequency bands. In most cases the transitions which are responsible for microwave absorption lines occur between closely spaced levels whose separation is due to differences in the quantised angular momentum of the molecule as a whole. These are called rotational transitions, and the microwave rotational spectra of many gases have already been studied. Of the non-rotational spectra, one of particular interest—and the first to be studied experimentally—is that of ammonia. Ammonia shows strong absorption around a wavelength of 1·25 centimetres due to a property of the NH_3 molecule known as inversion. Its spectrum, which appears as a number of lines, has been examined in detail and many of the lines have been identified with those predicted by theory. Microwave spectra also provide one of the most powerful methods known for investigating quadrupole moments and nuclear spins.

Two gases whose spectra are of interest in radar because of their influence on the propagation of microwaves are oxygen and water vapour. The former has an electronic fine-structure transition in the vicinity of a wavelength of 5 millimetres, and the latter a rotational transition at 1·348 centimetres. The effect of these lines has been discussed in Chapter II.

As distinct from spectroscopic application, microwave absorption lines are now being used to control the frequency of oscillators in this region (where stabilisation by crystals is not possible) and a further application which promises to be of outstanding value is the use of such lines to provide a fundamental standard of frequency and, hence, of time.

Techniques in Use. A typical microwave spectroscope used for examining a small part of the spectrum at a time is shown in Fig. 349. The gas whose spectrum is to be examined is introduced into a sealed, evacuated section of the waveguide. The variable frequency oscillator is usually a reflex klystron, the frequency of which is swept over a small range by applying a sawtooth voltage to its reflector. The output from the crystal detector is amplified in a wide-band amplifier and applied to one pair of plates of a cathode ray oscilloscope. The frequency sweep, controlled by the sawtooth pulse, is applied to the other pair of plates. A microwave absorption line displayed by this type of equipment is shown in Fig. 350. Using standard methods of frequency measurement, measurements can be made with an accuracy better than one part in a million.

The most troublesome feature in this technique arises from mismatches in the waveguide which can cause both spurious signals in the spectrum and an unwanted wave pattern in the baseline. These effects can seriously obscure the absorption spectrum of the gas. One ingenious method of overcoming

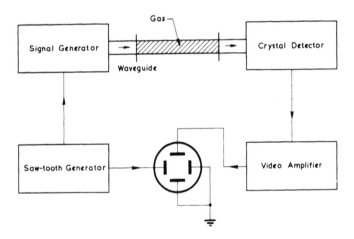

Figure 349.—Diagrammatic representation of a microwave spectroscope.

the difficulty is to apply to the gas a regular periodic electric field whose period is very much less than that of the frequency sweep. The periodic electric field has the effect of periodically splitting an absorption line into multiple lines (Stark effect), thereby modulating the amplitude of the original line. Spurious signals, on the other hand, remain unmodulated. An amplifier tuned to receive the modulated signals is incorporated in the detector circuit so that the spurious signals are eliminated. This refinement, known as Stark effect modulation, also improves the sensitivity of the system because the tuned amplifier eliminates low-frequency noise produced by the crystal.

In some microwave spectrometers the gas is contained in a relatively small resonant cavity instead of in the waveguide. Although the cavity type is more compact, the waveguide method has the advantage that the gas container requires no tuning.

The technique of microwave spectroscopy differs fundamentally from that of optical spectroscopy. The prism or grating is dispensed with because the microwave source (the

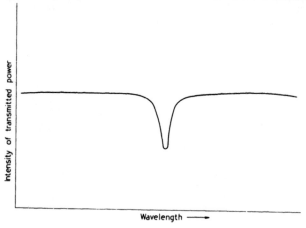

Figure 350.—The. appearance of a microwave absorption line on the oscilloscope of a microwave spectroscope of the type shown in Fig. 349.

signal generator) is monochromatic and may be varied in frequency at will. The absolute resolving power of a microwave spectrometer is about 10^5 times that of the best infrared grating spectrometer. Since the detector is a tuned device the sensitivity is very high.

The Velocity of Electromagnetic Waves

Several recent determinations of the velocity of electromagnetic waves, based on the use of radar techniques, yield a value some 16 kilometres per second greater than that which has previously been accepted. As one of the fundamental physical constants the velocity of electromagnetic waves has been the subject of study for many years. The experimental results have been analysed and weighted, and the statistical best value for the free-space velocity given by Birge[26] in 1941 was 299,776 ± 4 kilometres per second. This was supported by later reviewers. Early experiments involved a measurement of the velocity of light and, with few exceptions, this remained the case for all accurate work up to the end of World War II.

The development of radar navigational aids during the war, however, created a practical need for an accurate value of the velocity of propagation of radio waves and, in addition, of the refractive index of the atmospheric medium. The optical free-space value was at first accepted without question, and indeed until the development of precision radar possible errors would have been unimportant. Since 1945, however, careful measurements have been made using the Oboe system in England[27] and the Shoran system in the United States,[28] Canada and Australia, to check the ultimate accuracy of the systems and their possible use in geodetic surveying. Series of distances, known from normal triangulation methods, were measured by radar by the method described in Chapter XVIII. Particularly in the case of the American measurements, systematic discrepancies have been observed consistent with a higher propagation velocity than that assumed. The consistency of the American results is very high and, when interpreted as a measurement of the velocity of propagation in free space, a value of 299,792 ± 2·4 kilometres per second is obtained.

Other recent work tends to substantiate the view that a higher value of the velocity of propagation should be used. Essen,[29] at the National Physical Laboratory in England, has obtained a value from a study of radio waves in an evacuated

---

[26] R. T. Birge, "The general physical constants," *Reports on Progress in Physics*, Vol. 8, 1941, pp. 90-134.

[27] F. E. Jones, "The measurement of the velocity of propagation of centimetre radio waves as a function of height above the Earth," *J.I.E.E.*, Vol. 94, Part III, 1947, pp. 399-401.

[28] C. I. Aslakson, "Can the velocity of propagation of radio waves be measured by Shoran?" *Trans. Amer. Geophys. Union*, Vol. 30, 1949, pp. 475-87.

[29] L. Essen, "The velocity of propagation of electromagnetic waves derived from the resonant frequencies of a cylindrical cavity resonator," *Proc. Roy. Soc. A*, Vol. 204, 1950, pp. 260-77.

hollow metal cylinder. It is known that the resonant frequencies of a length of cylindrical waveguide closed by metal discs at both ends depend on the internal dimensions and the free-space velocity c according to the relationship

$$c = \frac{f'_{l, m, n}\left(1 + \dfrac{1}{2Q}\right)}{\left[\left(\dfrac{\tau}{\pi D}\right)^2 + \left(\dfrac{n}{2L}\right)^2\right]^{\frac{1}{2}}},$$

where $f'_{l, m, n}$ is the measured frequency of the mode of resonance denoted by the suffix l, m, n,

D and L are the diameter and length of the resonator,

Q is the sharpness of resonance,

τ is a constant for a series of particular l, m modes,

n is the number of half wavelengths in the length L.

Hence from measured values of $f'_{l, m, n}$, D, L and Q a value of c can be obtained. Essen used a silver-plated mild steel cylinder 6·5 centimetres in diameter, the electrical length of which could be varied by means of a piston. Measurements were made with the cylinder evacuated and the piston in a series of positions $\lambda/2$ apart such that the resonant frequency in various $H_{0, 1, n}$ modes was either 9,500, 9,000 or 5,960 megacycles per second. The resonator was very carefully constructed and allowances were made for the depth of penetration of the waves into the inside surface of the cylinder. The scatter in the individual measurements was equivalent to a proportional error of only 3×10^{-6} and the final result is quoted as a velocity of propagation of 299,792·5 $\pm$ 3 kilometres per second, the error here being the maximum of a single measurement.

Thus there are two recent determinations using radio techniques which give a value for the velocity of electromagnetic waves 16 kilometres per second greater than the previously accepted value.

New optical measurements by Bergstrand[30] in Sweden give support for this larger value. The method he used is a development of that used by earlier workers. A light source, modulated by a Kerr cell at about 8 megacycles per second, is reflected from a distant plane mirror and received by a photo-cell modulated at the same frequency as the source. When the distance bears a fixed relationship to the modulation frequency the light is received at a period of maximum sensitivity. Bergstrand's final value for the velocity of light is 299,792·7 ± 0·25 kilometres per second.

The following table gives particulars of a number of the more important measurements. It includes two early determinations using electrical techniques, the two optical measurements thought until recently to be most reliable, and values obtained by any method since 1940.

Date of publication	Author	Method	Velocity in vacuo (kilometres per second)
1907	Rosa, Dorsey	Ratio of units	299,784 ± 10
1923	Mercier	Standing waves on wires	299,782 ± 30
1935	Michelson, Pease, Pearson	Rotating mirror	299,774 ± 11
1941	Anderson	Kerr cell	299,776 ± 6
1947	Smith, Franklin, Whiting	Pulse radar	299,775 ± 50
1947	Jones	Pulse radar	299,767 ± 25
1948	Essen, Gordon-Smith	Cavity resonator	299,792 ± 9
1949	Bergstrand	Kerr cell	299,793 ± 2
1949	Aslakson	Pulse radar	299,792 ± 2·4
1950	Bergstrand	Kerr cell	299,792·7 ± 0·25
1950	Essen	Cavity resonator	299,792·5 ± 3

The consistency of results obtained over the last few years by such diverse methods is very high and there is little doubt that radar has contributed the motive and some of the tools to increase the accuracy with which the velocity of light is known by a further order of magnitude.

[30] E. Bergstrand, "A determination of the velocity of light," *Ark. fys.*, Vol. 2, 1950, pp. 119-50.

APPENDIX I

TABLE OF CONSTANTS

Permittivity of free space	$\epsilon_0 = 8\cdot854 \times 10^{-12}$ farads per metre
Permeability of free space	$\mu_0 = 4\pi \times 10^{-7}$ henries per metre
Velocity of light	$c = 2\cdot998 \times 10^8$ metres per second
Intrinsic impedance of free space	$\eta_0 = 377\cdot0$ ohms
Electronic charge	$e = 1\cdot590 \times 10^{-19}$ coulombs
Electronic mass	$m = 9\cdot038 \times 10^{-31}$ kilograms
Boltzmann's constant	$k = 1\cdot371 \times 10^{-23}$ joules per degree
Conductivity of copper	$5\cdot8 \times 10^7$ mhos per metre
Relative permittivity of polystyrene	$2\cdot5$
Q of polystyrene	2000
Refractive index of air	$n = 1 + \{79P_a/T + 380,000e/T^2\} \times 10^{-6}$
	(Chapter II)
Radar echo time	$= 10\cdot7$ microseconds per statute mile
Mean radius of curvature of earth in latitude 45 degrees	$= 6378$ kilometres
1 kilometre	$= 0\cdot6214$ statute miles
1 nautical mile	$= 1\cdot151$ statute miles

$e \qquad = 2\cdot7183$

$\pi \qquad = 3\cdot1416$

$\log 10 \quad = 2\cdot3026$

1 neper $= 8\cdot686$ decibels

1 radian $= 57\cdot296$ degrees

APPENDIX II

UNITS AND CONVERSION FACTORS

The rationalised metre-kilogram-second system of units is used throughout the book. The following table gives a list of MKS units, and factors for converting numerical magnitudes into unrationalised CGS electrostatic and electromagnetic units.

Rationalisation affects only three of the quantities listed, namely electric flux density, magnetomotive force, and magnetic field strength.

Quantity	Symbol	MKS Unit	Dimensions				Multiply by given factor to convert MKS Units to	
			M	L	T	Q	E.M.U.	E.S.U.
Force	F	newton	1	1	-2	0	10^5	10^5
Energy	W	joule	1	2	-2	0	10^7	10^7
Power	P	watt	1	2	-3	0	10^7	10^7
Charge	Q	coulomb	0	0	0	1	10^{-1}	3×10^9
Current	I	ampere	0	0	-1	1	10^{-1}	3×10^9
Resistance	R	ohm	1	2	-1	-2	10^9	$(1/9) \times 10^{-11}$
Conductivity	g	mho per metre	-1	-3	1	2	10^{-11}	9×10^9
Potential difference	V	volt	1	2	-2	-1	10^8	$(1/3) \times 10^{-2}$
Electric field strength	E	volt per metre	1	1	-2	-1	10^6	$(1/3) \times 10^{-4}$
Capacity	C	farad	-1	-2	2	2	10^{-9}	9×10^{11}
Electric flux density	D	coulomb per sq. metre	0	-2	0	1	$4\pi \times 10^{-5}$	$12\pi \times 10^5$
Permittivity	ϵ	farad per metre	-1	-3	2	2	10^{-11}	9×10^9
Magnetic flux	ϕ	weber	1	2	-1	-1	10^8	$(1/3) \times 10^{-2}$
Magnetic flux density	B	weber per sq. metre	1	0	-1	-1	10^4	$(1/3) \times 10^{-4}$
Magnetomotive force		ampere-turn	0	0	-1	1	$4\pi \times 10^{-1}$	$12\pi \times 10^9$
Magnetic field strength	H	ampere per metre	0	-1	-1	1	$4\pi \times 10^{-3}$	$12\pi \times 10^7$
Inductance	L	henry	1	2	0	-2	10^9	$(1/9) \times 10^{-11}$
Permeability	μ	henry per metre	1	1	0	-2	10^7	$(1/9) \times 10^{-13}$

APPENDIX III

CHARTS OF THE COMPLEX HYPERBOLIC TANGENT

The Kennelly and Smith charts which are reproduced below on pp. 602–5 are extensively used for carrying out numerical calculations with transmission lines. A typical problem is to find the input impedance to a length of transmission line terminated with a general impedance. The charts are described in Chapter VI (p. 132) and examples of their use are given in Chapter VII.

Each chart represents the formula

$$\frac{R}{Z_0} + j\frac{X}{Z_0} = \tanh\,(\text{artanh}\,S + j\theta),$$

where $R + jX$ is the impedance at a point on a transmission line of characteristic impedance Z_0, S is the standing wave ratio, and θ the space phase of the standing wave pattern (p. 130). The Kennelly chart differs slightly from that published by Kennelly in that the angle θ is marked in revolutions (or fractions of a wavelength), not in quadrants, and the circles of constant standing wave ratio are marked directly with values of S not of artanh S. An auxiliary scale of the latter quantity is, however, included. It is required in problems where the attenuation in the transmission line is not negligible, and reads directly in nepers.

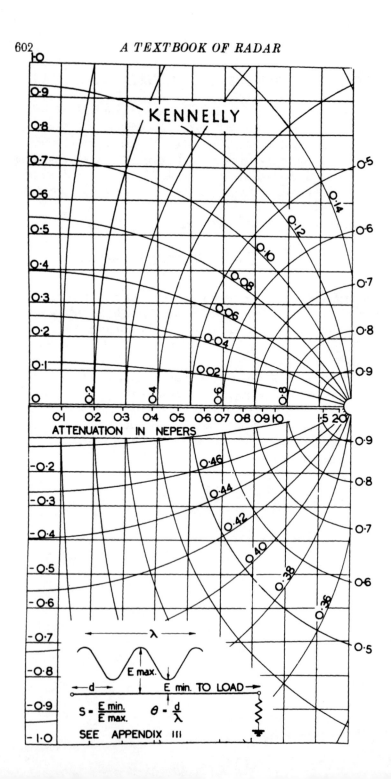

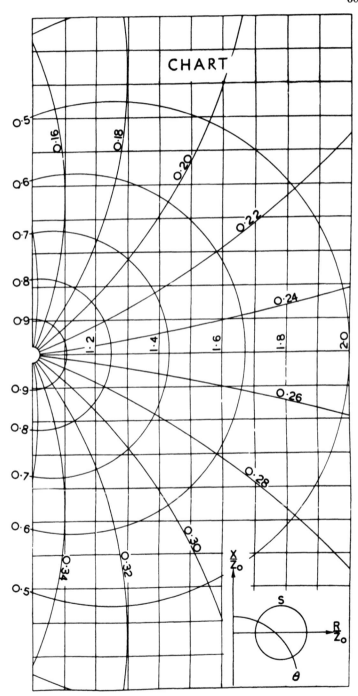

CHART

SMITH

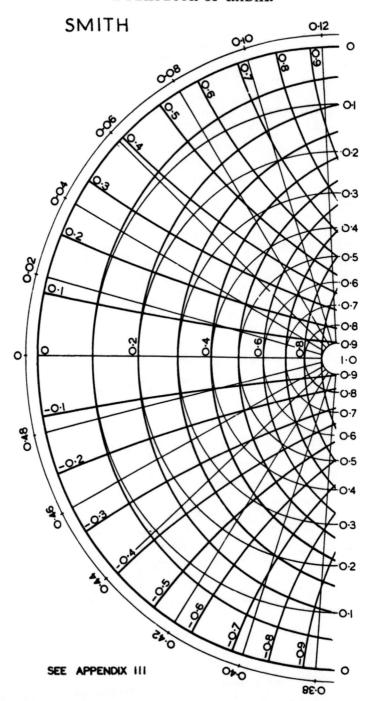

CHART

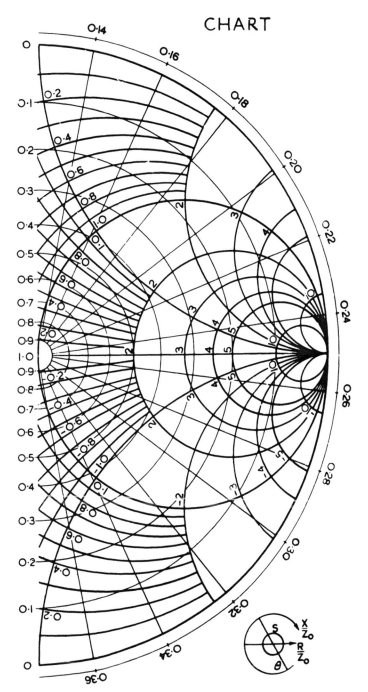

INDEX

Frequency (continued)
  stability of klystron, 384
  of magnetron, 54
  spectrum of magnetron, 54
Frequencies used for radar, 7

Gain control of IF amplifier, 433
Gain
  effect on valve characteristic, 433
  instantaneous automatic, 433
Galactic centre, determination of, 585
'Galactic noise', 573, 584-7
Gap
  iris type, 308
Gate, 330
Gate generator, 446
  in auto-ranging circuit, 489-90
  in display circuit, 446
  pulse length in auto-ranging circuit, 490
G.C.A. (Ground Controlled Approach)
  radar, 557
G.C.I. (Ground Controlled Interception) radar, 524
Gee, 542-4
Generator
  gate, 446, 489, 490
  pulse, 465
  range mark, 467
  sawtooth, 451
  sweep, 467, 494, 497

H radar system, 537, 538, 562, 563
H₂K radar, 536
H₂S (Home to Station) radar, 536
H₂X radar, 537
Hard tube modulator, 74, 75, 106
Harmonic energy from magnetrons, 308
'Hartree diagram', 49
Height-Finding radar, 517, 522-4
Helical scan, 511
Hertz, 1
'Hesitation' search circuit, 497
Horns, 271
  biconical, 276
  conical, 276
  pyramidal, 275
  sectoral, 272
Hyperbolic navigational systems, 542-7, 551-4
  Gee, 542-5
  Loran, 545-7
  M.T.R. 552 4

IAGC, 433
IFF (Identification Friend or Foe), 539

Impedance characteristic, 130, 135, 163
  charts, 132, 165
  gear, slotted line, 164
  intrinsic, 115
  measurement, 164, 213
  mixer, 399, 411
  normalized, 130
  transformer, 165-70, 194, 298, 303-9
  wave, 135
Inductance
  for DC charging, 94
  of cathode lead, 330, 416
Integrating circuit, 445
Interferometers, 574-7
Ionospheric research, 565-6, 568
Iris discharge gap, 308
  in waveguide, 193

Jaeger, J. C., 23
Johnson noise, 335-7, 412
Joints
  choke, 189
  coaxial, 185
  rotating, 185, 198
Junction
  magic tee, 196, 311, 407
  tee, 195, 297, 311

Keep-alive electrode in gas switch, 302, 308
Kennelly chart, 132, 165, 171, 601
Kerr cell, for velocity of light determination, 597
Klystron
  AFC, 385
  bunching, 359, 371
  construction, 356, 381
  current distribution, 359, 362, 375
  double resonator, 352, 368
  electronic tuning, 384
  frequency stability, 384
  limitations, 367
  mixer, 395
  oscillator, 351
  power output, 377
  reflex, 352, 371, 374
  second order effects in, 366

Leakage power for TR switches, 299, 314, 317, 398, 400
Lens aerial, 269
Lighthouse valves, 344, 394
Limiting in video amplifier, 324, 438
Linear accelerators, 587-91
Loading effect on magnetron, 54
Lobe due to ground reflection, 22

CPSIA information can be obtained
at www.ICGtesting.com
Printed in the USA
LVOW10s2347180518
577697LV00001B/18/P